W9-BBP-148

SPATIAL ANALYSIS AND PLANNING UNDER IMPRECISION

Studies in Regional Science and Urban Economics

Series Editors

ÅKE E. ANDERSSON
WALTER ISARD
PETER NIJKAMP

Volume 17

NORTH-HOLLAND – AMSTERDAM • NEW YORK • OXFORD • TOKYO

Spatial Analysis and Planning under Imprecision

YEE LEUNG
Department of Geography
The Chinese University of Hong Kong
Shatin, N.T.
Hong Kong

1988

NORTH-HOLLAND – AMSTERDAM • NEW YORK • OXFORD • TOKYO

© ELSEVIER SCIENCE PUBLISHERS B.V., 1988

All rights reserved. No part of this publication may be reproduced, stored in a retrieval system, or transmitted, in any form or by any means, electronic, mechanical, photocopying, recording or otherwise, without the prior permission of the copyright owner.

ISBN: 0 444 70390 x

Publisher:
ELSEVIER SCIENCE PUBLISHERS B.V.
P.O. Box 1991
1000 BZ Amsterdam
The Netherlands

Sole distributors for the U.S.A. and Canada:
ELSEVIER SCIENCE PUBLISHING COMPANY, INC.
52 Vanderbilt Avenue
New York, N.Y. 10017
U.S.A.

LIBRARY OF CONGRESS
Library of Congress Cataloging-in-Publication Data

Leung, Yee.
Spatial analysis and planning under imprecision / Yee Leung.
p. cm. -- (Studies in regional science and urban economics ; v. 17)
Bibliography: p.
Includes index.
ISBN 0-444-70390-X (U.S.)
1. Spatial analysis (Statistics) 2. Fuzzy sets. I. Title. II. Series.
QA278.2.L483 1988
519.5'36--dc19 87-36007
CIP

PRINTED IN THE NETHERLANDS

To My Family

PREFACE

A baffling issue in the analysis of human behavior in space and time is the seemingly incompatible nature of objectivity and subjectivity, precision and imprecision, simplicity and complexity, certainty and uncertainty, as well as abstraction and realism. The attempt to analyze the intertwining of subjectivity, imprecision, complexity, and uncertainty (within human systems in general and spatial systems in particular) with objective, precise, simple, and deterministic mathematical systems gives rise to a certain sense of dissatisfaction.

Over the years, significant spatial theories and models under certainty have been quite successfully advanced through the use of classical mathematics. Probability and stochastic processes have further extended our abilities in analyzing spatial behavior under randomness. Though our achievements have been gratifying in terms of formalism, we are still quite remote from an adequate description in terms of realism.

In our quest for objectivity, simplicity, and precision, we customarily fit spatial behavior to rigid mathematical models which make no provision for systems complexity and the imprecision in our cognition, perception, valuation, and decisionmaking processes. Human subjectivity and imprecision have conventionally been regarded as absurd in scientific investigations. Being precise has been a virtue of science. Valuation is almost a forbidden word in formal models. Our unabating effort in achieving higher level of precision has been the instrument of spectacular advancements in the physical sciences.

Following the path of the physical sciences, social scientists have come to believe that human systems are also precise in nature and can be efficiently analyzed by classical mathematics. To be precise, we have attempted to force artificial precision on imprecise phenomena and processes, and in so doing have lost the intrinsic imprecision in human systems in search of precision as a goal. In addition, we have failed to realize that our ability to be precise diminishes as the system becomes more complex. In cases of extreme complexity precision is usually an impossibility. Spatial models which neglect these intrinsic characteristics tend to be over-simplified, too mechanical, and too inflexible to give an adequate description of the complex and elastic real world.

To have a closer approximation to and control of our spatial systems, it is essential to restore human values and to treat imprecision with rigor in theory and model constructions. Appropriate operational models should be able to mimic human behavior in space and time. The behavioral approach for example is developed for the purpose of making formal models more appropriate to the human condition. Most behavioral models, however, appear to be either too loose or too superficial in terms of formalism. The lack of a cohesive mathematical system which enables formal analysis of imprecise spatial behavior is perhaps a major obstacle for behavioral models to be regarded as elegant formalizations of spatial systems with operational values. It is therefore imperative to have a system which renders mathematical representations of imprecise human behavior.

Among existing methods fuzzy set theory appears to be a mathematical system which is instrumental in constructing formal models of imprecise spatial behavior. It allows us to restate the importance of treating human subjectivity, albeit imprecise, in model formulations. It can also provide a bridge between verbal and formal models. Under fuzzy set theory, the human system is a flexible system within which cognition, perception, thought, reasoning, and behavior are a matter of degree.

Since the inception of fuzzy set theory in 1965, its applications in information science, computer science, control engineering, management science, operations research, and formal logic have been phenomenal. Similar developments nevertheless have not been taking place in social science in general and spatial analysis and planning in particular, although human behavior is the subject of analysis in these disciplines. Fragmentary research and the relatively slow dissemination of fuzzy set theory in these fields naturally lead to confusion, ambivalence, and misunderstanding.

Though there are a few books on fuzzy set theory, they do not address the issues of imprecision in spatial analysis and planning. The undertaking of the present monograph was triggered by my perception of such a void in our literature. My purpose is to put the fuzzy set approach in perspective. It involves the reexamination of conventional concepts and the laying of a foundation for fuzzy set theoretic spatial analysis and planning under imprecision.

This monograph is written for geographers, regional scientists, spatial economists, urban and regional planners, fuzzy set researchers, and other interested social scientists. It is of course difficult to satisfy such a variety of readers. I hope the monograph can serve as a stimulus for a more comprehensive effort in analyzing human behavior in space and time.

I would like to thank Dr. Walter Isard for his open-mindedness and continual encouragement of this project. Dr. Peter Nijkamp's comments and his enthusiasm in having the monograph refereed for publication in the present series is deeply appreciated. I am also grateful for Dr. Peizhuang Wang's detailed analysis and Dr. Manfred Fischer's comments on the earlier draft. To Dr. Claude Ponsard goes my deepest appreciation for his in-depth comments and suggestions on the early and final drafts. I remain of course responsible for all errors and deficiencies.

My deepest gratitude goes to Mrs. Jane Wan for typing and re-typing the entire monograph with patience and excellence, and to Mr. See-lou Too for his professionally done illustrations. I also thank Miss Ho-kwan Lo for her professional indexing.

Y. Leung

CONTENTS

CHAPTER 1

INTRODUCTION

1.1. ON SYSTEMS COMPLEXITY AND IMPRECISION

Human systems in general and spatial systems in particular are complex systems with omnipresence of subjectivity, incompleteness, and imprecision. No matter what we are studying, an individual, a firm, a group, a city, a region, a nation, or the whole world, encountering awesome complexities is ordinarily the rule rather than the exception. In general, spatial systems at the micro, meso, and macro scales are too complex and valuative for precision to be effective and deterministic statements to be realistic.

At the micro scale, we are usually concerned with decisions of an individual. Job search, residential choice, location and operation of a firm are typical examples. Taking an individual's job search as an example, to be completely certain, he needs to amass precise information on all aspects of the job market in the context of space and time. Based on the availabilities of jobs, differences in wages, job stabilities, prospects, living standards, and the living environments, he then decides on the one which somehow can optimize or satisfy a set of criteria or a composite criterion which he set for a desirable job.

With limited amount of time, financial resources, and mental capacity, acquiring complete and precise information on these aspects within a complicated job market is impossible. Therefore, information is never complete and is generally imprecise. The job-search process is often based on approximated data.

In choosing a job, the criteria are usually imprecise.

Under the utility maximizing principle, for example, "utility" is a term with imprecise denotation. Even if the concept makes sense to the individual, it is extremely difficuilt for him to fathom what "utility" precisely means to him. Thus, utility maximization is actually a fuzzy criterion guiding the decisionmaking process.

Instead of employing a composite criterion such as utility, a job seeker may want to base his decision on how well the job satisfies a set of criteria such as wage, location, work hours, job security, and prospect. A valuative decision rule such as "*high* wage, *convenient* location, *relatively short* work hours, *secure*, and *good* prospect" might be formulated as a standard to evaluate plausible jobs. Again, the rule is fuzzy because *high*, *convenient*, *relatively short*, *secure*, and *good* are imprecise subjective valuations. They, however, carry significant meanings and play an important role in the decisionmaking process.

To elaborate my argument, residential choice is another common micro-scale spatial decision in which complexity, valuation, and imprecision are rampant. The system is complex because residential choice, similar to job search, is not independent of other basic living preferences such as those on job, education for children, life style, and amenities. Within the system, they are highly interdependent. To be completely certain about a decision, obtaining precise information on the availabilities of housing units, differences of rents or sale prices, housing qualities, public service provisions, amenities, and crime rates in all localities is necessary. With time and money constraints, information gathered is again imperfect. For example, an individual may only know the approximate housing stock, approximate rents or sale prices in the market, rough descriptions of public service provisions and crime rates, and subjective valuations of housing qualities and amenities.

In making a decision, a sought after house has to satisfy a value-based standard such as "*short* distance from home, *high* quality, *moderate* rent or sale price, *good* public services, *quiet* and *clean* neighborhood with *low* crime rate". Instead of making a precise judgement of distance, quality, rent or sale price, public services, amenities, and crime rate, we conceptualize them by fuzzy categories such as *short*, *high*, *moderate*, *good*, *quiet* and *clean*, and *low* which human brains can process effectively. Again, we can cope with these complexities and imprecisions and are able to make a decision.

The behavior of an organization such as a firm, is again a micro-scale behavior of high complexity. In the locational analysis of a firm, for example, complexity is intertwined with imprecision and subjectivity. To find a location which can maximize its profit, a firm needs to obtain perfect information on the locations and availabilities of resources, locations of labors with respect to their costs, skills, and work morales, locations of markets, consumer preferences, as well as transport systems in terms of modes, routes, and costs. Our experiences have shown that such a task is impossible. While complete and precise information is beyond reach, incomplete and imprecise information is what a firm has to settle for.

If the objective is to maximize profit or minimize cost, it is quite a clear-cut locational rule provided that unit profits or unit costs can be precisely determined. If a firm prefers the satisfaction principle to the optimization principle, a fuzzy goal such as "Total profit should be *approximately* 10 million dollars", or "Total costs should be below 2 million dollars *as much as possible*" is specified. Apparently, subjectivity and flexibility exist in human thought and locational decisions are based on imprecise rules.

Expanding our spatial context to the behavior of an international trade firm, the decisionmaking environment is even more complex. To be certain, a firm needs to have perfect information on the national and international markets in terms of productions, consumptions, price differentials, and means of transportation. However, such a perfection seldom exists in the real world. Partial and imprecise information is the reality international trade firms have to accept. Their decisionmaking process is embedded with imprecision and subjectivity. It is again due to our ability to reckon, think, reason, and communicate globally with imprecise information, that a rational decision can always be made.

Therefore, spatial behavior at the micro scale is generally imprecise. The reason, to recapitulate, is mainly twofold. First, complexity and precision are mutually conflicting. A higher level of precision can only be achieved when a system is simple. The level of precision, however, decreases as our spatial systems become more complex. Imprecision and complexity are thus an integral phenomenon. Second, human thought, language, and perception, as discussed previously, are usually imprecise.

Thus, the complexity of a spatial system and human subjectivity often make precision impossible or even meaningless at the extremes. We, however, can cope with complexity and take global information in a granular and imprecise scale. We are able to imagine and think with loose concepts and to deduce in a logical manner. Thus, our decision is rational, albeit imprecise sometimes.

The same phenomenon can also be observed at the meso scale. In the study of urban and regional structures, information on the transitions of land uses or interzonal flows, for example, is seldom complete or precise. Complexities of demands and supplies among zones, imprecision on the perception of the attractiveness and distance of a place, and the imprecise preference-utility-choice structures of a person or a group all make precision an impossible task. Compounded by multiple objectives and multiple decisionmakers, we have, to be realistic, to take imprecision as a piece of fact in our analysis of such spatial systems. Our ability in handling complexity and imprecision is thus crucial.

Similar to the micro and meso scale, spatial systems at the macro scale are highly complex. Though we investigate systems with a small number of variables, as a way to cope with complexity, systems behavior is still imprecise. In the study of urban hierarchy, city orders may be treated as a function of population size and per capita income. Precisely how the two control variables influence the transitions between levels of the urban hierarchy is usually difficult to determine. Unless very simple assumptions are employed, threshold values for urban growth and decline cannot be precisely determined. We may observe that for *large* population and *high* income, there corresponds a *high* order in the hierarchy. We may further observe that for *sufficiently high* incomes, the transition between different levels may be smooth. Due to the complexities of the urban system, we are vague on exactly where to draw the line.

In the study of regional welfare, the effect of agglomeration and negative externalities on welfare is imprecise. In the analysis of urban population growth, precisely how people perceive and react to the differentials of urban and rural utilities is again difficult to determine.

Therefore, spatial systems at the micro, meso, and macro scales have built-in complexities and imprecision. It is however possible to perform formal analysis on these systems in such a way that imprecision can be catered for and directly dealt with.

1.2. ON UNCERTAINTY IN SPATIAL SYSTEMS

Following the successful applications of classical mathematics to physical systems under certainty, spatial scientists, in the past three decades or so, have constructed deterministic models of human behavior over space and time. Though the models are instrumental in explicating the basic mechanisms of spatial systems, a very important step towards the understanding of complex systems, they however fall short in capturing the uncertainty of systems behavior. The models tend to be oversimplistic, inflexible, and too mechanical. Empirical evidence often shows that too much of the human imprecisions and spatial systems complexities has been sacrificed for the simplicity and analytical elegance of the deterministic models.

To relax the assumptions of deterministic models, the notion of uncertainty is often injected in the model building process. A common presumption is that uncertainty exists because there are random processes within spatial systems. We are uncertain about a spatial behavior because there is a chance factor influencing human thinking processes and the operations of the whole system. We cannot be deterministic and precise because random occurrences of events cannot be completely and precisely predicted. Therefore, uncertainty is equated with randomness, albeit undefined. To account for uncertainty, probability models have been extensively constructed.

However, there is another aspect of uncertainty in human systems which cannot be attributed to randomness. Often, uncertainty is due to systems complexity, incomplete information, and imprecise human thought, language, and perception discussed in section 1.1. This type of uncertainty has nothing to do with randomness.

Take the previously discussed residential choice as an example, the system is too complex to warrant any kind of precise information. Consider the sale price of a house, we may only have a piece of approximate information such as "The sale price of a three-bedroom house is *approximately* 80,000 dollars." Our uncertainty is over the imprecise denotation of the rough figure "*approximately* 80,000 dollars". It differs from the probabilistic information such as "There is an eighty percent chance that the sale price of this house will be 80,000 dollars next month." Here, our uncertainty is over the likelihood of the exact figure, 80,000 dollars, occurred as the sale price. Therefore, uncertainty due to imprecision is not a random phenomenon. It is strictly a state of impreciseness. Modelling it as a random process is philosophically and theoretically inappropriate. (See section 2.5 for a more thorough discussion)

The same argument can also be applied to the decision rule in residential choice. The uncertainty encountered is completely a problem of imprecision over the valuative terms "*short* distance from home", "*high* quality", "*moderate* rent or sale price", "*good* public services", "*quiet* and *clean* neighborhood", and "*low* crime rate".

Therefore, to model a simple spatial decision such as residential choice, imprecision of information and human valuation should be treated directly and rigorously. The implication then is imprecision and valuation should not be considered as absurd in spatial analysis. On the contrary, they should be regarded as basic characteristics pertinent to the construction of spatial models which mimic human perception, thought, and behavior.

To solve imprecision problems, conventional methods force precision into otherwise imprecise objects. The two-valued logic is a typical example. By choosing an arbitrary cut-off value, a precise boundary can be imposed on a fuzzy object with gradual rather than abrupt transition from membership to non-membership. The cut-off value then serves as a precise boundary separating the object and the non-object. For example, the term "*short*" is a valuative term defined on a measurable physical distance. To characterize the term "*short*", an arbitrary physical distance, say 2 km., can be employed as the cut-off point so that any distance shorter than or equal to 2 km. is considered as *short*, and any distance longer than 2 km. is not considered as *short*.

Mathematically, "*short*" is a set defined by a characteristic function

$$
\begin{aligned}
f: X &\longrightarrow \{0, 1\}, \\
x &\longrightarrow f_{short}(x),
\end{aligned}
\tag{1-1}
$$

such that

$$
f_{short}(x) = \begin{cases} 1, & \text{if } x \leq 2 \text{ km.}, \\ 0, & \text{otherwise.} \end{cases}
\tag{1-2}
$$

(see Fig. 1.1)

Employing such an artificial characterization, there is no imprecision associated with the term "*short*". However, "*short*" is in fact a term with imprecise denotation. It lacks a clear-cut boundary. The progression from "*short*" to "*not short*" is gradual. A given physical distance only belongs to the term "*short*" to a certain degree. The law of the excluded middle thus nullifies the intrinsic vagueness of the term "*short*". In general, human beings think in terms of sets with fuzzy boundaries. Our uncertainty rests on whether or not an element belongs to a set. The two-valued logic is thus inappropriate

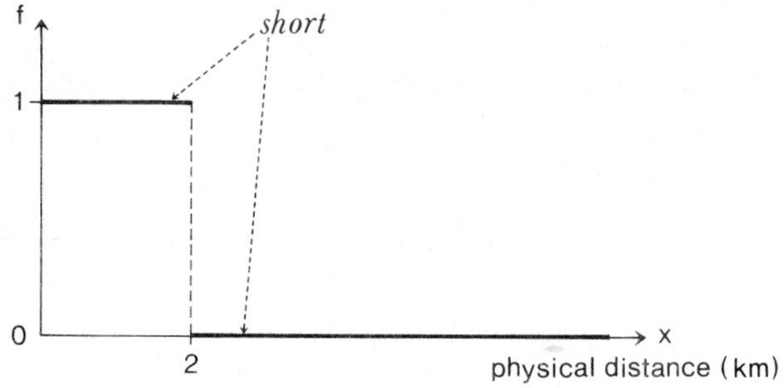

Fig. 1.1 Characteristic function of "*short*" under the two-valued logic

in modelling imprecise concepts and processes. The more we exclude fuzziness, the less realistic classical mathematics are in modelling real world problems.

Imprecision in our language, perception, thought and reasoning processes has long been discussed in human history. The Greeks posed the paradoxes of *sōrites* (the heap) and *falakros* (the bald man). To make the long story short, the problem of *sōrites* asks the question: "How many seeds constitute a heap?", and the problem of *falakros* asks the question: "How many strands of hair should there be for a man to be considered as *bald*?". Since the transition from not being a heap to being a heap, and from not being bald to being bald is gradual, the paradoxes then pose the problem of where to draw the boundary. A problem to which a natural and satisfactory answer cannot be provided by the two-valued logic.

In recent decades, the problem of imprecision has also received attention in philosophical and psychological research (see for example Peirce, 1902; Rusell, 1923; Black, 1937, 1963, 1970; Copilowish, 1939; Hempel, 1939; and Korner, 1957). However, a systematic theory had not been advanced until Zadeh (1965) proposed a mathematical concept named fuzzy sets.

Within the framework of fuzzy set theory, the term "*short*" can be conceptualized as a fuzzy subset with imprecise boundary. The transition from membership to nonmembership is gradual rather than abrupt. Its characteristic function should exhibit such a gradation as depicted in Fig. 1.2. The basic idea is that the degree of belongingness to the term "*short*" should only change slightly if the physical distance increases or decreases slightly. That is, a slight change in physical distance should not lead to an abrupt change in class, such as "*short*". Such a conceptualization is a more natural reflection of imprecision intrinsically embedded in the fuzzy term.

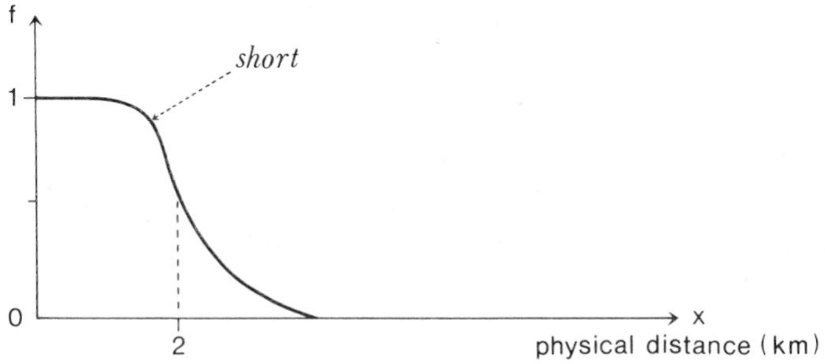

Fig. 1.2 Characteristic function of "*short*" under the concept of a fuzzy subset

Fuzzy set theory is thus a mathematical theory for imprecision. As long as there is imprecision, there remains a margin of uncertainty. The theory would certainly contribute to a more comprehensive modelling of uncertainty in spatial systems. Fig. 1.3 depicts where in spatial systems research would fuzzy set theory be resourceful. When the system is under certainty, classical mathematics are instrumental in analysis and model formulation. When the system is under uncertainty and is due to randomness, probability theory is a useful method of analysis. When the system is under uncertainty and is due to imprecision, fuzzy set theory appears to be an appropriate analytical framework. In some situations, we may need both.

While spatial systems under certainty and randomization related uncertainty have been extensively studied over the

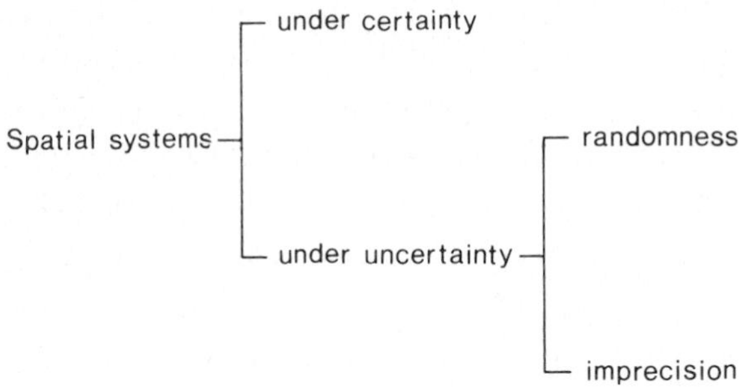

Fig. 1.3 A classification of spatial systems

years, uncertainty due to imprecision, though overwhelming, has been largely unexplored, misdirected, or evaded (see Gale 1972; Leung 1983c, 1985b for some general discussions). Since fuzzy set theory is still in its early stage of development, suspicion, misunderstanding, and ambivalence towards the approach is common. Unfamilarity with the theory and incomplete probing of its usefulness in spatial systems research are perhaps the major contributing factors.

1.3. PURPOSE AND STRUCTURE OF THE MONOGRAPH

The purpose of this monograph is to make an attempt to piece together fragmentary research and to lay a foundation of fuzzy set approach to spatial analysis and planning under imprecision. Emphasis is placed on its theoretical arguments and empirical relevance. It is demonstrated that analytical elegance can still be achieved in the face of imprecision. I hope that this modest effort can put the approach in perspectives and stimulate further research.

Since the word "fuzzy" may carry a negative connotation, a word of caution should be made here. Fuzzy set theory is a mathematical theory which deals with sets having fuzzy boundaries. There are no implications that the mathematical system is fuzzy and there are no intentions to make precise systems imprecise. It is, however, intended to make mathematics work in the sphere of imprecision. By the same token, when I say, for example, fuzzy decisionmaking or fuzzy optimization, I do not mean that the decision or optimization process is necessarily fuzzy. I mean that rational decisionmaking or optimization with various degrees of imprecision is performed with fuzzy information, performance criteria, and constraints. In general, we want to defuzzify uncalled for fuzziness but to treat intrinsic fuzziness with formality.

To make the presentation less abstract, fuzzy set theoretic analysis via a higher level of generalization, e.g. the theory of categories, is avoided. To maintain the rigor of analysis but to lessen the burden of formalism in the text, proofs of the basic theoretic results are given in the appendices. Numerical examples are employed to illustrate the theoretical counterparts and applications are provided to show their empirical relevance. Mathematically sophisticated readers may find the mathematical arguments a little superfluous. I hope the content and style of presentation are useful to spatial scientists and planners in general.

To facilitate our discussion, I first outline the plan for the remainder of the monograph. Chapter 2 is prepared for the purpose of providing a basic working knowledge of fuzzy set theory. Fundamental concepts and operations of fuzzy subsets are discussed. Connections with ordinary set theory are drawn wherever appropriate.

In chapter 3, regional concepts and regionalization, a basis of spatial analysis and planning, are first investigated within the framework of imprecision The two-valued logic of spatial classification is challenged and a new philosophical and mathematical framework is constructed. Concept of a region, core of a region, and boundary of a region are defined. Methods of regionalization are also proposed. It is demonstrated that the traditional framework does not fully capture the imprecise and valuative aspects in regional classification and regionalization. The fuzzy set framework turns out to be more natural, informative, and realistic. Furthermore, it embodies the traditional framework as a special case.

In chapter 4, I focus the discussion on the imprecision of human perception, preference, utility, and choice over space. Differing from the conventional conceptualization, the preference-utility-choice structure is treated as imprecise. Individuals' behavior is built on such premises. Producers' and consumers' spatial partial equilibriums are formulated as an optimization problem involving fuzzy utility and fuzzy technological or budget constraint. Spatial general equilibrium under imprecision is also scrutinized.

Moving away from the individual level, decisionmaking in the context of systems planning is discussed in chapter 5. Emphasis is placed on the single objective fuzzy optimization problem. Notions of a fuzzy objective, fuzzy constraint, fuzzy decision space, and fuzzy optimization are proposed. Situations involving a precise objecive and fuzzy constraints, a fuzzy objective and precise constraints, and a fuzzy objective and fuzzy constraints are analyzed. Robust programming involving fuzzy coefficients is also discussed. Spatial planning within the fuzzy optimization framework is demonstrated to be more realistic, flexible, and algorithmically solvable.

To have a better approximation to the complex decision-making environment, fuzzy mathematical programming is extended to solve multiobjective spatial planning problems in chapter 6. Conflict resolutions involving precise objectives and precise constraints, fuzzy objectives and precise constraints, fuzzy objectives and fuzzy constraints, and precise objectives and fuzzy constraints are scrutinized. Imprecision over objective and constraint specifications turns out to be instrumental in the conflict resolution process.

In chapter 7, dynamics and controls of fuzzy spatial systems are discussed. A general framework of fuzzy dynamic spatial systems is formulated. To enable the analysis of optimization over time, multistage spatial decision problems are investigated via a discrete-time fuzzy optimal control process. The chapter is rounded up with a comment on various analytical aspects of fuzzy dynamic spatial systems.

To conclude our discussion, a critical evaluation of

spatial analysis and planning under imprecision is made in chapter 8. Empirical issues of the fuzzy set concepts are discussed. Directions for further research are proposed.

Though the monograph is intended to be read as a whole, for readers who are interested in specific areas of the monograph, the following are suggested combinations for reading:

(a) chapters 1, 2, 3, and 8 (for regionalization problems);
(b) chapters 1, 2, 4, and 8 (for spatial economics problems);
(c) chapters 1, 2, 5, and 8 (for single objective spatial planning problems);
(d) chapters 1, 2, 5, 6, and 8 (for single and multiple objective spatial planning problems);
(e) chapters 1, 2, 7, and 8 (for fuzzy spatial systems and optimal control problems);
(f) the whole monograph.

CHAPTER 2

BASIC CONCEPTS OF FUZZY SET THEORY

2.1. INTRODUCTION

The purpose of this chapter is to provide a basic knowledge of fuzzy set theory which is relevant to the discussions in the subsequent chapters. Due to the rapid development in its still early stage, new concepts and operations are bound to surface in the future. Therefore, the word "basic" here does not imply an exhaustive account of all fundamental concepts of fuzzy set theory. My intention is to make the discussion in this monograph self-contained as much as possible.

To include more relevant items, rigor is traded off for quantity to a certain extent. I try to skimp on theorems but to enlist more concepts. Concepts related to specific spatial context are introduced in relevant places in other chapters. A more extensive exposure of the theory can be found in Kaufmann (1975), Negoita and Ralescu (1975), Dubois and Prade (1980), Wang (1983), and Zimmermann (1985).

To relate fuzzy set theory to ordinary set theory, some basic notions of ordinary set theory are briefly reviewed in section 2.2. Concepts and operations of fuzzy subsets are then examined in section 2.3. In section 2.4, our discussion is extended to a notion of a linguistic variable and a theory of possibility. To differentiate fuzzy set theory and probability theory, differences between the two approaches are investigated in section 2.5.

To distinguish symbolically between ordinary sets and fuzzy sets, the lower bar under capital letters is employed to denote ordinary sets throughout. For example, $\underline{A}$, indicates an ordinary set and, A, indicates a fuzzy set.

2.2. A BRIEF REVIEW OF ORDINARY SET THEORY

Roughly speaking, a set is a collection of certain well-defined objects of our perception or thought. The word "well-defined" means that it is possible to tell without ambiguity an object from a non-object.

A set $\underline{A}$ is defined on a universe of discourse $\underline{X}$. Let x be an element of $\underline{X}$. If x is an element of $\underline{A}$, we write $x \in \underline{A}$. If x is not an element of $\underline{A}$, we write $x \notin \underline{A}$. Therefore, to determine $\underline{A}$ in $\underline{X}$, we only need to decide whether $x \in \underline{A}$ or $x \notin \underline{A}$ for any $x \in \underline{X}$.

In general, a set can be defined by listing its elements as

$$\underline{A} = \{x_1, x_2, \ldots, x_n\}. \tag{2-1}$$

For example, $\underline{A} = \{a, e, i, o, u\}$ is the set of all vowel sound letters of the English alphabet. That is, we list all elements in $\underline{X}$ which satisfies the condition $x \in \underline{A}$. Alternatively, especially when a set is infinite, we can define a set by a given proposition P(x) which may be true or false. A set $\underline{A}$ then contains all elements in $\underline{X}$ for which P(x) is true. We denote it by

$$\underline{A} = \{x | P(x)\}. \tag{2-2}$$

For example, the set of rational numbers between 0 and 1 can be denoted as $\underline{A} = \{x | x \text{ is rational and } 0 < x < 1\}$. A set which contains no elements is called an empty set, denoted as ϕ.

Let $\underline{A}$, $\underline{B}$ be two sets in $\underline{X}$, the set $\underline{A}$ is said to be a subset of $\underline{B}$, we say $\underline{A}$ is included in $\underline{B}$ and denote it as $\underline{A} \subseteq \underline{B}$, if all elements of $\underline{A}$ are also elements of $\underline{B}$, i.e.

$$\underline{A} \subseteq \underline{B} \iff (x \in \underline{A} \Longrightarrow x \in \underline{B}). \tag{2-3}$$

Thus, the set $\underline{A}$ in $\underline{X}$ can be denoted as $\underline{A} \subseteq \underline{X}$. A proper subset of $\underline{B}$, denoted as $\underline{A} \subset \underline{B}$, means that there exists at least one $x \in \underline{B}$ but $x \notin \underline{A}$.

Two sets $\underline{A}$ and $\underline{B}$ are equal, denoted as $\underline{A} = \underline{B}$, if and only if $\underline{A}$ is included in $\underline{B}$ and $\underline{B}$ is included in $\underline{A}$, i.e.

$$\underline{A} = \underline{B} \iff \underline{A} \subseteq \underline{B} \text{ and } \underline{B} \subseteq \underline{A}. \tag{2-4}$$

Let $\mathscr{C}(\underline{X})$ be the power set of $\underline{X}$ whose elements are sets in $\underline{X}$. Then the set $\underline{A}$ in $\underline{X}$ can be denoted as $\underline{A} \subset \underline{X}$ or $\underline{A} \in \mathscr{C}(\underline{X})$.

Let $\underline{A}, \underline{B} \in \mathscr{C}(\underline{X})$, then the union of $\underline{A}$ and $\underline{B}$, denoted as $\underline{A} \cup \underline{B}$, is defined by

$$\underline{A} \cup \underline{B} = \{x | x \in \underline{A} \text{ or } x \in \underline{B}\}, \tag{2-5}$$

or in the logical symbol "or", $\vee$,

$$\underline{A} \cup \underline{B} = \{x | x \in \underline{A} \vee x \in \underline{B}\}. \tag{2-6}$$

The intersection of $\underline{A}$ and $\underline{B}$, denoted as $\underline{A} \cap \underline{B}$, is defined by

$$\underline{A} \cap \underline{B} = \{x | x \in \underline{A} \text{ and } x \in \underline{B}\}, \tag{2-7}$$

or in the logical symbol "and", $\wedge$,

$$\underline{A} \cap \underline{B} = \{x | x \in \underline{A} \wedge x \in \underline{B}\}. \tag{2-8}$$

The complement of a set $\underline{A}$, denoted as $\overline{\underline{A}}$, is defined by

$$\overline{\underline{A}} = \{x | x \notin \underline{A}\}. \tag{2-9}$$

Since the complement of a set is always defined with respect to the universe of discourse, then $\overline{\underline{A}}$ should technically be expressed as $\overline{\underline{A}}_{\underline{X}}$. I, however, choose the former expression for simplicity purpose.

The difference of sets $\underline{A}$ and $\underline{B}$, denoted as $\underline{A} - \underline{B}$, is defined by

$$\underline{A} - \underline{B} = \underline{A} \cap \overline{\underline{B}} = \{x | x \in \underline{A} \text{ and } x \in \overline{\underline{B}}\}. \tag{2-10}$$

Let $\underline{A}, \underline{B}, \underline{C} \in \mathscr{C}(\underline{X})$, with respect to the operations $\cup$, $\cap$, and $^{-}$, the fundamental algebraic properties are:

(a) Law of idempotence

$$\underline{A} \cup \underline{A} = \underline{A}, \quad \underline{A} \cap \underline{A} = \underline{A}; \tag{2-11}$$

(b) Law of commutativity

$$\underline{A} \cup \underline{B} = \underline{B} \cup \underline{A}, \quad \underline{A} \cap \underline{B} = \underline{B} \cap \underline{A}; \tag{2-12}$$

(c) Law of associativity

$$(\underline{A} \cup \underline{B}) \cup \underline{C} = \underline{A} \cup (\underline{B} \cup \underline{C}),$$

$$(\underline{A} \cap \underline{B}) \cap \underline{C} = \underline{A} \cap (\underline{B} \cap \underline{C}); \qquad (2\text{-}13)$$

(d) Law of distributivity

$$\underline{A} \cup (\underline{B} \cap \underline{C}) = (\underline{A} \cup \underline{B}) \cap (\underline{A} \cup \underline{C}),$$

$$\underline{A} \cap (\underline{B} \cup \underline{C}) = (\underline{A} \cap \underline{B}) \cup (\underline{A} \cap \underline{C}); \qquad (2\text{-}14)$$

(e) Law of two extremes

$$\underline{A} \cup \phi = \underline{A}, \ \underline{A} \cap \underline{X} = \underline{A} \text{ (identity)},$$

$$\underline{A} \cup \underline{X} = \underline{X}, \ \underline{A} \cap \phi = \phi; \qquad (2\text{-}15)$$

(f) Law of the excluded-middle

$$\underline{A} \cup \overline{\underline{A}} = \underline{X}, \ \underline{A} \cap \overline{\underline{A}} = \phi; \qquad (2\text{-}16)$$

(g) Law of involution

$$\overline{\overline{\underline{A}}} = \underline{A}; \qquad (2\text{-}17)$$

(h) De Morgan's Laws

$$\overline{\underline{A} \cup \underline{B}} = \overline{\underline{A}} \cap \overline{\underline{B}},$$

$$\overline{\underline{A} \cap \underline{B}} = \overline{\underline{A}} \cup \overline{\underline{B}}. \qquad (2\text{-}18)$$

Let $\underline{A}$, $\underline{B} \in \mathscr{C}(\underline{X})$. The Cartesian product of $\underline{A}$ and $\underline{B}$, denoted as $\underline{A} \times \underline{B}$, is the set of all ordered pairs (x, y) defined by

$$\underline{A} \times \underline{B} = \{(x, y) \mid x \in \underline{A} \text{ and } y \in \underline{B}\}. \qquad (2\text{-}19)$$

The most familiar example of a Cartesian product is the Cartesian coordinates, $\mathbb{R} \times \mathbb{R}$, in the plane of analytic geometry.

In addition to the characterizations in (2-1) and (2-2), the concept of a set can be formulated through the idea of a function. Let $\underline{A} \in \mathscr{C}(\underline{X})$. It, in fact, determines a mapping f from $\underline{X}$ into $\{0, 1\}$ which maps $x \in X$ into $f_{\underline{A}}(x)$ in $\{0, 1\}$, i.e.

$$f: X \longrightarrow \{0, 1\},$$

$$x \longrightarrow f_{\underline{A}}(x). \qquad (2\text{-}20)$$

The mapping f is called the characteristic function of $\underline{A}$ and is defined by

$$f_{\underline{A}}(x) = \begin{cases} 1, & \text{if } x \in \underline{A}, \\ 0, & \text{if } x \notin \underline{A}. \end{cases} \tag{2-21}$$

The value $f_{\underline{A}}(x)$ can be interpreted as the degree of belonging of x to $\underline{A}$. If $x \in \underline{A}$, then its degree of belonging to $\underline{A}$ is 1, a 100%. If $x \notin A$, then its degree of belonging to A is 0, a 0%. Therefore, no ambiguity exists under the concept of an ordinary set. We are certain whether or not an element belongs to a set. Thus, a set $\underline{A}$ can be defined as a set of ordered pairs:

$$\underline{A} = \{x, f_{\underline{A}}(x)\}, \ \forall \ x \in \underline{X}, \tag{2-22}$$

with $f_{\underline{A}}$ being its defining function and $f_{\underline{A}}(x) = 1$.

Based on the idea of a characteristic function, the operations $\cup$, $\cap$, and $^-$ satisfy respectively the following conditions:

$$f_{\underline{A} \cup \underline{B}}(x) = \max[f_{\underline{A}}(x), f_{\underline{B}}(x)], \tag{2-23}$$

$$f_{\underline{A} \cap \underline{B}}(x) = \min[f_{\underline{A}}(x), f_{\underline{B}}(x)], \tag{2-24}$$

$$f_{\bar{\underline{A}}}(x) = 1 - f_{\underline{A}}(x). \tag{2-25}$$

Therefore, the notion of an ordinary set is for characterizing well-defined concepts whose elements can be precisely determined. In section 2.3, the concept of a fuzzy subset can be seen as a generalization of the concept of an ordinary set and is constructed for the purpose of characterizing imprecise concepts.

2.3. CONCEPTS AND OPERATIONS OF FUZZY SUBSETS

Our thought and perception ordinarily involve concepts whose meanings are imprecise. If such concepts are defined as sets, it is difficult if not impossible to determine whether or not an object belongs to a set. Nevertheless, it is possible to say to what degree an object belongs to a set. The concept of a set described in section 2.2, henceforth referred to as ordinary set, is thus inappropriate and the concept of a fuzzy subset appears to be suitable.

In subsection 2.3.1, fundamental concepts of a fuzzy subset are introduced. Basic operations are then discussed in subsection 2.3.2. Decompositions of fuzzy subsets and the principles of extention, a bridge between ordinary sets and and fuzzy subsets, are examined in subsections 2.3.3 and 2.3.4 respectively. In subsection 2.3.5, concepts of and operatioins on fuzzy relations are presented.

2.3.1. The Concept of a Fuzzy Subset

Definition 2.1. Let $\underline{X}$ be a universe of discourse. Let $x \in \underline{X}$. Then a fuzzy subset A in $\underline{X}$ is a set of ordered pairs

$$\{[x, \mu_A(x)]\}, \quad \forall\, x \in \underline{X}, \tag{2-26}$$

where $\mu_A: \underline{X} \longrightarrow \underline{M}$ is a membership function which maps $x \in \underline{X}$ into $\mu_A(x)$ in a totally ordered set $\underline{M}$, called the membership set, and $\mu_A(x)$ indicates the grade of membership (degree of belonging) of x in A. The membership set $\underline{M}$ can take on a general structure such as a lattice. Throughout this monograph, the membership set is restricted to the closed interval [0, 1].

A is called a fuzzy subset because the universe of discourse (the reference set) is always an ordinary set of well-defined objects. It is the subset which may be fuzzy.

Employing the concepts in (2-20) and (2-21), a fuzzy subset is completely determined by its membership function. If $\mu_A(x) = 1$, then x completely belongs to A. If $\mu_A(x) = 0$, then x completely does not belong to A. Thus, no imprecision is involved in both cases. However, for $0 < \mu_A(x) < 1$, x belongs to A only to a certain degree. We are ambiguous in determining whether or not x belongs to A. Nonetheless, the closer $\mu_A(x)$ is to 1, the more likely it is considered to be an element of A. The physical meaning is that a gradual instead of an abrupt transition from membership to nonmembership should be experienced when imprecise concepts are encountered.

When the membership set is restricted to the set $\{0, 1\}$,

μ_A becomes the characteristic function $f_{\underline{A}}$ in (2-20) and (2-21) and A becomes an ordinary set. Therefore, an ordinary set is only a special case of a fuzzy subset.

When $\underline{X}$ is countable, say $\underline{X} = \{x_1, x_2, \ldots, x_n\}$, the following formats are equivalent in expressing a fuzzy subset:

(a) as union of fuzzy singletons

$$A = \sum_{i=1}^{n} \mu_A(x_i)/x_i$$

$$= \mu_A(x_1)/x_1 + \mu_A(x_2)/x_2 + \ldots + \mu_A(x_n)/x_n, \qquad (2\text{-}27)$$

where "+", i.e. "$\sum$", stands for the union, not "sum", of the fuzzy singletons $\mu_A(x_i)/x_i$, and "/" denotes that $\mu_A(x_i)$ and x_i are related;

(b) as ordered pairs

$$A = \{[x_1, \mu_A(x_1)], [x_2, \mu_A(x_2)], \ldots, [x_n, \mu_A(x_n)]\}; \qquad (2\text{-}28)$$

(c) as finite dimensional vectors

$$A = [\mu_A(x_1), \mu_A(x_2), \ldots, \mu_A(x_n)]. \qquad (2\text{-}29)$$

When $\underline{X}$ is a continuum, A can be expressed as

$$A = \int_{\underline{X}} \mu_A(x)/x. \qquad (2\text{-}30)$$

Again, "$\int$" denotes union, not "integration".

Of course, a fuzzy subset in both cases can always be represented by its membership function μ_A, if available.

Example 2.1. Let X = {5, 11, 20, 50, 100} be a collection of physical distances measured in kilometers. Let A $\underline{\underline{\Delta}}$ "*approximately* 10 kilometers". (The symbol "$\underline{\underline{\Delta}}$" means defined as throughout). Then A is a fuzzy subset which may be expressed as:

A = 0.6/5 + 0.9/11 + 0.3/20 + 0.1/50 + 0/100, by (2-27),

or

$A = \{(5, 0.6), (11, 0.9), (20, 0.3), (50, 0.1), (100, 0)\}$,
by (2-28),

or

$A = (0.6, 0.9, 0.3, 0.1, 0)$, by (2-29).

The grades of membership here are determined subjectively. Hereupon, for simplicity of presentation, the format in (2-29), unless stated otherwise, will be employed to write a fuzzy subset in a finite universe of discourse.

Example 2.2. Let $\underline{X} = [0, \infty)$ be a measure of physical distance. Let $A \triangleq$ "*long*" and is defined by

$$\mu_{long}(x) = \begin{cases} 0, & \text{if } x \leq a, \\ 1 - e^{-\left(\frac{x - a}{a}\right)^2}, & \text{if } x \geq a. \end{cases} \tag{2-31}$$

Then, by (2-30), "*long*" can be expressed as

$$= \int_{x \leq a} 0/x + \int_{x \geq a} \left[1 - e^{-\left(\frac{x - a}{a}\right)^2}\right] / x. \tag{2-32}$$

(see Fig. 2.1)

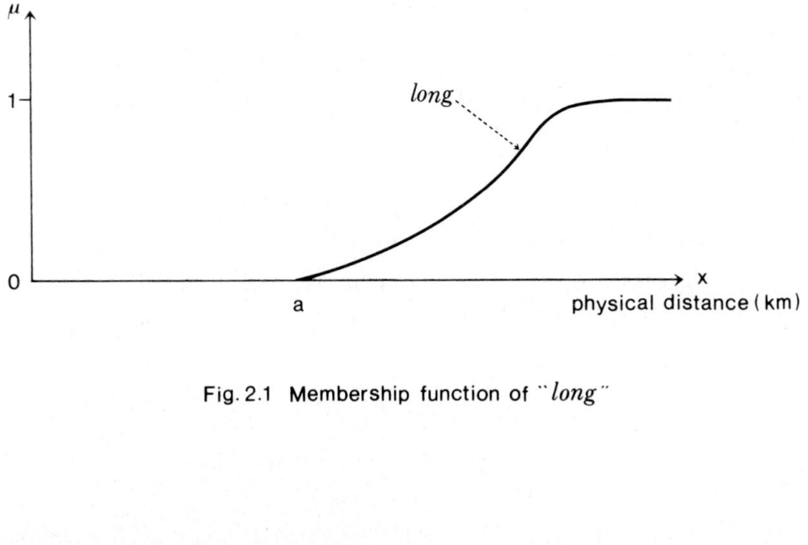

Fig. 2.1 Membership function of "*long*"

Definition 2.2. A fuzzy subset is empty, denoted as ϕ, if $\mu_A(x) = 0, \ \forall\, x \in \underline{X}$.

Definition 2.3. A fuzzy subset A is included in a fuzzy subset B, denoted as $A \subseteq B$, if and only if $\mu_A(x) \leq \mu_B(x), \ \forall\, x \in \underline{X}$.

Definition 2.4. Fuzzy subsets A and B are equal, denoted as $A = B$, if and only if $\mu_A(x) = \mu_B(x), \ \forall\, x \in \underline{X}$.

Definition 2.5. The support of a fuzzy subset A is the ordinary subset of $\underline{X}$ defined by

$$\text{supp}(A) = \{x \mid \mu_A(x) > 0\}. \tag{2-33}$$

Definition 2.6. The height of a fuzzy subset A is the least upper bound of A defined by

$$\text{hgt}(A) = \sup_x \mu_A(x). \tag{2-34}$$

Definition 2.7. A fuzzy subset A is said to be normalized if there exists an $x \in \underline{X}$ such that $\mu_A(x) = 1$.

2.3.2. Fundamental Operations on Fuzzy Subsets

Parallel to the theory of ordinary sets, basic operations such as union, $\cup$, intersection, $\cap$, and complementation, $^{-}$, on fuzzy subsets can also be defined.

Let

$$\mathscr{F}(\underline{X}) = \{A \mid \mu_A : \underline{X} \longrightarrow [0, 1]\} \tag{2-35}$$

be the fuzzy power set of $\underline{X}$ whose elements are fuzzy subsets. The operations $\cup$, $\cap$, and $^{-}$ on $\underline{X}$ are defined in definitions 2.8, 2.9, and 2.10 respectively.

Definition 2.8. Let $A, B \in \mathscr{F}(\underline{X})$. The union of A and B, denoted as $A \cup B$, is defined by

$$\mu_{A \cup B}(x) = \max[\mu_A(x), \mu_B(x)], \quad \forall\, x \in \underline{X}, \qquad (2\text{-}36)$$

or in the logical symbol "or", $\vee$,

$$\mu_{A \cup B}(x) = \mu_A(x) \vee \mu_B(x), \quad \forall\, x \in \underline{X}. \qquad (2\text{-}37)$$

Definition 2.9. Let $A, B \in \mathscr{F}(\underline{X})$. The intersection of A and B, denoted as $A \cap B$, is defined by

$$\mu_{A \cap B}(x) = \min[\mu_A(x), \mu_B(x)], \quad \forall\, x \in \underline{X}, \qquad (2\text{-}38)$$

or in the logical symbol "and", $\wedge$,

$$\mu_{A \cap B}(x) = \mu_A(x) \wedge \mu_B(x), \quad \forall\, x \in \underline{X}. \qquad (2\text{-}39)$$

Definition 2.10. Let $A \in \mathscr{F}(\underline{X})$. The complement of A, denoted as $\overline{A}$, is defined by

$$\mu_{\overline{A}}(x) = 1 - \mu_A(x), \quad \forall\, x \in \underline{X}. \qquad (2\text{-}40)$$

When $M = \{0, 1\}$, Definitions 2.8, 2.9, and 2.10 become the union, intersection, and complementation of ordinary sets and satisfy respectively (2-23), (2-24), and (2-25).

Definition 2.11. Let $A, B \in \mathscr{F}(\underline{X})$. The difference of A and B, denoted as $A - B$, is defined by

$$\mu_{A - B}(x) = \mu_{A \cap \overline{B}}(x) = \min[\mu_A(x), \mu_{\overline{B}}(x)], \quad \forall\, x \in \underline{X}. \qquad (2\text{-}41)$$

Example 2.3. Let $\underline{X} = \{x_1, x_2, x_3, x_4, x_5\}$. Let $A = (1, 0.9, 0.7, 0.5, 0.4)$ and $B = (0, 0.3, 0.6, 0.8, 1)$. Then

$$A \cup B = (1, 0.9, 0.7, 0.8, 1),$$

$$A \cap B = (0, 0.3, 0.6, 0.5, 0.4),$$

$$\overline{A} = (0, 0.1, 0.3, 0.5, 0.6).$$

Example 2.4. Let $\underline{X} = [0, 24]$ be a measure of time in hours.

Let A $\triangleq$ "*a lot longer than* 5 hours", and B $\triangleq$ "*much shorter than* 8 hours", and are defined respectively by the following membership functions:

$$\mu_A(x) = \begin{cases} 0, & \text{if } x \leq 5, \\ \dfrac{x-5}{4}, & \text{if } 5 < x \leq 9, \\ 1, & \text{if } x > 9; \end{cases} \tag{2-42}$$

$$\mu_B(x) = \begin{cases} 1, & \text{if } x \leq 6, \\ \dfrac{8-x}{2}, & \text{if } 6 < x \leq 8, \\ 0, & \text{if } x > 8. \end{cases} \tag{2-43}$$

Then, $A \cup B \triangleq$ "*a lot longer than* 5 hours" or "*much shorter than* 8 hours", $A \cap B \triangleq$ "*a lot longer than* 5 hours" and "*much shorter than* 8 hours", and $\bar{A} \triangleq$ "*not a lot longer than* 5 hours", are defined respectively by the following membership functions:

$$\mu_{A \cup B}(x) = \begin{cases} 1, & \text{if } x \leq 6, \text{ or } x > 9, \\ \dfrac{8-x}{2}, & \text{if } 6 < x \leq 7, \\ \dfrac{x-5}{4}, & \text{if } 7 < x \leq 9; \end{cases} \tag{2-44}$$

$$\mu_{A \cap B}(x) = \begin{cases} 0, & \text{if } x \leq 5, \text{ or } x > 8, \\ \dfrac{x-5}{4}, & \text{if } 5 < x \leq 7, \\ \dfrac{8-x}{2}, & \text{if } 7 < x \leq 8; \end{cases} \tag{2-45}$$

$$\mu_{\bar{A}}(x) = \begin{cases} 1, & \text{if } x \leq 5, \\ \dfrac{9-x}{4}, & \text{if } 5 < x \leq 9, \\ 0, & \text{if } x > 9. \end{cases} \tag{2-46}$$

(see Fig. 2.2a, b, and c).

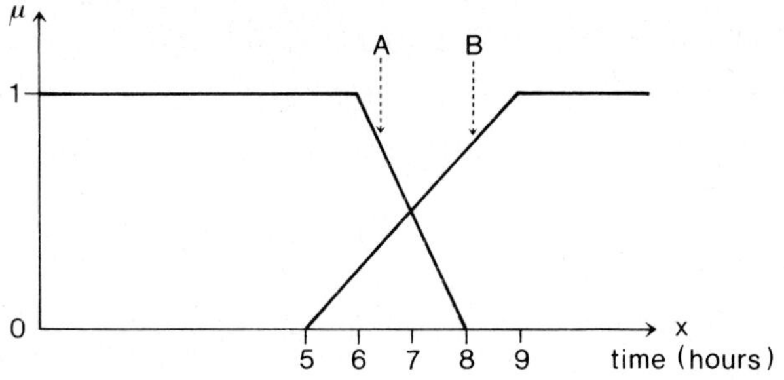

Fig. 2.2a Membership functions of A≙"*a lot longer than* 5 hours" and B≙"*much shorter than* 8 hours"

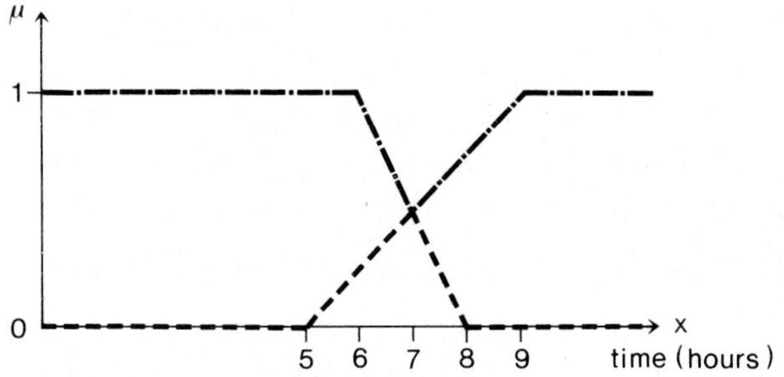

Fig. 2.2b Membership functions of A∪B (—·—·—·—) and A∩B (- - - - - - -)

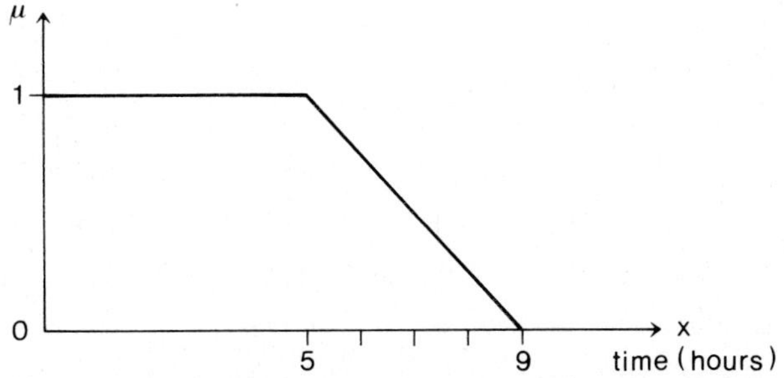

Fig. 2.2c Membership function of $\overline{A}$≙"*not a lot longer than* 5 hours"

In defining disjunction and conjunction of fuzzy subsets, "max" and "min" are not the only operators. Bellman and Giertz (1973), however, show that they are the only operators which satisfy the following conditions:

(a) The membership value of x in a compound fuzzy subset depends only on the membership values of x in the attributing elementary fuzzy subsets. That is

$$\mu_{A \cup B}(x) = f(\mu_A(x), \mu_B(x)), \quad \forall x \in \underline{X},$$

$$\mu_{A \cap B}(x) = g(\mu_A(x), \mu_B(x)), \quad \forall x \in \underline{X};$$

(b) f and g are commutative, associative, and mutually distributive operators;

(c) f and g are continuous and non-decreasing in each of their arguments. That is, $\mu_{A \cup B}(x)$ or $\mu_{A \cap B}(x)$ cannot decrease if $\mu_A(x)$ or $\mu_B(x)$ increases, and a small increase of $\mu_A(x)$ or $\mu_B(x)$ cannot induce large increase of $\mu_{A \cup B}(x)$ or $\mu_{A \cap B}(x)$;

(d) f(x, x) and g(x, x) are strictly increasing. That is, if $\mu_A(x_1) = \mu_B(x_1) > \mu_A(x_2) = \mu_B(x_2)$, then $\mu_{A \cup B}(x_1) > \mu_{A \cup B}(x_2)$ or $\mu_{A \cap B}(x_1) > \mu_{A \cap B}(x_2)$;

(e) $\mu_{A \cup B}(x) \geq \max[\mu_A(x), \mu_B(x)], \quad \forall x \in \underline{X},$
$\mu_{A \cap B}(x) \leq \min[\mu_A(x), \mu_B(x)], \quad \forall x \in \underline{X};$

(f) $\mu_A(x) = 0$ and $\mu_B(x) = 0 \Longrightarrow \mu_{A \cup B}(x) = 0$, i.e. $f(0, 0) = 0$,
$\mu_A(x) = 1$ and $\mu_B(x) = 1 \Longrightarrow \mu_{A \cap B}(x) = 1$, i.e. $g(1, 1) = 1$.

Conditions (a) to (f) are consistent and sufficient to guarantee the unique choice of "max" and "min" as the operators for union and intersection. Based on a set of slightly different assumptions, Fung and Fu (1975) also showed that "max" and "min" are the only operators. The implication then is that different operators can be defined if any of the above conditions is relaxed or a new set of conditions are imposed.

In justifying negation defined in (2-40), Bellman and Giertz (1973) impose the following conditions:

(a) $\mu_{\bar{A}}(x)$ depends only on $\mu_A(x)$, i.e. $\mu_{\bar{A}}(x) = h(\mu_A(x))$;

(b) $h(0) = 1$ and $h(1) = 0$. This is to retain the complementation properties of ordinary sets;

(c) h is continuous and strictly monotonically decreasing;

(d) h is involutive, i.e. $h(h(\mu_A(x))) = \mu_A(x)$;

(e) $\mu_A(x_1) - \mu_A(x_2) = \mu_A(x_2) - \mu_A(x_1)$. That is, a certain change in the membership value in A should have the same effect on the membership value in $\bar{A}$, or

(e') If $\mu_A(x_1) + \mu_A(x_2) = 1$, then $\mu_{\bar{A}}(x_1) + \mu_{\bar{A}}(x_2) = 1$, $\forall\, x_1, x_2 \in \underline{X}$. [A less restrictive condition suggested by Gaines (1976)].

Again, different operators can be employed if any of the conditions from (a) to (e) is relaxed.

Let union, intersection, and negation be defined by (2-36), (2-38), and (2-40) respectively. Then the operations $\cup$, $\cap$, and $^-$ retain all the algebraic properties in (2-11) to (2-18) except (2-16) in ordinary-set operations. Thus, the law of the excluded-middle no longer holds in the operations of fuzzy subsets. In general,

$$A \cup \bar{A} \neq \underline{X}, \quad A \cap \bar{A} \neq \phi. \tag{2-47}$$

Due to the fuzziness of the boundary of A and consequently $\bar{A}$, the overlapping property of A and $\bar{A}$ in (2-47) is then natural. However, when A is an ordinary set, the law of the excluded-middle in (2-16) is retained.

Remark. As discussed above, though "max" and "min" are employed to define the operations $\cup$ and $\cap$, they are by no means the only operators. The union formed by "max" however is the smallest fuzzy subset that contains both A and B, and the intersection formed by "min" is the largest fuzzy subset that is contained in both A and B. These two operators are regarded as non-interactive because a high value of one cannot compensate for a low value of another. Thus, a modification of A or B does not necessarily lead to a change in $A \cup B$ or $A \cap B$. In defining $\cup$ and $\cap$, the following are some other operators which may be employed if circumstances arise.

Definition 2.12. Let $A, B \in \mathscr{F}(\underline{X})$. The algebraic sum of A and B, denoted as $A \hat{+} B$, is defined by

$$\mu_{A \hat{+} B}(x) = \mu_A(x) + \mu_B(x) - \mu_A(x) \cdot \mu_B(x), \ \forall x \in \underline{X}. \qquad (2\text{-}48)$$

Definition 2.13. Let A, B $\in \mathscr{F}(\underline{X})$. The algebraic product of A and B, denoted as $A \cdot B$, is defined by

$$\mu_{A \cdot B}(x) = \mu_A(x) \cdot \mu_B(x), \ \forall x \in \underline{X}. \qquad (2\text{-}49)$$

Remark. $\hat{+}$ and $\cdot$ reflect trade-offs between A and B. They can be regarded as interactive operators in defining $\cup$ and $\cap$. However, $\hat{+}$ and . satisfy only the law of commutativity, associativity, two-extremes, and De Morgan's Law.

Definition 2.14. Let A, B $\in \mathscr{F}(\underline{X})$. The bounded sum of A and B, denoted as $A \boxplus B$, is defined by

$$\mu_{A \boxplus B}(x) = \min[1, \mu_A(x) + \mu_B(x)], \ \forall x \in \underline{X}. \qquad (2\text{-}50)$$

Definition 2.15. Let A, B $\in \mathscr{F}(\underline{X})$. The bounded product of A and B, denoted as $A \boxdot B$, is defined by

$$\mu_{A \boxdot B}(x) = \max[0, \mu_A(x) + \mu_B(x) - 1], \ \forall x \in \underline{X}. \qquad (2\text{-}51)$$

Remark. $\boxplus$ and $\boxdot$ can be employed to define $\cup$ and $\cap$. They satisfy the law of commutativity, associativity, two-extremes, De Morgan's Laws, and the law of the excluded-middle.

Based on the properties of "max", "min"; $\hat{+}$, $\cdot$; and $\boxplus$, $\boxdot$, for all A, B $\in \mathscr{F}(\underline{X})$, the following conditions are satisfied:

$$\cup: \max[\mu_A(x), \mu_B(x)] \le \mu_{A \hat{+} B}(x) \le \mu_{A \boxplus B}(x), \ \forall x \in \underline{X}, \qquad (2\text{-}52)$$

$$\cap: \min[\mu_A(x), \mu_B(x)] \ge \mu_{A \cdot B}(x) \le \mu_{A \boxdot B}(x), \ \forall x \in \underline{X}. \qquad (2\text{-}53)$$

In addition to "max", "min"; $\hat{+}$, $\cdot$; and $\boxplus$, $\boxdot$, the following are some other operators, presented in pairs, by which $\cup$ and $\cap$ can be defined:

(a) $$\cup: \mu_{A \overset{+}{\varepsilon} B}(x) = \frac{\mu_A(x) + \mu_B(x)}{1 + \mu_A(x) \cdot \mu_B(x)}, \quad \forall\, x \in \underline{X}, \tag{2-54}$$

$$\cap: \mu_{A \overset{\cdot}{\varepsilon} B}(x) = \frac{\mu_A(x) \cdot \mu_B(x)}{1 + (1 - \mu_A(x))(1 - \mu_B(x))}, \quad \forall\, x \in \underline{X}; \tag{2-55}$$

(b) $$\cup: \mu_{A \overset{+}{\gamma} B}(x) = \frac{\mu_{A \hat{+} B}(x) - (1 - \gamma)\mu_A(x) \cdot \mu_B(x)}{\gamma + (1 - \gamma)(1 - \mu_A(x) \cdot \mu_B(x))},$$

$$\gamma \in [0, \infty), \quad \forall\, x \in \underline{X}, \tag{2-56}$$

$$\cap: \mu_{A \overset{\cdot}{\gamma} B}(x) = \frac{\mu_A(x) \cdot \mu_B(x)}{\gamma + (1 - \gamma)(\mu_{A \hat{+} B}(x))},$$

$$\gamma \in [0, \infty), \quad \forall\, x \in \underline{X}; \tag{2-57}$$

(c) $$\cup: \mu_{A \overset{\wedge}{\nu} B}(x) = \min\left[1, (\mu_A^{\nu}(x) + \mu_B^{\nu}(x))^{\frac{1}{\nu}}\right],$$

$$\nu \in [1, \infty), \quad \forall\, x \in \underline{X}, \tag{2-58}$$

$$\cap: \mu_{A \overset{\vee}{\nu} B}(x) = 1 - \min\left\{1, \left[(1 - \mu_A(x))^{\nu} + (1 - \mu_B(x))^{\nu}\right]^{\frac{1}{\nu}}\right\},$$

$$\nu \in [1, \infty), \quad \forall\, x \in \underline{X}. \tag{2-59}$$

Of course, other operators can also be formulated to define $\cup$ and $\cap$. Nevertheless, all the above sets of operators generalize that of the ordinary set. That is, when $M = \{0, 1\}$, $\cup$ and $\cap$ thus defined become the operations in ordinary set theory.

Unlike ordinary set theory, we do not have an unique definition of union and intersection of fuzzy sets. Such a variety of operators makes the theory more flexible but in the same time more confusing in model formulations. Zimmermann (1985) proposes some criteria on which the selection of appropriate operators may be based. Some of them are however conflicting and certain degree of compromising need to be made.

(a) Axiomatic strength. Everything being equal, the less restricting the axioms are, the better are the operators.
(b) Empirical fit. The operators, besides satisfying some basic axioms, should capture systems behavior and meet empirical tests.
(c) Adaptability. To better model a variety of systems behavior, operators should be adaptable to specific context. The max- and min-operators are very rigid while those with adjustable parameters, see for example (2-58) and (2-59), are quite adaptable.
(d) Numerical efficiency. To be able to solve large scale problems, operators should be computationally efficient. Thus, max- and min-operators have a high level of numerical efficiency while those defined in (2-56) to (2-59), for example, are more cumbersome.
(e) Compensation. Depending on the situations, compensatory (interactive) operators may be necessary. Obviously, max- and min-operators are non-compensatory while most of the others are. The larger are the range of compensation, the better are the operators.
(f) Aggregating behavior. There are no definitive rules for aggregating fuzzy subsets. The algebraic product, for example, would decrease the aggregated degree of membership as additional fuzzy subsets are being aggregated.
(g) Scale. For membership information, scale can be nominal, ordinal, and interval. Different operators require different scales. Other things being equal, operators which require the lowest scale level is more functional in information gathering.

Therefore, there are no definitive criteria by which connective operators should be selected. It is impossible and unrealistic to find a so-to-speak "ideal operator" which can meet, for example, all of the above criteria. To decide on a particular set of operators, the problem to be analyzed has to be considered. We may only be able to require the selected operators to meet some of the criteria above. My opinion is max- and min-, unless believed otherwise, are natural and simple operators for defining union and intersection on the theoretical and practical basis. They are employed for discussions in the remainder of the monograph. Other operators, however, are also discussed in appropriate places. Empirical relevance of the operators are discussed in chapter 8.

2.3.3. Level Sets and Decompositions

Since a fuzzy subset A lacks a precise boundary, then the identification of its elements can be problematic. Except for those elements whose grades of membership are unity or zero, all other elements belong to a fuzzy subset to a certain degree. Therefore, to identify the ordinary subset corresponding to a fuzzy subset, we need to impose a rule for determining unambiguous membership. One way is to require the elements of the corresponding ordinary subset to possess a certain grade of membership in the fuzzy subset. The concept of α-level sets is thus formulated. It serves as an important transfer between ordinary sets and fuzzy subsets. It also plays an important role in the construction of fuzzy subsets by ordinary sets.

Definition 2.16. Let $A \in \mathscr{F}(\underline{X})$. The α-level set of A is the ordinary subset $\underline{A}_\alpha \in \mathscr{C}(\underline{X})$ such that

$$\underline{A}_\alpha = \{x \mid \mu_A(x) \geq \alpha\}, \quad \alpha \in [0, 1], \tag{2-60}$$

and is defined by the characteristic function

$$f_{\underline{A}_\alpha}(x) = \begin{cases} 1, & \text{if } \mu_A(x) \geq \alpha, \\ 0, & \text{if } \mu_A(x) < \alpha. \end{cases} \tag{2-61}$$

(see Fig. 2.3)

Example 2.5. Let $\underline{X} = \{x_1, x_2, x_3, x_4, x_5\}$. Let A = (0.7, 0.4, 0, 0.2, 1). Then $\underline{A}_1 = \{x_5\}$, $\underline{A}_{0.7} = \{x_1, x_5\}$, $\underline{A}_{0.4} = \{x_1, x_2, x_5\}$, $\underline{A}_{0.2} = \{x_1, x_2, x_4, x_5\}$, $\underline{A}_0 = \underline{X}$.

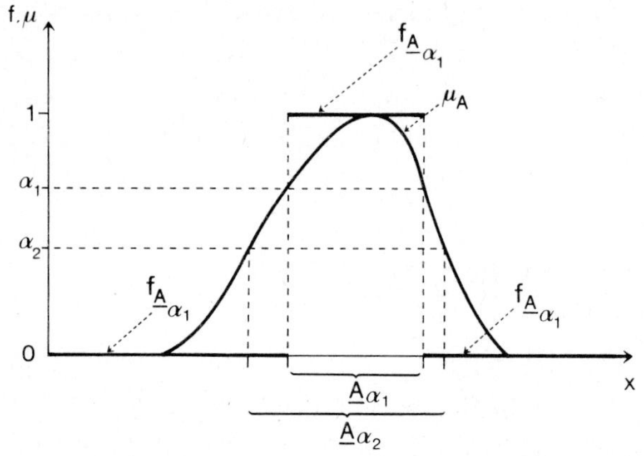

Fig. 2.3 α-level sets of a fuzzy subset A

Definition 2.17. A fuzzy subset is bounded if and only if the α-level sets are bounded for all $\alpha > 0$.

Therefore, a α-level set is in fact an ordinary subset whose elements belong to the corresponding fuzzy subset to a certain degree. The α-level sets exhibit the following basic properties:

Let A, B ε $\mathscr{F}(\underline{X})$. Then

(a) $(\underline{A \cup B})_\alpha = \underline{A}_\alpha \cup \underline{B}_\alpha$; (2-62)

(b) $(\underline{A \cap B})_\alpha = \underline{A}_\alpha \cap \underline{B}_\alpha$; (2-63)

(c) $(\underline{\bar{A}})_\alpha \neq \overline{\underline{A}_\alpha}$; (2-64)

(d) $\alpha_1 \geq \alpha_2 \Longrightarrow \underline{A}_{\alpha_1} \subseteq \underline{A}_{\alpha_2}$. (2-65)

Employing the concept of α-level sets, the connection between ordinary subsets and fuzzy subsets can be established by the following theorem:

Theorem 2.1. (Decomposition Theorem)

Let $A \in \mathscr{F}(\underline{X})$. Then

$$A = \bigcup_{\alpha \in [0,\ 1]} \alpha \cdot \underline{A}_\alpha . \qquad (2\text{-}66)$$

Theorem 2.1 states that fuzzy subset A can be decomposed into a series of α-level sets $\underline{A}_\alpha$ by which A can be reconstructed. It can be seen from Fig. 2.4 that with respect to x, μ_A is in fact the graph which connects the "sup" of $\mu_{\alpha \cdot \underline{A}_\alpha}(x)$ for all $\alpha \in [0,\ 1]$. Therefore, any fuzzy subset is a family of ordinary subsets.

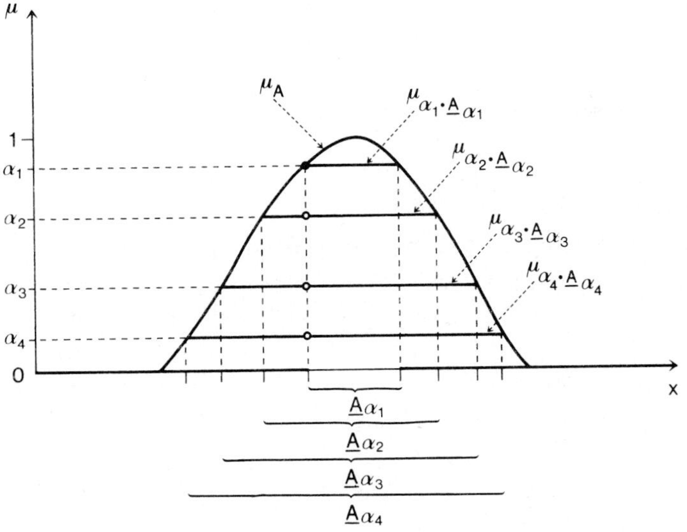

Fig. 2.4 Decomposition of a fuzzy subset A

Example 2.6. Let $\underline{X} = \{x_1, x_2, x_3, x_4, x_5\}$ and A = (0.7, 0.4, 0, 0.2, 1). Based on (2-60), (2-61), and theorem 2.1,

$$A = \bigcup_{\alpha \in [0,\ 1]} \alpha \cdot \underline{A}_\alpha = 1 \cdot \underline{A}_1 \cup 0.7 \cdot \underline{A}_{0.7} \cup 0.4 \cdot \underline{A}_{0.4} \cup 0.2 \cdot \underline{A}_{0.2} \cup 0 \cdot \underline{A}_0$$

$= 1(0, 0, 0, 0, 1) \cup 0.7(1, 0, 0, 0, 1) \cup 0.4(1, 1, 0, 0, 1) \cup 0.2(1, 1, 0, 1, 1) \cup 0(1, 1, 1, 1, 1)$

$= [\max(0.7, 0.4, 0.2), \max(0.4, 0.2), 0, 0.2, \max(1, 0.7, 0.4, 0.2)]$

$= (0.7, 0.4, 0, 0.2, 1).$

Example 2.7. Let $A \in \mathscr{F}(\underline{X})$ and is defined by

$$\mu_A(x) = 1 - \frac{1}{1 + x^2}, \quad x \in \mathbb{R}^+. \tag{2-67}$$

For $\alpha \in [0, 1]$, the characteristic function of the α-level set is

$$f_{\underline{A}_\alpha}(x) = \begin{cases} 1, & \text{if } \mu_A(x) \in [\alpha, 1], \\ 0, & \text{if } \mu_A(x) \notin [\alpha, 1]. \end{cases} \tag{2-68}$$

Therefore,

$$f_{\underline{A}_\alpha}(x) = \begin{cases} 1, & \text{if } \mu_A(x) \geq \alpha \Longrightarrow x \geq \left(\dfrac{\alpha}{1 - \alpha}\right)^{\frac{1}{2}}, \\ 0, & \text{if } \mu_A(x) < \alpha \Longrightarrow x < \left(\dfrac{\alpha}{1 - \alpha}\right)^{\frac{1}{2}}. \end{cases} \tag{2-69}$$

Thus, for all values of α in $[\alpha, 1]$, $\mu_A(x)$ can be decomposed.

The decomposition theorem can take on another format which is stated in theorem 2.2.

Theorem 2.2. Let $A \in \mathscr{F}(\underline{X})$ and is defined by μ_A. Then

$$\mu_A(x) = \sup_{\alpha \in [0, 1]} \min[\alpha, f_{\underline{A}_\alpha}(x)]. \tag{2-70}$$

2.3.4. Extension Principles and Fuzzy Numbers

In subsection 2.3.3, α-level sets and the theorem of decomposition are demonstrated to be fundamental in relating fuzzy subsets to ordinary subsets and vice versa. The extension principle to be discussed in this subsection is important in extending mathematical concepts and operations of precise systems to fuzzy systems. A natural application is to extend algebraic operations in $\mathbb{R}$ to operate on fuzzy numbers (fuzzy subsets) in $\mathbb{R}$.

To examine the extension principle of fuzzy subsets, let us review first the extension principles of ordinary sets in definitions 2.18 and 2.19.

Definition 2.18. (Extension principle of ordinary sets)

Let $f: \underline{X} \longrightarrow \underline{Y}$. Let $\mathscr{C}(\underline{X})$ and $\mathscr{C}(\underline{Y})$ be power sets in $\underline{X}$ and $\underline{Y}$ respectively. Let $\underline{A} \in \mathscr{C}(\underline{X})$ and $\underline{B} \in \mathscr{C}(\underline{Y})$. Then f induces two functions f and f^{-1} such that f induces a set $\underline{B}$ in $\underline{Y}$ from a set $\underline{A}$ in $\underline{X}$ and f^{-1} induces a set $\underline{A}$ in $\underline{X}$ from a set $\underline{B}$ in $\underline{Y}$ as follows:

$$f: \mathscr{C}(\underline{X}) \longrightarrow \mathscr{C}(\underline{Y}),$$

$$A \longrightarrow f(\underline{A}); \tag{2-71}$$

$$f^{-1}: \mathscr{C}(\underline{Y}) \longrightarrow \mathscr{C}(\underline{X}),$$

$$B \longrightarrow f^{-1}(\underline{B}); \tag{2-72}$$

where

$$f(A) = \{y \mid y = f(x),\ x \in \underline{A}\} \text{ and } f^{-1}(\underline{B}) = \{x \mid x \in \underline{X},\ f(x) \in \underline{B}\}.$$

Let $g_{\underline{A}}$, $g_{\underline{B}}$, $g_{f(\underline{A})}$, and $g_{f^{-1}(\underline{B})}$ be the characteristic functions of A, $\underline{B}$, $f(\underline{A})$, and $f^{-1}(\underline{B})$ respectively. Then

$$g_{\underline{B}}(y) = g_{f(\underline{A})}(y) = \sup_{y=f(x)} g_{\underline{A}}(x),\ \forall\, y \in \underline{Y}; \tag{2-73}$$

$$g_{\underline{A}}(x) = g_{f^{-1}(\underline{B})}(x) = g_{\underline{B}}(f(x)),\ \forall\, x \in \underline{X}. \tag{2-74}$$

Definition 2.19. (Extension principle of Cartesian products of ordinary sets)

Let $f: \underline{X} \times \underline{Y} \longrightarrow \underline{Z}$. Let $\underline{A} \in \mathscr{C}(\underline{X})$, $\underline{B} \in \mathscr{C}(\underline{Y})$, and $\underline{C} \in \mathscr{C}(\underline{Z})$. Then f induces a set $\underline{C}$ in $\mathscr{C}(\underline{Z})$ from a set $\underline{A}$ in $\mathscr{C}(\underline{X})$ and a set

$\underline{B}$ in $\mathscr{C}(\underline{Y})$ as follows:

$$f: \ \mathscr{C}(\underline{X}) \times \mathscr{C}(\underline{Y}) \longrightarrow \mathscr{C}(\underline{Z}),$$

$$\underline{A} \times \underline{B} \longrightarrow f(\underline{A} \times \underline{B}), \tag{2-75}$$

where

$$f(\underline{A} \times \underline{B}) = \{ z \mid z = f(x, y), (x, y) \ \varepsilon \ \underline{A} \times \underline{B} \}.$$

Let $g_{\underline{A}}$, $g_{\underline{B}}$, $g_{\underline{A} \times \underline{B}}$, and $g_{\underline{C}}$ be the characteristic functions of A, $\underline{B}$, $\underline{A} \times \underline{B}$, and $\underline{C}$ respectively. Then

$$g_{\underline{A} \times \underline{B}}(x, y) = \min[g_{\underline{A}}(x), g_{\underline{B}}(y)], \ \forall \, x \, \varepsilon \, \underline{X}, \ \forall \, y \, \varepsilon \, \underline{Y}; \tag{2-76}$$

$$g_{\underline{C}}(z) = g_{f(\underline{A} \times \underline{B})}(z) = \sup_{z=f(x,y)} \min[g_{\underline{A}}(x), g_{\underline{B}}(y)]. \tag{2-77}$$

The extension principles in definitons 2.18 and 2.19 are generalized respectively by their fuzzy counterparts depicted in definitions 2.20 and 2.21.

Definition 2. 20. (Extension principle of fuzzy subsets)

Let f: $\underline{X} \longrightarrow \underline{Y}$. Let $\mathscr{F}(\underline{X})$ and $\mathscr{F}(\underline{Y})$ be fuzzy power sets in $\underline{X}$ and $\underline{Y}$ respectively. Then f induces two functions f and f^{-1} such that f induces a fuzzy subset B in $\underline{Y}$ from a fuzzy subset A in $\underline{X}$ and f^{-1} induces a fuzzy subset A in $\underline{X}$ from a fuzzy subset B in $\underline{Y}$ as follows:

$$f: \ \mathscr{F}(\underline{X}) \longrightarrow \mathscr{F}(\underline{Y}),$$

$$A \longrightarrow f(A); \tag{2-78}$$

$$f^{-1}: \ \mathscr{F}(\underline{Y}) \longrightarrow \mathscr{F}(\underline{X}),$$

$$B \longrightarrow f^{-1}(B). \tag{2-79}$$

Let μ_A, μ_B, $\mu_{f(A)}$, and $\mu_{f^{-1}(B)}$ be the corresponding membership functions of A, B, f(A), and $f^{-1}(B)$. Then

$$\mu_B(y) = \mu_{f(A)}(y) = \begin{cases} \sup\limits_{y=f(x)} \mu_A(x), \\ 0, \quad \text{if } f^{-1}(y) = \phi; \end{cases} \tag{2-80}$$

$$\mu_A(x) = \mu_{f^{-1}(B)}(x) = \mu_B(f(x)), \quad \forall\, x \in f^{-1}(\underline{Y}). \qquad (2\text{-}81)$$

(see Fig. 2.5)

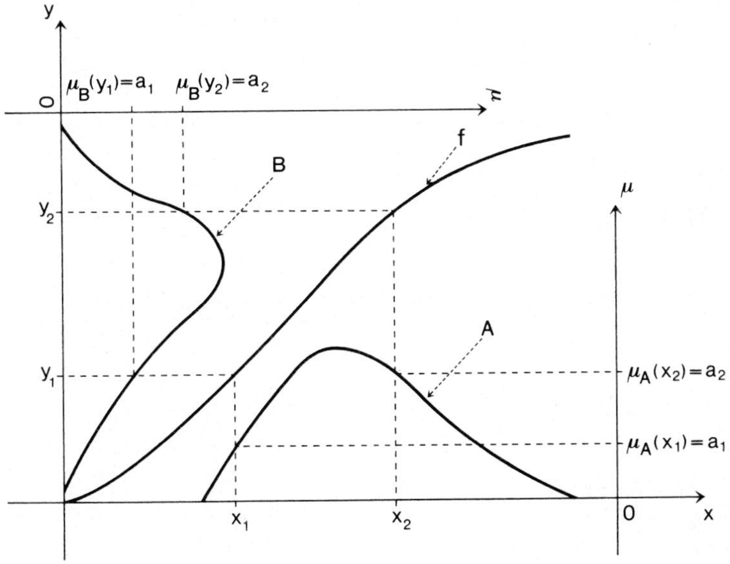

Fig. 2.5 A simple representation of the extension principle

Example 2.8. Let $\underline{X} = \{x_1, x_2, x_3, x_4, x_5, x_6\}$. Let $\underline{Y} = \{y_1, y_2, y_3, y_4\}$. Let $f: \underline{X} \longrightarrow \underline{Y}$ be defined by

$$f(x) = \begin{cases} y_1, & \text{if } x = x_1, x_3, x_4, \\ y_2, & \text{if } x = x_2, \\ y_4, & \text{if } x = x_5, x_6. \end{cases}$$

Let $A \in \mathscr{F}(\underline{X})$ be defined by

$$A = (0,\ 0.5,\ 0.2,\ 1,\ 1,\ 0.6).$$

Then by (2-80), $B = [\max(0, 0.2, 1), 0.5, 0, \max(1, 0.6)] = (1, 0.5, 0, 1)$.

(see Fig. 2.6)

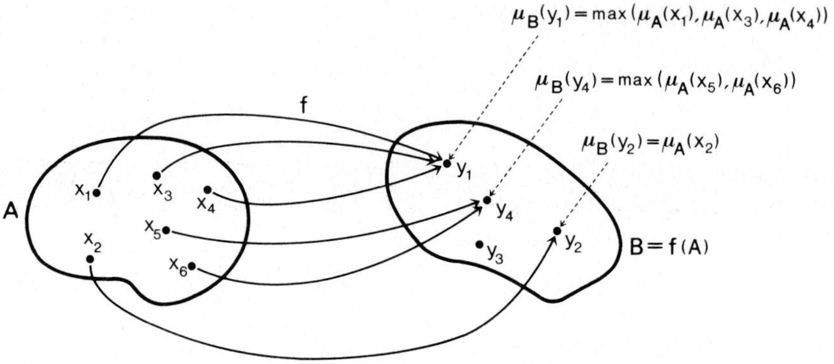

Fig. 2.6 Fuzzy subsets through the extension principle

Example 2.9. Let $\underline{X} = \{0, 1, 2, 3, 4\}$ be waiting hours. Let $\underline{Y} = \{0, 1, 4, 9, 16\}$. Let $f: \underline{X} \longrightarrow \underline{Y}$ be defined by $f(x) = x^2$. Let $A \triangleq$ "*short*" $= 1/0 + 0.5/1 + 0.3/2 + 0.1/3 + 0/4$. Then $B \triangleq$ "*very short*" $= 1/0 + 0.25/1 + 0.09/4 + 0.01/9 + 0/16$.

Definition 2.21. (Extension Principle of Cartesian products of fuzzy subsets)

Let $f: \underline{X}_1 \times \underline{X}_2 \times \ldots \times \underline{X}_n \longrightarrow \underline{Y}$. Let $A_i \in \mathscr{F}(\underline{X}_i)$, $i = 1, \ldots, n$. Let μ_{A_i}, $i = 1, \ldots, n$, and μ_B be respectively membership functions of A_i, $i = 1, \ldots, n$, and B. Let the Cartesian product of A_i's be defined by

$$\mu_{A_1 \times A_2 \times \ldots \times A_n}(x_1, x_2, \ldots, x_n)$$

$$= \min[\mu_{A_1}(x_1), \mu_{A_2}(x_2), \ldots, \mu_{A_n}(x_n)],$$

$$\forall x_1 \in \underline{X}_1, \forall x_2 \in \underline{X}_2, \ldots, \forall x_n \in \underline{X}_n. \qquad (2\text{-}82)$$

Then f induces a fuzzy subset B in Y from the fuzzy subsets A_i's in $\mathscr{F}(\underline{X}_i)$'s as follows:

$$f:\ \mathscr{F}(\underline{X}_1) \times \mathscr{F}(\underline{X}_2) \times \ldots \times \mathscr{F}(\underline{X}_n) \longrightarrow \mathscr{F}(\underline{Y}),$$

$$A_1 \times A_2 \times \ldots \times A_n \longrightarrow f(A_1 \times A_2 \times \ldots \times A_n). \qquad (2\text{-}83)$$

That is,

$$\mu_B(y) = \begin{cases} \sup\limits_{y=f(x_1, x_2, \ldots, x_3)} \min[\mu_{A_1}(x_1), \mu_{A_2}(x_2), \ldots, \mu_{A_n}(x_n)], \\ 0, \quad \text{if } f^{-1}(y) = \phi. \end{cases} \qquad (2\text{-}84)$$

Thus, B is the largest fuzzy subset of $A_1 \times A_2 \times \ldots \times A_n$ through f, and $A_1 \times A_2 \times \ldots \times A_n$ is the inverse image of B through f^{-1}.

Example 2.10. Let $\underline{X}_1 \triangleq$ distance (in km) = {10, 20, 30, 40}. Let $\underline{X}_2 \triangleq$ transportation cost (in \$) = {1, 2, 2.5, 2.8}. Let $A_1 \triangleq$ "*long*" = 0.2/10 + 0.5/20 + 0.7/30 + 0.9/40. Let $A_2 \triangleq$ "*high*" = 0.1/1 + 0.5/2 + 0.8/2.5 + 0.9/2.8. Then by (2-84),

B = $A_1 \times A_2 \triangleq$ "*long* distance" and "*high* cost" = 0.1/(10, 1) + 0.2/(10, 2) + 0.2/(10, 2.5) + 0.2/(10, 2.8) + 0.1/(20, 1) + 0.5/(20, 2) + 0.5/(20, 2.5) + 0.5/(20, 2.8) + 0.1/(30, 1) + 0.5/(30, 2) + 0.7/(30, 2.5) + 0.7/(30, 2.8) + 0.1/(40, 1) + 0.5/(40, 2) + 0.8/(40, 2.5) + 0.9/(40, 2.8).

Observe that the operator "sup" does not play any role here since there are no identical elements in $A_1 \times A_2$.

One of the applications of the extension principle in fuzzy set theory is the extension of the algebraic operations +, −, •, and ÷ in the ordinary number system to fuzzy subsets. Such an extension is made possible by theorem 2.3.

Theorem 2.3. Let $*: \underline{X} \times \underline{X} \longrightarrow \underline{X}$ be a binary operation defined on $\underline{X}$. Let $\mathscr{F}(\underline{X})$ be the fuzzy power set in $\underline{X}$. Let A, B, C $\in \mathscr{F}(\underline{X})$. Then, by the extension principle, * can be extended to operate on fuzzy subsets as

$$*:\ \mathscr{F}(\underline{X}) \times \mathscr{F}(\underline{X}) \longrightarrow \mathscr{F}(\underline{X}),$$

$$A \times B \longrightarrow C; \qquad (2\text{-}85)$$

such that

$$\mu_C(x) = \mu_{A*B}(x) = \sup_{x=x_1*x_2} \min[\mu_A(x_1), \mu_B(x_2)]. \qquad (2\text{-}86)$$

Based on theorem 2.3, *: +, −, •, and ÷ in $\mathbb{R}$ can be respectively extended to operate on fuzzy subsets in $\mathbb{R}$ as follows:

(a) Addition, ⊕ :

$$\mu_C(x) = \mu_{A \oplus B}(x) = \sup_{x=x_1+x_2} \min[\mu_A(x_1), \mu_B(x_2)]$$

$$= \sup_{x_1 \in \mathbb{R}} \min[\mu_A(x_1), \mu_B(x - x_1)]. \qquad (2\text{-}87)$$

(b) Subtraction, ⊖ :

$$\mu_C(x) = \mu_{A \ominus B}(x) = \sup_{x=x_1-x_2} \min[\mu_A(x_1), \mu_B(x_2)]$$

$$= \sup_{x_1 \in \mathbb{R}} \min[\mu_A(x_1), \mu_B(x_1 - x)]. \qquad (2\text{-}88)$$

(c) Multiplication, ⊙ :

$$\mu_C(x) = \mu_{A \odot B}(x) = \sup_{x=x_1 \cdot x_2} \min[\mu_A(x_1), \mu_B(x_2)]$$

$$= \begin{cases} \sup_{x_1 \in \mathbb{R}} \min\left[\mu_A(x_1), \mu_B\left(\frac{x}{x_1}\right)\right], & \text{if } x \neq 0, \\ \max\left\{ \sup_{x_1 \in \mathbb{R}} \min[\mu_A(x_1), \mu_B(0)], \sup_{x_2 \in \mathbb{R}} \min[\mu_A(0), \mu_B(x_2)] \right\}, & \text{if } x = 0; \end{cases}$$

(2-89)

(d) Division, ⊕ :

$$\mu_C(x) = \mu_{A \oplus B}(x) = \sup_{x=x_1 \div x_2} \min[\mu_A(x_1), \mu_B(x_2)]$$

$$= \sup_{x_1 \in \mathbb{R}} \min\left[\mu_A(x), \mu_B\left(\frac{x_1}{x}\right)\right]$$

$$= \sup_{x_2 \in \operatorname{supp}(B)} \min[\mu_A(x \cdot x_2),\ \mu_B(x_2)],$$

$$0 \notin \operatorname{supp}(B). \tag{2-90}$$

Example 2.11. Let X = ℝ. Let A ≜ "*approximately* 3" = 0.5/2 + 1/3 + 0.6/4. Let B ≜ "*approximately* 5" = 0.6/4 + 1/5 + 0.7/6. Then

A ⊕ B ≜ "*approximately* 3" ⊕ "*approximately* 5"
= 0.5/6 + max(0.5/7, 0.6/7) + max(0.5/8, 1/8, 0.6/8) + max(0.7/9, 0.6/9) + 0.6/10
= 0.5/6 + 0.6/7 + 1/8 + 0.7/9 + 0.6/10.

A ⊙ B ≜ "*approximately* 3" ⊙ "*approximately* 5"
= 0.5/8 + 0.5/10 + 0.6/12[= max(0.5/12, 0.6/12)] + 1/15 + 0.6/16 + 0.7/18 + 0.6/20 + 0.6/24.

Example 2.12. (Mizumoto & Tanaka, 1979)

Let $\underline{X}$ = ℝ. Let

$$A = B \triangleq \text{"}\mathit{approximately}\ 2\text{"} = \int_1^2 (x - 1)/x + \int_2^3 (3 - x)/x.$$

Then

A ⊕ B ≜ "*approximately* 2" ⊕ "*approximately* 2"

$$= \int_2^4 \left(\frac{x}{2} - 1\right)\Big/x + \int_4^6 \left(3 - \frac{x}{2}\right)\Big/x;$$

A ⊖ B ≜ "*approximately* 2" ⊖ "*approximately* 2"

$$= \int_{-2}^0 \left(\frac{x}{2} - 1\right)\Big/x + \int_0^2 \left(1 - \frac{x}{2}\right)\Big/x;$$

A ⊙ B ≜ "*approximately* 2" ⊙ "*approximately* 2"

$$= \int_1^4 (\sqrt{x} - 1)/x + \int_4^9 (3 - \sqrt{x})/x;$$

A $\oplus$ B $\underline{\triangle}$ "*approximately* 2" $\oplus$ "*approximately* 2"

$$= \int_{\frac{1}{3}}^{1} \left(3 - \frac{4}{x+1}\right) x + \int_{1}^{3} \left(\frac{4}{x+1} - 1\right) x.$$

(See Fig. 2.7)

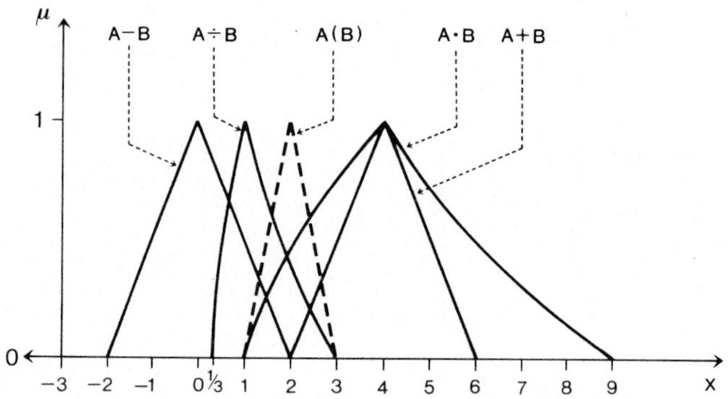

Fig. 2.7   Membership functions of A=B $\overset{\triangle}{=}$ " *approximately* 2 ";
A+B; A−B; A·B; and A÷B

Therefore, mathematical operations of ordinary real numbers can be extended to the operations of fuzzy subsets. Since fuzzy numbers are fuzzy subsets (see definitions 2.22 and 2.23), the extension principle can then be applied to operate on fuzzy numbers.

Definition 2.22   Let A ϵ $\mathscr{F}(\underline{X})$. Let x_1, x_2 ϵ $\underline{X}$. Then A is a convex fuzzy subset if

$$\mu_A(\lambda x_1 + (1 - \lambda)x_2) \geq \min[\mu_A(x_1), \mu_A(x_2)], \forall \lambda \epsilon [0, 1]. \quad (2\text{-}91)$$

(see Fig. 2.8a and b)

Therefore, all fuzzy subsets whose membership functions are quasi-concave are convex.

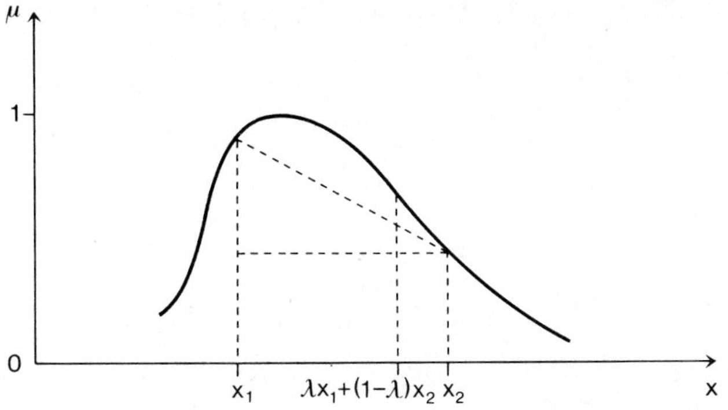

Fig. 2.8a A convex fuzzy subset

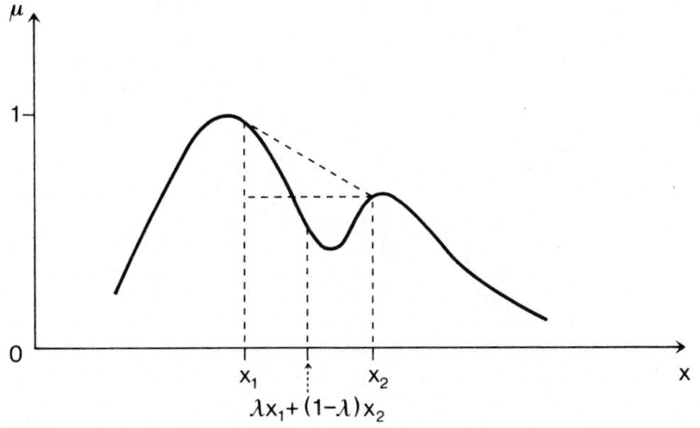

Fig. 2.8b A non-convex fuzzy subset

Definition 2.23. A fuzzy number is a normalized and bounded convex fuzzy subset in the real space.

A special type of fuzzy number is the L-R fuzzy number (Dubois and Prade, 1978). To define the L-R fuzzy number, we first define a reference function in definition 2.24.

Definition 2.24. A reference function ψ in $\mathbb{R}$ is a normalized convex function, $\psi: \mathbb{R} \longrightarrow [0, 1]$, defined by

$$\psi(x) = \begin{cases} F_L(x), & \text{if } -\infty < x < 0, \\ 1, & \text{if } x = 0, \\ F_R(x), & \text{if } 0 < x < \infty, \end{cases} \qquad (2\text{-}92)$$

where $F_L(x)$ is the monotonically increasing part and $F_R(x)$, not necessarily symmetric to $F_L(x)$, is the monotonically decreasing part.

(see Fig. 2.9)

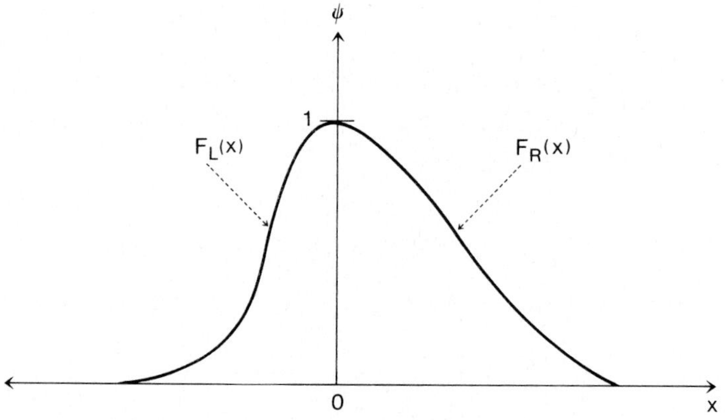

Fig. 2.9 A L-R reference function

Definition 2.25. A fuzzy number A is said to be a L-R fuzzy number if it is defined by

$$\mu_A(x) = \begin{cases} F_L\left(\frac{x - m}{\alpha}\right), & \text{if } -\infty < x < m, \ \alpha > 0, \\ 1, & \text{if } x = m, \\ F_R\left(\frac{x - m}{\delta}\right), & \text{if } m < x < +\infty, \ \delta > 0. \end{cases} \qquad (2\text{-}93)$$

(see Fig. 2.10)

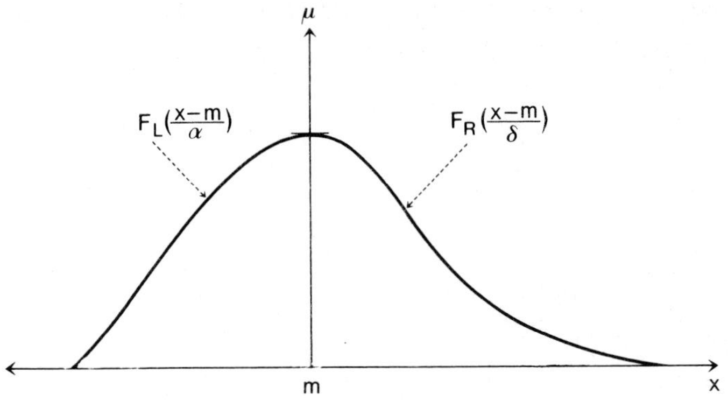

Fig. 2.10 A L–R fuzzy number

The L-R fuzzy number can, in short, be expressed as $A = (m, \alpha, \delta)$, with m, α, and δ being the mean value, the left, and right spreads respectively.

By the extension principle, it can be shown that the operations on L-R fuzzy numbers have the following properties:

Let $A = (m_A, \alpha_A, \delta_A)$ and $B = (m_B, \alpha_B, \delta_B)$. Then

(a) Addition, ⊕ .

$$A \oplus B = (m_A + m_B, \alpha_A + \alpha_B, \delta_A + \delta_B). \tag{2-94}$$

Example 2.13. (Kaufmann and Gupta, 1985, pp. 64-65)

Let the following be a reference function:

$$F_L(x) = \begin{cases} 0, & \text{if } x < -1, \\ \sqrt{1 + x}, & \text{if } -1 \le x \le 0; \end{cases} \tag{2-95}$$

$$F_R(x) = \begin{cases} 1 - x^2, & \text{if } 0 \le x < 1, \\ 0, & \text{if } x \ge 1. \end{cases} \tag{2-96}$$

Let A and B be two L-R fuzzy numbers such that $A = (m_A, \alpha_A, \delta_A) = (4, 2, 3)$ and $B = (m_B, \alpha_B, \delta_B) = (8, 3, 5)$. Then A and B are defined respectively by:

$$\mu_A(x) = \begin{cases} 0, & \text{if } x < 2, \\ \sqrt{1 + \dfrac{x-4}{2}}, & \text{if } 2 \leq x < 4, \\ 1, & \text{if } x = 4, \\ 1 - \left(\dfrac{x-4}{3}\right)^2, & \text{if } 4 < x \leq 7, \\ 0, & \text{if } x > 7. \end{cases} \tag{2-97}$$

$$\mu_B(x) = \begin{cases} 0, & \text{if } x < 5, \\ 1 + \dfrac{x-8}{3}, & \text{if } 5 \leq x < 8, \\ 1, & \text{if } x = 8, \\ 1 - \left(\dfrac{x-8}{5}\right)^2, & \text{if } 8 < x \leq 13, \\ 0, & \text{if } x > 13. \end{cases} \tag{2-98}$$

By (2-94), we have the addition $A \oplus B = (m_A + m_B, \alpha_A + \alpha_B, \delta_A + \delta_B) = (12, 5, 8)$ defined by

$$\mu_{A \oplus B}(x) = \begin{cases} 0, & \text{if } x < 7, \\ 1 + \dfrac{x-12}{5}, & \text{if } 7 \leq x < 12, \\ 1, & \text{if } x = 12, \\ 1 - \left(\dfrac{x-12}{8}\right)^2, & \text{if } 12 \leq x \leq 20, \\ 0, & \text{if } x > 20. \end{cases} \tag{2-99}$$

(see Fig. 2.11)

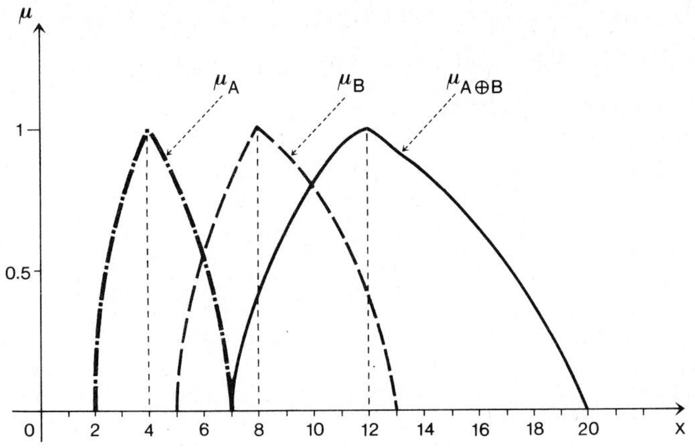

Fig. 2.11 Addition of two L−R fuzzy numbers A and B

(b) Multiplication by a real number k.

$$k \cdot A = \begin{cases} (km_A,\ k\alpha_A,\ k\delta_A), & \text{if } k > 0, \\ (km_A,\ -k\delta_A,\ -k\alpha_A), & \text{if } k < 0. \end{cases} \tag{2-100}$$

(c) Subtraction, $\ominus$.

$$A \ominus B = A + (-1)B = (m_A - m_B,\ \alpha_A + \delta_B,\ \delta_A + \alpha_B). \tag{2-101}$$

(If $F_L \neq F_R$, then $A \ominus B$ is not a L-R fuzzy number)

(d) Multiplication, $\odot$.

$$A \odot B \sim (m_A m_B,\ m_A\alpha_B + m_B\alpha_A,\ m_A\delta_B + m_B\delta_A). \tag{2-102}$$

(" $\sim$ " means approximately equals to)

(e) Division, $\oplus\!\!\!\!\div$.

$$A \,\hat{\div}\, B \sim \left(\frac{m_A}{m_B},\ \frac{m_A\delta_B + m_B\alpha_A}{m_B^2},\ \frac{m_A\alpha_B + m_B\delta_A}{m_B^2}\right),\ m_B \neq 0. \tag{2-103}$$

(f) Fuzzy max and min, $\underset{\sim}{\max}$, $\underset{\sim}{\min}$.

$\underset{\sim}{\max}(A,\ B) \sim [\max(m_A,\ m_B),\ \min(\alpha_A,\ \alpha_B),\ \max(\delta_A,\ \delta_B)];$

$$\underset{\sim}{\min}(A, B) \sim [\min(m_A, m_B), \max(\alpha_A, \alpha_B), \max(\delta_A, \delta_B)]. \quad (2\text{-}104)$$

(Except for special cases, $A \odot B$, $A \oplus B$, $\underset{\sim}{\max}$, and $\underset{\sim}{\min}$ are not L-R fuzzy numbers)

(g) Ordering.

$$A \leq B \Longleftrightarrow m_A \leq m_B,\ \alpha_A \geq \alpha_B,\ \delta_A \leq \delta_B;$$

$$A \subseteq B \Longleftrightarrow m_A + \delta_A \leq m_B - \delta_B \text{ or } A = B. \quad (2\text{-}105)$$

2.3.5. Fuzzy Relations

The concept of a fuzzy relation is formulated for the purpose of studying imprecise relations. In comparing distances, we may say that "distance x is *much longer* than distance y". In describing a region, we may say that "region S is *poor* and *highly populated*". In grouping spatial units, we may say that "place A is *similar* to place B". In characterizing friendship, we may say that "person A *likes* person B" (Leung 1980b). All these relations are imprecise because the words in italic have imprecise denotations. To have a formal analysis of such imprecise relations, a formal representation of fuzzy relations is imperative.

Definition 2.26. Let $\underline{X}$ and $\underline{Y}$ be two universes of discourse. A fuzzy binary relation R in the Cartesian product $\underline{X} \times \underline{Y}$ is a fuzzy subset in $\underline{X} \times \underline{Y}$ defined by the membership function

$$\mu_R: \underline{X} \times \underline{Y} \longrightarrow [0, 1],$$

$$(x, y) \longrightarrow \mu_R(x, y),\ x \in \underline{X},\ y \in \underline{Y}, \quad (2\text{-}106)$$

where the grade of membership $\mu_R(x, y)$ indicates the degree of relationship between x and y.

Example 2.14. Let $\underline{X} = \underline{Y} = \mathbb{R}$. Then $R \triangleq$ "x and y are *very similar*" is a fuzzy binary relation which may be defined by.

$$\mu_R(x, y) = e^{-k(x - y)^2}, \quad k > 1. \tag{2-107}$$

(see Fig. 2.12)

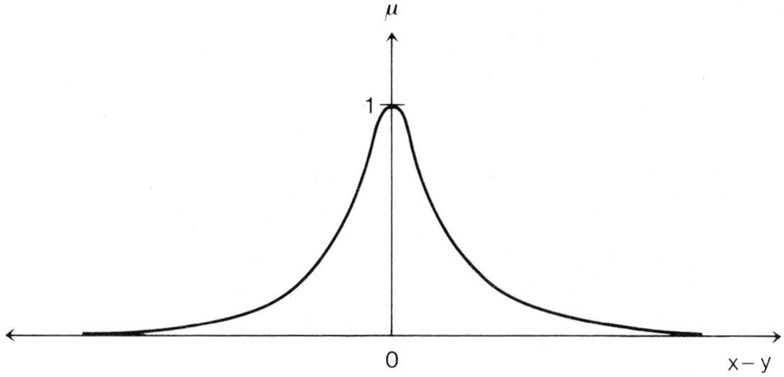

Fig. 2.12 Membership functioı of the fuzzy relation $R \triangleq$ "x and y are *very similar*"

Example 2.15. Let $\underline{X} = \underline{Y} = \mathbb{R}$. Then $R \triangleq$ "x is *much larger than* y" is a fuzzy binary relation which may be defined by

$$\mu_R(x, y) = \begin{cases} [1 + (x - y)^{-2}]^{-1}, & \text{if } x > y, \\ 0, & \text{if } x \leq y. \end{cases} \tag{2-108}$$

(see Fig. 2.13)

Definition 2.27. A fuzzy n-ary relation R in the universe of discourse $\underline{X}_1 \times \underline{X}_2 \times \ldots \times \underline{X}_n$ is a fuzzy subset in $\underline{X}_1 \times \underline{X}_2 \times \ldots \times \underline{X}_n$ defined by the membership function

$$\mu_R: \underline{X}_1 \times \underline{X}_2 \times \ldots \times \underline{X}_n \longrightarrow [0, 1],$$

$$(x_1, x_2, \ldots, x_n) \longrightarrow \mu_R(x_1, x_2, \ldots, x_n),$$

$$x_1 \in \underline{X}_1, \; x_2 \in \underline{X}_2, \; \ldots, \; x_n \in \underline{X}_n. \tag{2-109}$$

If the universes of discourse in (2-106) are finite, then

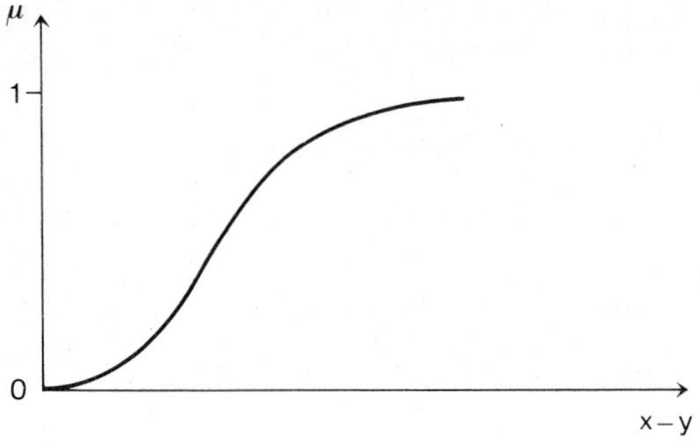

Fig. 2.13 Membership function of the fuzzy relation R ≜ "x is *much larger* than y"

the fuzzy binary relation can be represented as a fuzzy matrix, i.e. $R \triangleq (r_{ij})$, $i = 1, \ldots, m$; $j = 1, \ldots, n$; $0 \le r_{ij} = \mu_R(x_i, y_j) \le 1$. Of course, it can also be viewed as a fuzzy graph.

Example 2.16. Let $\underline{X} = \{a, b, c, d\}$ be a set of four cities. Let $R \triangleq$ "***resemblance***". Then, R can be represented by a fuzzy matrix

$$R = \begin{array}{c} \\ a \\ b \\ c \\ d \end{array} \begin{array}{c} \begin{array}{cccc} a & b & c & d \end{array} \\ \begin{bmatrix} 1 & 0.9 & 0.1 & 0.5 \\ 0.9 & 1 & 0.3 & 0.8 \\ 0.1 & 0.3 & 1 & 0.6 \\ 0.5 & 0.8 & 0.6 & 1 \end{bmatrix} \end{array},$$

where $\mu_R(a, d) = 0.5$ indicates the degree of resemblance of a and d.

Since fuzzy relations are fuzzy subsets, then the concepts and operations of fuzzy subsets can be parallelly applied to fuzzy relations. For example, the operations $\cup$, $\cap$, and $^-$ can be respectively defined as follows:

Let $\mathscr{F}(\underline{X} \times \underline{Y})$ be the fuzzy power set. Let R, S ε $\mathscr{F}(\underline{X} \times \underline{Y})$. Then the union of R and S, denoted as $R \cup S$, is defined by

$$\mu_{R \cup S}(x, y) = \max[\mu_R(x, y), \mu_S(x, y)], \quad \forall (x, y) \in \underline{X} \times \underline{Y}. \qquad (2\text{-}110)$$

The intersection, denoted as $R \cap S$, is defined by

$$\mu_{R \cap S}(x, y) = \min[\mu_R(x, y), \mu_S(x, y)], \quad \forall (x, y) \in \underline{X} \times \underline{Y}. \qquad (2\text{-}111)$$

The complement, denoted as $\bar{R}$, is defined by

$$\mu_{\bar{R}}(x, y) = 1 - \mu_R(x, y), \quad \forall (x, y) \in \underline{X} \times \underline{Y}. \qquad (2\text{-}112)$$

Example 2. 17. Let $R = \begin{bmatrix} 0.6 & 0.2 \\ 0.8 & 0.4 \end{bmatrix}$, $S = \begin{bmatrix} 0.2 & 0.5 \\ 0.3 & 0.6 \end{bmatrix}$. Then $R \cup S = \begin{bmatrix} 0.6 & 0.5 \\ 0.8 & 0.6 \end{bmatrix}$, $R \cap S = \begin{bmatrix} 0.2 & 0.2 \\ 0.3 & 0.4 \end{bmatrix}$, and $\bar{R} = \begin{bmatrix} 0.4 & 0.8 \\ 0.2 & 0.6 \end{bmatrix}$.

Definition 2. 28. Let R ε $\mathscr{F}(\underline{X} \times \underline{Y})$. Then the projection of $\underline{Y}$ on $\underline{X}$ through R, denoted as $R(\underline{X})$, is defined by

$$\mu_{R(\underline{X})}(x) = \sup_y \mu_R(x, y); \qquad (2\text{-}113)$$

the projection of $\underline{X}$ on $\underline{Y}$, denoted as $R(\underline{Y})$, is defined by

$$\mu_{R(\underline{Y})}(y) = \sup_x \mu_R(x, y); \qquad (2\text{-}114)$$

and the global projection of R, denoted as g(R), is defined by

$$g(R) = \sup_x \sup_y \mu_R(x, y) = \sup_y \sup_x \mu_R(x, y). \qquad (2\text{-}115)$$

(see Fig. 2.14)

Example 2. 18. Numerically, the projections of a fuzzy relation R can be depicted as follows:

	y_1	y_2	$R(\underline{X})$
x_1	0.3	0.6	0.6
x_2	0.4	1	1
$R(\underline{Y})$	0.4	1	1 = g(R)

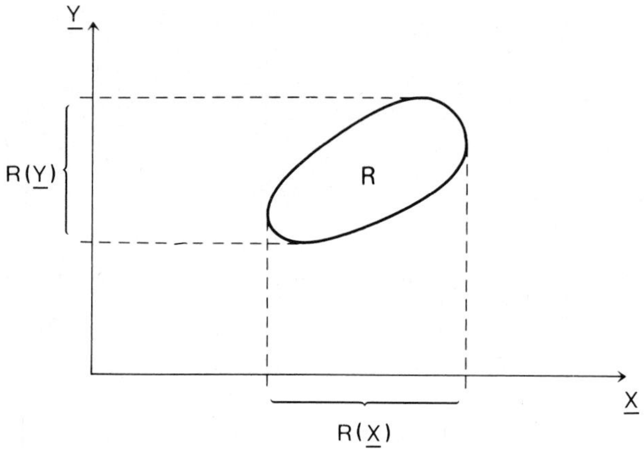

Fig. 2.14 Projections of R on X and Y

Definition 2.29. Let $R \in \mathscr{F}(\underline{X} \times \underline{Y})$, $S \in \mathscr{F}(\underline{Y} \times \underline{Z})$. The max-min composition of R and S, denoted as $R \circ S$, is defined by

$$\mu_{R \circ S}(x, z) = \max_{y} \min[\mu_R(x, y), \mu_S(y, z)],$$

$$x \in \underline{X}, \; y \in \underline{Y}, \; z \in \underline{Z}. \tag{2-116}$$

The compositions of fuzzy relations allow us to relay relationship in R to S and form a new fuzzy relation. Such a process is important in modelling relations and inferences in a variety of spatial problems.

Example 2.19. (Simplified example in Leung, 1980a)

Let $\underline{X} = \{x_1, x_2\}$ be two higher order objectives in a project evaluation process. Let $\underline{Y} = \{y_1, y_2, y_3\}$ be three lower order objectives. Let $\underline{Z} = \{z_1, z_2, z_3\}$ be three projects to be evaluated on their abilities in achieving the higher order objectives.

Let

$$R = \begin{matrix} & y_1 & y_2 & y_3 \\ x_1 & 0.7 & 0.5 & 0.1 \\ x_2 & 0.9 & 0.4 & 0.2 \end{matrix}, \quad S = \begin{matrix} & z_1 & z_2 & z_3 \\ y_1 & 0.8 & 0.4 & 0.1 \\ y_2 & 0.4 & 0.9 & 0.3 \\ y_3 & 0.1 & 0.2 & 0.9 \end{matrix}$$

be the respective achievement matrices with, for example $\mu_R(x_1, y_3) = 0.1$ indicates the degree of achieving the higher order objective x_1 by the lower objective y_3. Then

$$R \circ S = \begin{matrix} & z_1 & z_2 & z_3 \\ x_1 & 0.7 & 0.5 & 0.3 \\ x_2 & 0.8 & 0.4 & 0.3 \end{matrix}$$

is the new achievement matrix with $\mu_{R \circ S}(x_1, z_1) = 0.7$, for example, indicates the degree of achieving objective x_1 by project z_1.

Remark. To make the composition in (2-116) more general, "min" can be replaced by an operator * as long as * is associative and monotonically nondecreasing in each argument. Furthermore, composition such as "min-max" can also be defined if circumstances require (see section 3.4.4 for an application in regionalization).

Definition 2.30. Let $A \in \mathscr{F}(\underline{X})$ and $B \in \mathscr{F}(\underline{Y})$. The fuzzy subset B is said to be conditioned on $\underline{X}$ if its membership function depends on $x \in \underline{X}$ as a parameter. The conditional membership function is given by:

$$\mu_B(y|x), \quad x \in \underline{X} \text{ and } y \in \underline{Y}. \tag{2-117}$$

If $A \in \mathscr{F}(\underline{X})$, then A induces a fuzzy subset $B \in \mathscr{F}(\underline{Y})$ defined by:

$$\mu_B(y) = \max_{x \in \underline{X}} \min[\mu_B(y|x), \mu_A(x)]. \tag{2-118}$$

Example 2.20. Let $\underline{X} = \{x_1, x_2\}$ and $\underline{Y} = \{y_1, y_2, y_3\}$. Let $A = (0.7, 0.4) \in \mathscr{F}(\underline{X})$. Let $R \in \mathscr{F}(\underline{X} \times \underline{Y})$ and

$$\mu_B(y|x) \triangleq R = \begin{array}{c} \\ x_1 \\ x_2 \end{array} \begin{array}{c} \begin{array}{ccc} y_1 & y_2 & y_3 \end{array} \\ \begin{bmatrix} 0.2 & 0.7 & 0.9 \\ 0.8 & 0.6 & 0.3 \end{bmatrix} \end{array} .$$

By (2-116) and (2-118), $B = R \circ A = (0.4, 0.7, 0.7)$.

2.4. LINGUISTIC VARIABLES AND POSSIBILITY DISTRIBUTION

In conventional mathematical systems, variables are assumed to be precise. That is, a variable can be precisely measured and have exact values. However, when we are dealing with our daily languages, imprecision usually prevails. Intrinsically, daily languages cannot be precisely characterized on either the syntactic or semantic level. Therefore, a word in our daily languages can technically be regarded as a fuzzy subset. Its imprecision is mostly possibilistic rather than propabilistic in nature. Since languages are usually encountered in spatial systems analysis, a formal representation procedure is then essential.

In general, a variable in a universe of discourse can be called a linguistic variable if its values are words or sentences in a daily or artificial language (see Zadeh 1975a, b, c and 1978 for details). For example, if distance is measured in terms of words such as *long* and *short* rather than physical units such as kilometers or hours, then it can be characterized as a linguistic variable whose spatial denotation is imprecise.

Definition 2.31. A linguistic variable is designated by a quintuple $(X, T(X), \underline{\mathscr{X}}, G, M)$. The symbol X is the name of the variable, for example "distance". T(X) is the term set of X, finite or infinite, which consists of terms F, linguistic values, such as *long*, *short*, *very long*, *rather short*, *not long*, and *long or short* in a universe of discourse $\underline{\mathcal{X}}$. In terms of physical distance, elements in $\underline{\mathcal{X}}$ may be unit distance in kilometers or hours. The terms in the term set are generated by a syntactic rule G. The semantic rule M associates each term its meaning, a fuzzy subset in $\underline{\mathcal{X}}$. Through M, a compatibility function $\mu: \underline{\mathcal{X}} \longrightarrow [0, 1]$ is constructed to associate each element in $\underline{\mathcal{X}}$ its compatibility with a linguistic value F.

Therefore, a linguistic variable is a fuzzy variable whose values are fuzzy subsets in a universe of discourse. The base variable of the linguistic variable is a precise variable which takes on individual values in its domain, i.e. the universe of discourse $\underline{\mathcal{X}}$. The domain of the linguistic variable is, however, the collection of all possible linguistic values, fuzzy subsets, defined in the same universe of discourse through the base variable.

To illustrate the use of linguistic variables in spatial analysis (see Leung, 1982a for details), let distance be a linguistic variable, denoted as (X, T(X), $\underline{\mathcal{X}}$, G, M), with

X $\triangleq$ distance.
T(X) = *short* + *long* + *not short* + *not long* + *very short* + *very long* + *somewhat short* + *somewhat long* + *somewhat short and not short* + *somewhat short or long* + ...

is the term set in which short and long are primary terms, linguistic values, defined on $\underline{\mathcal{X}} = [0, \infty)$ with $x \in \underline{\mathcal{X}}$ stands for physical distance in kilometers. Here, linguistic hedges such as *very* and *somewhat* are employed as operators to modify the primary terms to create new terms such as *very short* and *somewhat long*. Logical connectives such as "or", "and", and "not" can also be applied to combine existing terms to form new terms. All of these constitute the syntactic rule G.

To associate meaning to each term, a semantic rule M which serves as a fuzzy restriction on the base variable x imposed by the linguistic value F is required. The realization of a fuzzy restriction on the possible values in $\underline{\mathcal{X}}$ is based on a compatibility function specified for the term F.

If we denote R(X) as a fuzzy restriction, then the effect of F on X can be expressed as

$$R(X) = F. \tag{2-119}$$

Since X is a variable in $\underline{\mathscr{X}}$, F is then a fuzzy subset in $\underline{\mathscr{X}}$ and R(X) is a fuzzy restriction imposed by F. The expression in (2-119) is in fact equivalent to the linguistic proposition

$$P: X \text{ is } F, \tag{2-120}$$

which induces a possibility distribution

$$\Pi_X = F, \tag{2-121}$$

equivalent to the fuzzy restriction R(X).

Associated with the possibility distribution is a possibility distribution function such that

$$\text{Poss}(X = a) = \pi_X(a) \triangleq \mu_F(a). \tag{2-122}$$

Therefore, fuzzy restrictions, possibility distributions, and fuzzy subsets are closely related. For example, if the linguistic proposition is P: distance is *long*, then

$$\text{Poss(distance} = 10 \text{ km.)} = \pi_{\text{distance}}(10) = \mu_{long}(10) = 0.6,$$

indicates the possibility of 10 km. being considered as a long distance. That is, the possibility distribution indicates whether or not the proposition, P: distance is *long*, is possible. Equivalently, it denotes the degree of compatibility of 10 km. to the fuzzy subset *long*. Under this situation, *long* imposes a fuzzy restriction R(distance) on $\underline{\mathscr{X}}$.

Thus, the semantic rule M can be regarded as a linguistic proposition about the linguistic variable X. Consequently, linguistic values are characterized as fuzzy subsets.

The following shows how the semantic rule works:

Let

$$P: X \text{ is } F \triangleq \text{distance is } \textit{short}; \tag{2-123}$$

$$Q: X \text{ is } G \triangleq \text{distance is } \textit{long}, \tag{2-124}$$

be the linguistic propositions about the primary terms *short* and *long*. Then

$$\Pi_{\text{distance}} = \textit{short}; \tag{2-125}$$

$$\Pi_{\text{distance}} = \textit{long}, \tag{2-126}$$

are the respective possibility distributions such that

$$\pi_{\text{distance}}(x) = \mu_{\textit{short}}(x) = \begin{cases} 1, & \text{if } x \leq \alpha, \\ e^{-k\left(\frac{x-\alpha}{\alpha}\right)^2}, \quad k > 0, & \text{if } x > \alpha; \end{cases} \tag{2-127}$$

$$\pi_{\text{distance}}(x) = \mu_{\textit{long}}(x) = \begin{cases} 1, & \text{if } x \leq \beta, \\ 1 - e^{-k\left(\frac{x-\beta}{\beta}\right)^2}, \quad k > 0, & \text{if } x > \beta; \end{cases} \tag{2-128}$$

(see Fig. 2.15)

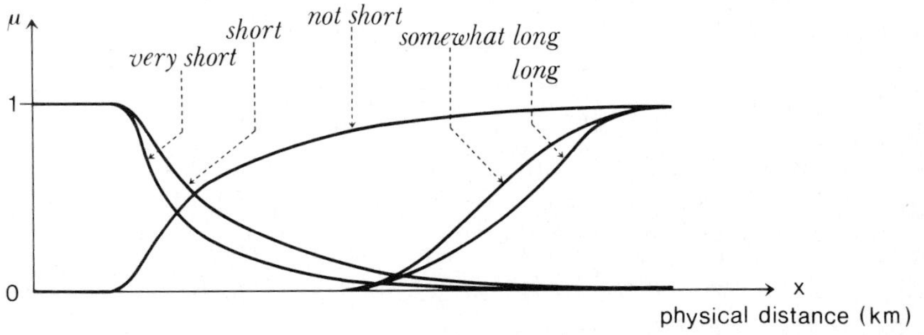

Fig. 2.15 Membership functions of some values of the linguistic variable "distance"

Based on (2-127) and (2-128), the terms "*not short*", "*very short*", and "*somewhat short*", for example, in the term set can be defined by applying respectively hedges "*not*" and "*very*" to *short* and "*somewhat*" to *long* to obtain the following:

$$\mu_{not\ short}(x) = \begin{cases} 0, & \text{if } x \leq \alpha, \\ 1 - e^{-k\left(\frac{x-\alpha}{\alpha}\right)^2}, & \text{if } x > \alpha; \end{cases} \qquad (2\text{-}129)$$

$$\mu_{very\ short}(x) = \begin{cases} 1, & \text{if } x \leq \alpha, \\ \left[e^{-k\left(\frac{x-\alpha}{\alpha}\right)^2}\right]^2, & \text{if } x > \alpha; \end{cases} \qquad (2\text{-}130)$$

$$\mu_{somewhat\ long}(x) = \begin{cases} 0, & \text{if } x \leq \beta, \\ \left[1 - e^{-k\left(\frac{x-\beta}{\beta}\right)^2}\right]^{\frac{1}{2}}, & \text{if } x > \beta. \end{cases} \qquad (2\text{-}131)$$

(see Fig. 2.15)

2.5. A REMARK ON FUZZY SET THEORY, PROBABILITY THEORY, AND POSSIBILITY THEORY

Since the inception of fuzzy set theory, doubts have been cast on the value of such a mathematical system. The superficial similarity of fuzzy set theory, especially the normalized membership function, and probability theory, especially the probability distribution function, further deepens such a suspicion. Those who have doubts usually ask the following questions: Are randomness and imprecision the same? What are the basic differences and similarities of probability theory and fuzzy set theory? Is fuzzy set theory in fact probability theory bearing a new name? What are the differences and similarities of probability theory and possibility theory? Can fuzzy set theory be a genuine field of scientific research? All these are legitimate questions which need to be clarified.

In subsection 2.5.1, the basic differences between randomness and imprecision are discussed. The arguments in turn constitute a strong support for separating probability theory and fuzzy set theory as independent mathematical systems which attempt to model randomness and imprecision respectively.

In subsection 2.5.2, probability theory and possibility theory are compared. Differences and similarities of probability measures and possibility measures are briefly investigated.

In subsection 2.5.3, the concept of a fuzzy event is employed to unite randomness and imprecision. It is demonstrated that probability theory and fuzzy set theory actually complement each other under some uncertain situations.

2.5.1. Randomness versus Imprecision

Classical mathematics, such as differential equations, are very effective in analyzing causal relationships under certainty. Their effectiveness, however, decreases as systems become more complex and causal relationships start to disintegrate. Given a cause, we are no longer certain about the outcome under uncertainty. Randomness is an uncertainty which arises from imperfect cause-effect relationships. It attempts to grasp general cause-effect relationship through probability laws.

Thus, the birth of probability theory was triggered by the conflict between certainty and randomness. Realizing that any study of randomness would be meaningless if randomness cannot be constructed by nonrandom means, realizing that employing random means to study randomness would lead us to an unproductive circular logic, realizing that randomness cannot be replaced by certainty, we attempt to approximate randomness by certainty instead. We attempt to determine the occurrence of an event by determining the likelihood of its occurrence. We attempt to search for laws out of randomness.

Probability is in fact an abstraction of random events but not the events themselves. Its success rests on our ability in demonstrating the stability of the probability laws, i.e. stability of the frequency of occurrence.

As pointed out in section 1.2, randomness only depicts a

partial picture of uncertainty. Imprecision or fuzziness, on the other hand, is an uncertainty which is induced by the break down of the law of the excluded-middle. It attempts to grasp the tendency of extremities through transitions in the middle grounds.

The birth of fuzzy set theory was triggered by the conflict between precision and fuzziness. The study of imprecision is meaningless if our analysis is based on imprecise means. Employing imprecise means to study imprecision would again lead us to a circular logic. Our approach then is to approximate imprecision by precision. In place of an all-or-nothing determination of an object, we study its degree of belongingness, grade of membership. We attempt to search for laws out of imprecision.

Thus, grade of membership is actually a precise concept. It is an abstraction of imprecise objects but not the objects themselves. Similar to probability, its success rest on whether or not such an abstraction has an objective existence, and whether or not we can establish stability of the membership function.

To recapitulate, randomness is an uncertainty resulting directly from the breakdown of deterministic cause-effect relationships, and imprecision is an uncertainty resulting directly from the breakdown of the law of the excluded-middle. Probability theory and fuzzy set theory are a means to study randomness and imprecision respectively. They have fundamental differences but are in a way closely connected.

2.5.2. Probability Theory versus Possibility Theory

The superfical resemblance of the probability distribution function and the membership function have caused some researchers to claim that fuzzy set theory is indeed probability theory with a new mask. To provide a counter argument of the claim is, however, not straight forward. To be convincing, we have to show that the two theories are in fact different philosophically and mathematically.

Except for the clear-cut contrast of their philosophical basis in subsection 2.5.1, direct comparison of the two theories in terms of their mathematical structures is in general difficult. Up to now, we do not have a unique definition of probability. If probability is defined as the degree of truth of a statement, i.e. subjective probability, its mathematical structure is a Boolean ring within the system of two-valued logic (Koopman, 1940). If it is defined by the concept of statistical frequency, its mathematical structure is a σ-algebra (Kolmogoroff, 1950).

In a parallel way, there is no unique definition of imprecision. The mathematical structure of fuzzy sets is no longer a Boolean algebra if union and intersection are defined by the max- and min-operators. Varying with different operators defining union and intersection (see for example those defined in subsection 2.3.2), we have a family of mathematical structures instead.

Thus, probability theory and fuzzy set theory cannot be directly compared. However, partial comparisons can be made if membership functions defining fuzzy subsets are treated to be numerically equal to possibility distribution functions discussed in section 2.4, and probability is interpreted as a statistical frequency. Their basic differences are discussed as follows:

The axiomatic foundation of Kolmogorov's system is a probability space $(\underline{\Omega}, \underline{\mathscr{A}}, P)$. The set $\underline{\Omega}$ is the outcome space and $\underline{\mathscr{A}}$ is a system of subsets of $\underline{\Omega}$, a Borel field. The sets of $\underline{\mathscr{A}}$ are called the events. The set $\underline{\Omega}$ and the system $\underline{\mathscr{A}}$ satisfies the following conditions:

(a) $\underline{\Omega} \in \underline{\mathscr{A}}$; (2-132)

(b) If $\underline{A} \in \underline{\mathscr{A}}$, then $\overline{\underline{A}} \in \underline{\mathscr{A}}$; (2-133)

(c) Let $\{\underline{A}_j\}$ be a sequence of sets. If $\underline{A}_j \in \underline{\mathscr{A}}$,

then $\bigcup_{j=1}^{\infty} A_j \in \underline{\mathscr{A}}$ and $\bigcap_{j=1}^{\infty} A_j \in \underline{\mathscr{A}}$, $\forall$ j. (2-134)

Thus, the system of subsets $\underline{\mathscr{A}}$ is a σ-algebra.

Let P(•) be a probability defined in $\underline{\mathscr{A}}$ such that

(a) $P: \underline{\mathscr{A}} \longrightarrow [0, 1]$; (2-135)

(b) $P(\underline{\Omega}) = 1$; (2-136)

(c) Let $\{\underline{A}_j\}$ be a sequence of events such that $\underline{A}_i \cap \underline{A}_j = \phi$, $i \neq j$, then $P(\bigcup_{j=1}^{\infty} \underline{A}_j) = \sum_{j=1}^{\infty} P(\underline{A}_j)$, $\forall j$, $\underline{A}_j \in \underline{\mathscr{A}}$. (2-137)

Thus, the set function P is σ-additive.

A possibility distribution Π, on the other hand, is a function defined in the fuzzy power set $\mathscr{F}(\underline{X})$ such that

(a) $\Pi : \mathscr{F}(\underline{X}) \longrightarrow [0, 1]$; (2-138)

(b) $\Pi(\phi) = 0$; (2-139)

(c) If $A, B \in \mathscr{F}(\underline{X})$ and $A \subseteq B$, then $\Pi(A) \leq \Pi(B)$; (2-140)

(d) Let $\{A_j\}$ be a sequence of fuzzy subsets. If $A_j \in \mathscr{F}(\underline{X})$, then $\Pi(\bigcup_j A_j) = \sup_j \Pi(A_j)$, $\forall j$, $A_j \in \mathscr{F}(\underline{X})$. (2-141)

Thus, Π is not σ-additive. For $\underline{X}$ finite, it is in fact a special case of the fuzzy measure proposed by Sugeno (for a review, see Sugeno, 1977).

Let $\mathscr{B}$ be a Borel field of $\underline{X}$. A set function f(•) defined in $\mathscr{B}$ is a fuzzy measure if it satisfies:

(a) $f(\phi) = 0$, $f(\underline{X}) = 1$; (2-142)

(b) If $A, B \in \mathscr{B}$ and $A \subseteq B$, then $f(A) \leq f(B)$; (2-143)

(c) Let $\{A_j\}$ be a sequence of fuzzy subsets. If $A_j \in \mathscr{B}$, then $\lim_{j\to\infty} f(A_j) = f(\lim_{j\to\infty} A_j)$. (2-144)

Therefore, possibility theory refers specifically to fuzzy subsets whose union and intersection are obtained through the max- and min-operators respectively. Given a possibility distribution, Π_X, associated with a variable X in $\underline{\mathscr{X}}$, and a fuzzy subset A in $\underline{\mathscr{X}}$, the possibility measure, $\pi(A)$, of A is defined by

$$\text{Poss}(X \text{ is } A) = \pi(A) = \sup_{x \in \mathscr{X}} \min[\mu_A(x), \pi_X(x)]. \tag{2-145}$$

In terms of union and intersection, the basic differences between probability and possibility are:

$$P(\underline{A} \cup \underline{B}) = P(\underline{A}) + P(\underline{B}), \text{ if } \underline{A} \cap \underline{B} = \phi, \tag{2-146}$$

$$P(\underline{A} \cap \underline{B}) = P(\underline{A})P(\underline{B}), \quad \text{if } \underline{A} \text{ and } \underline{B} \text{ are independent;} \tag{2-147}$$

versus,

$$\Pi(A \cup B) = \max[\Pi(A), \Pi(B)], \text{ if A and B are non-interactive,} \tag{2-148}$$

$$\Pi(A \cap B) = \min[\Pi(A), \Pi(B)], \text{ if A and B are non-interactive.} \tag{2-149}$$

Nevertheless, probability and possibility may coexist in some situations. Given one, we may be able to obtain the other. Leung (1981d), for example, shows how subjective probabilities can be derived through the knowledge of a possibility distribution within a maximum entropy framework.

2.5.3. Probability of Fuzzy Events

Up to this point, we have differentiated two basic types of uncertainty, namely the uncertainty due to randomness and the uncertainty due to imprecision. Nevertheless, some uncertainties may possess simultaneously the components of randomness and imprecision. For example, "there is a seventy percent chance that the per capita income of region S would be ***around*** a next year" is a statement about the occurrence of a fuzzy event. It is a fuzzy event because "***around*** a" is a fuzzy subset having random occurrence.

Consider the probability space ($\underline{\Omega}$, $\mathscr{A}$, P). If A is a fuzzy event in $\underline{\Omega}$, then the probability of A is defined by

$$P(A) = \sum_{i=1}^{n} \mu_A(x_i)p(x_i) = E[\mu_A], \text{ if } \underline{\Omega} \text{ is finite,} \tag{2-150}$$

and,

$$P(A) = \int_{\underline{\Omega}} \mu_A(x) dP = E[\mu_A], \text{ if } \underline{\Omega} \text{ is infinite.} \tag{2-151}$$

Obviously, if A is an ordinary set, then μ_A becomes the 0-1 characteristic function of a precise event and P(A) in (2-150) and (2-151) becomes the respective conventional definitions of probability.

Example 2. 21. Let per capita income be depicted by the fuzzy subset A $\triangleq$ "*around* a" and is defined by

$$\mu_A(x) = e^{-(x - a)^2}, \; x \in \mathbb{R} \text{ (monetary unit).} \tag{2-152}$$

Let the probability distribution of per capita income be normal with mean m and σ^2:

$$p(x) = \frac{1}{\sqrt{2\pi}\,\sigma} e^{-\frac{1}{2}\left(\frac{x - m}{\sigma}\right)^2} . \tag{2-153}$$

Then, the probability of the fuzzy event A is given by

$$P(A) = \frac{1}{\sqrt{2\pi}\,\sigma} \int_{-\infty}^{+\infty} e^{-(x - a)^2} \cdot e^{-\frac{1}{2}\left(\frac{x - m}{\sigma}\right)^2} dx$$

$$= \frac{1}{\sqrt{2\pi}\,\sigma} \int_{-\infty}^{+\infty} e^{-\left[(x - a)^2 + \left(\frac{x - m}{2\sigma^2}\right)^2\right]} dx. \tag{2-154}$$

It can be demonstrated that the probability of the fuzzy event A, P(A), satisfies:

(a) $0 \le P(A) \le 1$; (2-155)

(b) $P(\phi) = 0, \; P(\underline{\Omega}) = 1$; (2-156)

(c) If $A \subseteq B$, then $P(A) \le P(B)$; (2-157)

(d) $P(\bar{A}) = 1 - P(A)$; (2-158)

(e) $P(A \cup B) = P(A) + P(B) - P(A \cap B)$; (2-159)

(f) If $A \cap B = \phi$, then $P(A \cup B) = P(A) + P(B)$. (2-160)

Without further elaboration, we can also analyze uncertainty involving fuzzy probabilities, probabilities which are words in our daily language, of precise or imprecise events. In place of a precise number, such as 0.8, it may be more natural and practical to describe or predict the occurrence of events by words, such as "*highly likely*", "*quite possible*" and "*slim chance*". Therefore, probability theory and fuzzy set theory can be effectively combined to study uncertain spatial phenomena.

CHAPTER 3

REGIONAL CONCEPTS AND REGIONALIZATION

3.1. A CRITIQUE OF THE TWO-VALUED LOGIC APPROACH

The concept of a region and the methods of regionalization are fundamental issues in spatial analysis and planning. They are crucial in organizing spatial information, in characterizing and identifying spatial structures and processes, and in constructing a suitable basis for the formulation of spatial theories and planning policies. Their importance is reflected in the voluminous literature in the field. (See for example the reviews by Berry, 1967; Grigg, 1967; and Fischer, 1983).

The purpose of this chapter is not to review but to reexamine the conventional logic of regional conceptualization and regionalization. I intend to demonstrate the drawbacks of the two-valued logic and the significance of treating imprecision in spatial classification. A new framework for regionalization is also proposed.

Conventionally, a region is treated as a theoretical construct which can be precisely identified and delimited. Boundaries separating regions and nonregions are clear-cut. With respect to a set of prespecified regions, the assignment of spatial units to regions is unambiguous, namely a all-or-nothing assignment. Likewise, a space can be subdivided with exactitude into a set of mutually exclusive and exhaustive regions, or equivalently, spatial units can be grouped into precisely delimited regions. Thus, regional characterization, regional assignment, and grouping are all based on a two-valued logic which makes no room for fuzziness in our spatial classification problems.

In reality, such a rigid system is unnatural and is

generally inadequate in handling imprecision in most regionalization problems. My basic argument is that two-valued logic fails to capture the gradation of transition from membership to nonmembership of a region. Moreover, it provides insufficient machinery to include or preserve information which is pertinent to a regionalization procedure.

Let us examine this argument from the theoretical and technical points of view. The theoretical argument is mainly threefold. First, the name of a region, the rudimentary element in any spatial classification problem, is ordinarily a word, a phrase, or a sentence in our daily language. For example, in climatic classification, "*hot*", "*warm*", and "*cold*" may be employed to characterize climatic regions with respect to temperature. "*Predominantly black*" may be employed to characterize neighborhoods within an urban area. "*Comparatively well-off* region" may be employed to characterize the area in space which is relatively wealthy. However, how high should a temperature be before it is considered as *hot*, how many blacks should an area has before it is considered as *predominantly black*, and how wealthy should a place be before it is classified as *comparatively well-off* have no clear-cut answers. The basic problem is that all these terms have imprecise denotations. With respect to temperature, for instance, it is difficult, except for extreme temperatures, for us to determine with exactitude which temperature signifies a hot weather. Consequently, it is difficult to decide unambiguously whether or not a spatial unit having a specific temperature belongs to a "*hot* region". Therefore, an exact delimitation is only possible if an arbitrary cut-off point or a threshold value is employed. However, a slight variation in temperature should have at most a small effect on our sensitivity towards it. That is, a temperature is considered as *hot* only to a certain degree. A small change of temperature should only lead to a small change in the degree of belongingness to "*hot*". This obviously does not correspond with the abrupt transition of classes imposed by a precise cut-off point. Therefore, the two-valued logic is too simplistic. It does not model closely the denotations of regional names coming from our daily language.

Second, most phenomena vary over space in a more or less continuous manner. Except for extreme physical environments or man-made situations, seldom do we find an abrupt change of phenomenon over space. Therefore, it is unreasonable to impose a precise boundary to separate a region and a non-region while the in-between transition is gradual. Gradation of membership is in general more realistic and should be preserved. The two-valued logic again falls short in achieving such an objective.

Third, similarity or difference is a matter of degree rather than kind. Two places are similar or different only to a certain degree. When we are dividing a space or grouping spatial units into regions, however, a common practice is to define similarity or difference by a two-valued system. Two spatial units are either similar or different. There are no in-betweens. By employing the two-valued logic, we are trading a certain amount of information, degree of similarity or difference, for artificial precisions. Regional patterns thus obtained do not show to what extent two regions are different and separable. Intermediate areas between regions, if they actually exist, would disappear in the regionalization process.

In addition to the theoretical problems discussed above, there are also technical difficulties in applying the two-valued logic to spatial classification. Besides precise numerical data, imprecise information such as words or sentences in our daily language or numbers with fuzzy numerical denotations ordinarily exist in the information base of a regionalization problem. For example, in measuring the level of industrial employment, the only measurement we may have may be a fuzzy number, "*approximately* 45%". This is especially common in countries with a poor data base. In some situations, classification may involve valuative information such as "*rather low*" in terms of a noise level. Since a fuzzy number or a linguistic term has fuzzy denotation, then the two-valued system tends to over-simplify or misrepresent its meaning.

To accommodate quantitative and qualitative information with varying degrees of imprecision, a finer system of representation and a more flexible regionalization procedure are thus necessary. Most importantly, human subjectivity in spatial classification should be closely modelled.

In this chapter, a fuzzy set theoretical framework for regionalization is presented. Basic components of regionalization, namely the characterization of a region, the assignment of spatial units to prespecified regions, and the grouping of spatial units into regions (not prespecified) are scrutinized.

In section 3.2, I discuss the characterization of a region through a theory of possibility. It is demonstrated that regions can be characterized by linguistic propositions with a mathematical translation. Concepts such as the core and boundary of a region are also examined.

In section 3.3, a pattern matching procedure is formulated to assign spatial units to prespecified regions. It is essentially a regional discrimination problem.

In section 3.4, a grouping framework is constructed so that grouping of spatial units into regions can be performed with precise or fuzzy information. Fuzzy set concepts of similarity and distance are employed.

3.2. CHARACTERIZATION OF A REGION

To model the gradual transition from membership to nonmembership of a region, Gale (1976) applies the notion of a three-valued fuzzy subset to conceptualize a region. Though it is a very limited version of a fuzzy subset, it points out the problem of imprecision in regional conceptualization.

Taking the classification of economic regions as an example, with respect to a base variable, say percent industrial employment, a *highly industrialized* region can be characterized as a fuzzy subset "*highly industrialized*" defined by a membership function $\mu_{highly\ industrialized}$ which maps the universe of discourse $\underline{\mathscr{X}}$, percent industrial employment, to the membership set, [0, 1]. The grade of membership, $\mu_{highly\ industrialized}(x)$ indicates the compatibility of x percent industrial employment with the term "*highly industrialized*". The larger the value of x, the higher is its compatibility with the linguistic term. Therefore, an area may be considered as a *highly industrialized* region to a certain degree.

Since region names such as "*highly industrialized*" come from words in our daily language, the formal characterization of a region should then reflect the denotation, fuzzy or not, of the name and retain the embedded information as much as possible. Leung (1985a, 1987a) employs a concept of a linguistic variable and a theory of possibility to formalize the properties of regions. Our discussion in this section is mainly based on that analysis.

In subsection 3.2.1, regions are characterized by linguistic propositions with precise mathematical translations. A bridge is constructed to relate the mathematical descriptions with its linguistic counterparts.

In subsection 3.2.2, cores and boundaries of regions are defined. Fuzziness of regional boundaries is investigated.

In subsection 3.2.3, the concept of a boundary is extended to the multiple-region case. Conditions under which regions can be separated are examined.

3.2.1. Representation of a Region

Let $\underline{\mathcal{Y}} = \{\underline{\mathcal{Y}}_k\}$, k = 1, ..., ℓ, be a set of ℓ differentiating characteristics by which a region is characterized. Let Z be a region which is identified as F. Then region Z can be characterized by the joint linguistic proposition P:

$$P\text{: } Y(Z) \text{ is } F, \tag{3-1}$$

where $Y = (Y_1, \ldots, Y_k, \ldots, Y_\ell)$ is a ℓ-ary linguistic variable; $F = F_1 \times \ldots \times F_k \times \ldots \times F_\ell$ is a fuzzy ℓ-ary relation in $\underline{\mathcal{Y}}_1 \times \ldots \times \underline{\mathcal{Y}}_k \times \ldots \times \underline{\mathcal{Y}}_\ell$; and "Y(Z) is F" reads Y of region Z is F.

The linguistic proposition in (3-1) induces a joint possibility distribution

$$\Pi_{Y(Z)} = F, \tag{3-2}$$

such that

$$\text{Poss}(Y(Z)=y) = \pi_Y(y) = \mu_F(y)$$
$$= \min[\mu_{F_1}(y_1), \ldots, \mu_{F_k}(y_k), \ldots, \mu_{F_\ell}(y_\ell)], \qquad (3\text{-}3)$$

where $y = (y_1, \ldots, y_k, \ldots, y_\ell) \varepsilon \underline{\mathscr{Y}}$ with $y_k \varepsilon \underline{\mathscr{Y}}_k$, $k = 1, \ldots, \ell$; π_Y is the corresponding joint possibility distribution function of Π_Y; and μ_F is the membership function of the fuzzy ℓ-ary relation F with F_k being a fuzzy subset in $\underline{\mathscr{Y}}_k$, $k = 1, \ldots, \ell$.

The min-operator in (3-3) implies the assumption of possibilistic independence of the differentiating characteristics. With suitable mathematical properties, other operators can be employed if assumed otherwise.

The linguistic proposition P in (3-1) is in fact a linguistic proposition which combines ℓ linguistic propositions characterizing region Z with respect to the individual characteristics $\underline{\mathscr{Y}}_k$'s. That is, for each characteristic $\underline{\mathscr{Y}}_k$, $k = 1, \ldots, \ell$, Z is characterized by

$$P_k\text{: } Y_k(Z) \text{ is } F_k, \qquad (3\text{-}4)$$

which induces a possibility distribution

$$\Pi_{Y_k(Z)} = F_k, \qquad (3\text{-}5)$$

such that

$$\text{Poss}(Y_k(Z)=y_k) = \pi_{Y_k}(y_k) = \mu_{F_k}(y_k), \qquad (3\text{-}6)$$

where F_k is a fuzzy subset in $\underline{\mathscr{Y}}_k$.

Thus, the linguistic proposition in (3-1) is in fact the following joint proposition:

$$P\text{: } Y_1(Z) \text{ is } F_1 \text{ and...and } Y_k(Z) \text{ is } F_k \text{ and...and } Y_\ell(Z) \text{ is } F_\ell, \qquad (3\text{-}7)$$

in condensed form. It means that if Z is characterized as F_k with respect to characteristic $\underline{\mathscr{Y}}_k$, $k = 1, \ldots, \ell$, then Z is characterized as "F_1 and ... and F_k and ... and F_ℓ" with respect to all characteristics.

Since the values of Y_k's are linguistic terms in our daily language, then Y_k's are linguistic variables. Regional characterization is in fact linguistically-based.

Example 3.1. Let $\mathcal{Y}$ = income (in montary unit) be the only characteristic by which economic regions are identified. Let Z be a region which is characterized as a "*moderate income* region". With respect to the measurable base variable income (in monetary unit), the denotation of *moderate* is inexact. It imposes a fuzzy restriction on the base variable income.

Formally, Z can be characterized by the linguistic proposition

$$P: \text{Income}(Z) \text{ is } \textit{moderate}, \tag{3-8}$$

which induces a possibility distribution

$$\Pi_{\text{income}(Z)} = \textit{moderate}, \tag{3-9}$$

such that

$$\text{Poss}(\text{Income}(Z)=a) = \pi_{\text{income}}(a) = \mu_{\textit{moderate}}(a). \tag{3-10}$$

Here, the linguistic variable in $\mathcal{Y}$ is Y $\triangleq$ "income".

Let "*moderate*" be defined in Fig. 3.1. If income(Z) = 19,000, then the possibility of Z being considered as a "*moderate* income region" is 0.6. The possibility becomes 1 when income(Z) equals 21,000. Within the interval (10,000, 32,000), a region is identified as a "*moderate* income region" to a certain degree.

Example 3.2. In Example 3.1, a region is characterized by a single characteristic. Let there be two characteristics $\mathcal{Y}_1$ and $\mathcal{Y}_2$, with $\mathcal{Y}_1$ = income (in monetary unit) and $\mathcal{Y}_2$ = percent industrial employment. Let Z be a region which is characterized as a "*moderate* income and *high* industrial employment region". With respect to income, let the linguistic proposition P_1 be defined by (3-8). With respect to industrial employment, let the linguistic proposition be

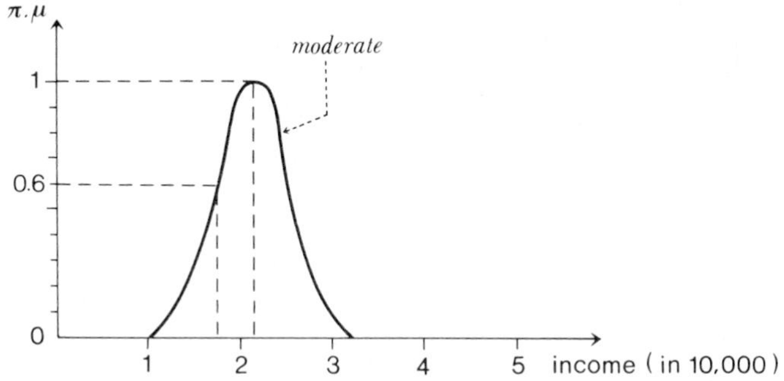

Fig. 3.1 Possibility distribution function of "*moderate*"

$$P_2\text{: Industrial employment(Z) is } \textit{high}, \tag{3-11}$$

which induces a possibility distribution

$$\Pi_{\text{industrial employment(Z)}} = \textit{high}, \tag{3-12}$$

such that

$$\text{Poss(industrial employment(Z)=b)}$$
$$= \pi_{\text{industrial employment}}(b) = \mu_{high}(b). \tag{3-13}$$

Combining P_1 and P_2, Z can be characterized by the joint linguistic proposition

$$\text{P: Income(Z) is } \textit{moderate} \text{ and industrial employment(Z) is } \textit{high}, \tag{3-14}$$

which induces the joint possibility distribution

$$\Pi_{[\text{Income(Z), industrial employment(Z)}]} = \textit{moderate} \times \textit{high}, \tag{3-15}$$

such that

$$\text{Poss(Income(Z)=a, Industrial employment(Z)=b)}$$
$$= \pi_{[\text{Income(Z), industrial employment(Z)}]}(a, b)$$
$$= \min[\mu_{moderate}(a), \mu_{high}(b)]. \tag{3-16}$$

Here, the linguistic variable in $\underline{\mathscr{Y}}_1 \times \underline{\mathscr{Y}}_2$ is $(Y_1, Y_2) \triangleq$ (income, industrial employment).

It states that if income is a and industrial employment is b, then the possibility of a region being considered as a "*moderate* income and *high* industrial employment region" equals the minimum of its grades of membership of being "*moderate*" with respect to income and being "*high*" with respect to industrial employment.

Let "*moderate*" be defined in Fig. 3.1 and "*high*" be defined in Fig. 3.2. Let Income(Z) = 19,000 and industrial employment(Z) = 60%. Then by (3-16), the possibility of Z being considered as a "*moderate* income and *high* industrial employment region" is: min(0.6, 0.4) = 0.4.

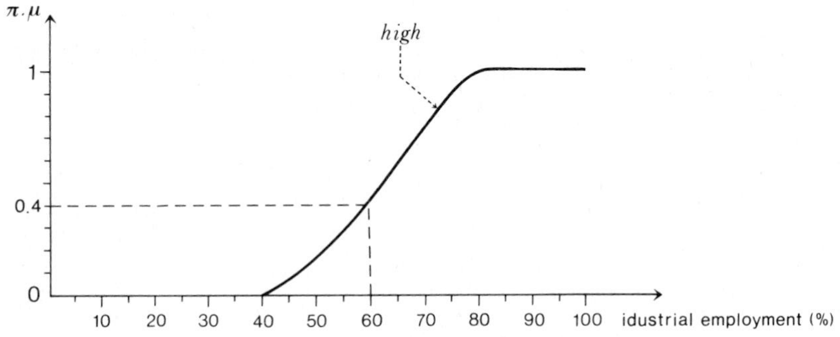

Fig. 3.2 Possibility distribution function of "*high*"

Compared to the two-valued logic, the present system of representation is more natural and informative. It provides a better approximation to our conceptualization of a region. Gradual transition from membership to nonmembership of a region is preserved. Should the transition be abrupt, it becomes a special case of the present system.

In the above discussion, the joint linguistic proposition of a region is constructed through the connective "and". Thus, individual possibility distribution functions are combined via the min-operator. That is, for a region to be considered as a prespecified region, it has to satisfy simultaneously the stated characterizations, precise or fuzzy, with respect to all characteristics. The possibility of a region being considered as a specific prespecified region depends on the degree to which the stated characterizations are jointly satisfied.

In some situations, however, we may only require a region to possess either one group of characteristics or that of the other. The joint linguistic proposition of a prespecified region should then be constructed with the connective "or". For instance, a joint linguistic proposition of a region Z with respect to two characteristics $\underline{\mathcal{Y}}_1$ and $\underline{\mathcal{Y}}_2$ may be stated as

$$P: Y_1(Z) \text{ is } F_1 \text{ or } Y_2(Z) \text{ is } F_2. \tag{3-17}$$

The induced joint possibility distribution becomes

$$\Pi_{[Y_1(Z),\ Y_2(Z)]} = F_1 \text{ or } F_2, \tag{3-18}$$

such that

$$\begin{aligned} &\text{Poss}(Y_1(Z)=a,\ Y_2(Z)=b) \\ &= \pi_{[Y_1(Z),\ Y_2(Z)]}(a,\ b) = \max[\mu_{F_1}(a),\ \mu_{F_2}(b)]. \end{aligned} \tag{3-19}$$

Under the assumption of possibilistic independence, the max-operator is employed to construct the joint possibility distribution function in (3-19).

In some cases, it may be necessary to formulate a joint linguistic proposition with both connectives "and" and "or". For example, a proposition may be stated as

$$P: [Y_1(Z) \text{ is } F_1 \text{ and } Y_2(Z) \text{ is } F_2] \text{ or } Y_3(Z) \text{ is } F_3, \tag{3-20}$$

which induces a possibility distribution

$$\Pi_{[Y_1(Z),\ Y_2(Z),\ Y_3(Z)]} = (F_1 \text{ and } F_2) \text{ or } F_3,$$

such that

$$\begin{aligned} &\text{Poss}(Y_1(Z)=a,\ Y_2(Z)=b,\ Y_3(Z)=c) \\ &= \pi_{[Y_1(Z),\ Y_2(Z),\ Y_3(Z)]}(a,\ b,\ c) \end{aligned}$$

$$= \max\left\{\min[\mu_{F_1}(a),\ \mu_{F_2}(b)],\ \mu_{F_3}(c)\right\}. \qquad (3\text{-}21)$$

By employing the connective "or", heterogeneity within a region becomes greater and the homogeneity among regions increases accordingly Under this situation, the separation of regions may be more difficult. It is, however, a natural consequence of the way a region is conceptualized under "or".

3.2.2. The Core and the Boundary of a Region

Core and boundary are basic but problematic notions in the conceptualization of regions. Though precisely delimited cores and boundaries may be preferrable, they are difficult to justify on the theoretical and empirical basis. Fuzziness often occurs at or around the cores and boundaries of regions. Due to the gradual variation of most phenomena over space, it is difficult to determine with exactitidue where the core is and where the region ends. Except for extreme cases, cores and boundaries can be considered as intrinsically fuzzy. This is especially so when regions are characterized linguistically. In general, the transition from the core to the edge is gradual. Such a gradation is oversimplified by the two-valued logic. However, a more natural delimitation can be achieved through the fuzzy set approach.

If core is treated as the most representative part of a region, then it should be the point or area in space whose characteristics are most compatible to the characterization of the region. Based on the characterization in (3-1), (3-2), and (3-3), the core of a region Z may be defined as the point or area in space having characteristics $y^* \in \underline{\mathcal{Y}}$.

Definition 3.1. The core of a region Z, denoted as CORE(Z), is defined by

$$\text{CORE}(Z) = \left\{ y^* \mid \mu_F(y^*) = \text{hgt}(F) = \sup_{y} \mu_F(y) \right.$$

$$= \min[\sup_{y_1} \mu_{F_1}(y_1),\ \ldots,\ \sup_{y_k} \mu_{F_k}(y_k),\ \ldots,$$

$$\sup_{y} \mu_{F_\ell}(y_\ell)]\Big\}. \tag{3-22}$$

Similar to the concept of a core, the boundary of a region can also be defined with respect to the possibility distribution characterizing a region. Since the linguistic characterization has imprecise spatial connotations, the corresponding boundary is naturally fuzzy. In general, it takes on a form of a gradient rather than a line. It is a zone which consists of all points in space whose characteristics are more or less compatible to the characterization of a region. That is, it is the transition from the core to the edge of a region. Based on (3-1), (3-2), and (3-3), the boundary of a region Z can be defined as:

Definition 3.2. The boundary of a region Z, denoted as BOUNDARY(Z), is defined by

$$\text{BOUNDARY(Z)} = \Big\{ y \,|\, 0 < \mu_F(y) < 1 \Longleftrightarrow 0 < \min[\mu_{F_1}(y_1), \ldots, \mu_{F_k}(y_k), \ldots, \mu_{F_\ell}(y_\ell)] < 1 \Big\}. \tag{3-23}$$

In the identification of a region, more tolerance may be required in some situations. Theoretically and pragmatically, it may be too rigid to consider only those points in space having complete compatibility to the regional characterization as the region. Furthermore, it may be inconvenient to have a zone instead of a line as the boundary. Under these circumstances, a precise boundary can always be delimited by restricting $\mu_F(y)$ in definition 3.2 to a specific value $\alpha \in (0, 1)$.

Definition 3.3. The α-boundary of a region Z, denoted as α-BOUNDARY(Z), is defined by

$$\alpha\text{-BOUNDARY(Z)} = \Big\{ y \,|\, \mu_F(y) = \alpha \Big\}, \quad 0 < \alpha < 1. \tag{3-24}$$

Therefore, the α-boundary of a region is a line connecting all points in space whose compatibility to the characterization of the region is α. Having imposed a precise boundary, the corresponding region can in turn be precisely delimited as the α-level set of F:

$$\underline{Z} = \left\{ y \mid \mu_F(y) \geq \alpha \right\}, \quad 0 < \alpha < 1, \tag{3-25}$$

with the characteristic function

$$f_F(y) = \begin{cases} 1, & \text{if } \mu_F(y) \geq \alpha, \\ 0, & \text{otherwise.} \end{cases} \tag{3-26}$$

That is, the α-boundary delimits a region whose elements are at least α-compatible to the characteristics of the region. The value serves as a qualifying threshold for the region. It can be viewed as a defuzzification mechanism out of necessity.

In addition to the core and boundary, it is useful to be able to delimit the edge of a region. By intuition, the edge should be those points in space which just fall short in meeting the characterization of a region. With respect to a region, they are adjacent points in space which possess none of the regional characteristics.

Definition 3.4. The edge of a region Z, denoted as EDGE(Z), is all those points $y \in \underline{\mathcal{Y}}$ such that y is neither the interior point not the exterior point of supp(F), the support of F.

Definition 3.4 only applies to $\underline{\mathcal{Y}} \subseteq \mathbb{R}^n$. In particular, let $\underline{\mathcal{Y}} \subseteq \mathbb{R}$. Then EDGE(Z) = inf[supp(F)] = $\inf\{y \mid \mu_F(y) > 0\}$ if μ_F is opened to the right. EDGE(Z) = sup[supp(F)] = $\sup\{y \mid \mu_F(y) > 0\}$ if μ_F is opened to the left. EDGE(Z) = inf[supp(F)] $\cup$ sup[supp(F)] if μ_F is closed at the right and left. If $\underline{\mathcal{Y}}$ is a discrete space, it is difficult to give the edge of a region a simple definition. Its identification, however, is possible once Y is known.

Example 3.3. In this example, regions are characterized by a single characteristic $\underline{\mathcal{Y}}$ = percent agricultural employment. Let Y $\underline{\underline{\Delta}}$ "agricultural employment". Three general types of membership functions are employed to illustrate the concepts discussed above.

(a) Case 1 (Membership function opened to the right)

Let Z be characterized as a region of *high* agricultural employment. Let the linguistic proposition be

$$P: Y(Z) \text{ is } high. \tag{3-27}$$

The induced possibility distribution is

$$\Pi_{Y(Z)} = high, \tag{3-28}$$

such that

$$\text{Poss}(Y(Z)=y) = \mu_Y(y) = \mu_{high}(y), \tag{3-29}$$

where

$$\mu_{high}(y) = \begin{cases} 1, & \text{if } y \geq 32, \\ \dfrac{y-24}{8}, & \text{if } 24 < y < 32, \\ 0, & \text{if } y \leq 24. \end{cases} \tag{3-30}$$

(see Fig. 3.3)

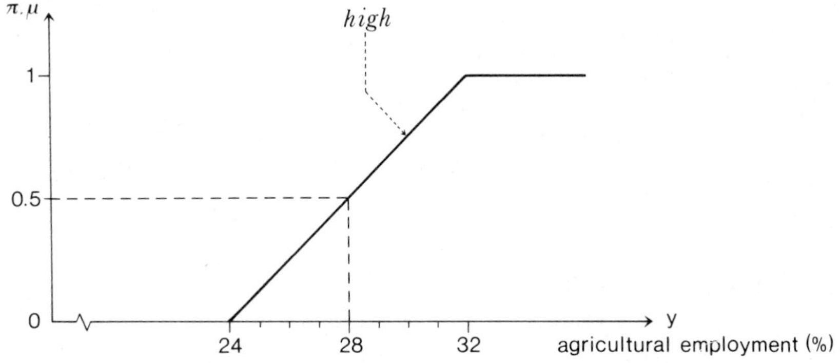

Fig. 3.3 Possibility distribution function of "*high*"

Then,

$$\text{CORE}(Z) = \{y \mid y \geq 32\};$$
$$\text{BOUNDARY}(Z) = \{y \mid 24 < y < 32\};$$
$$0.5\text{-BOUNDARY}(Z) = 28, \text{ and } \underline{Z} = \{y \mid y \geq 28\};$$
$$\text{EDGE}(Z) = 24.$$

(b) Case 2 (Membership function opened to the left)

Let Z be characterized as a region of *low* agricultural employment according to the format of (3-27) to (3-29). Let the membership function of "*low*" be

$$\mu_{low}(y) = \begin{cases} 1, & \text{if } y \leq 10, \\ \dfrac{20 - y}{10}, & \text{if } 10 < y < 20, \\ 0, & \text{if } y \geq 20. \end{cases} \qquad (3\text{-}31)$$

(see Fig. 3.4)

Then,

$$\text{CORE}(Z) = \{y \mid y \leq 10\};$$
$$\text{BOUNDARY}(Z) = \{y \mid 10 < y < 20\};$$
$$0.5\text{-BOUNDARY}(Z) = 15, \text{ and } \underline{Z} = \{y \mid y \geq 15\};$$
$$\text{EDGE}(Z) = 20. \qquad (3\text{-}32)$$

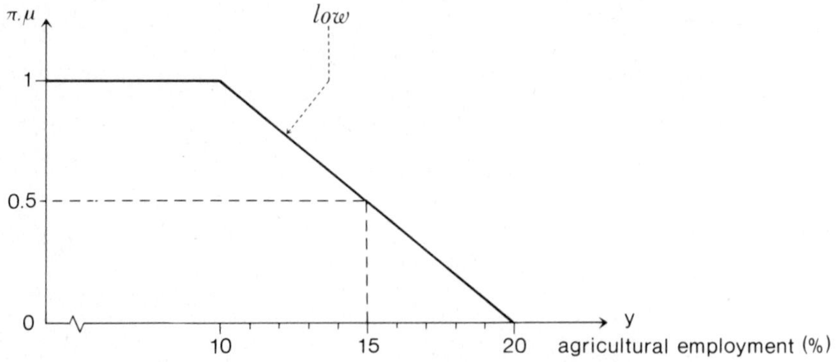

Fig. 3.4 Possibility distribution function of "*low*"

(c) Case 3 (Membership function closed at the right and left)

Let Z be characterized as a region of *moderate* agricultural employment. Let the membership function of "*moderate*" be

$$\mu_{moderate}(y) = \begin{cases} 0, & \text{if } y \le 16, \\ \dfrac{y-16}{4}, & \text{if } 16 < y < 20, \\ 1, & \text{if } 20 \le y \le 22, \\ \dfrac{26-y}{4}, & \text{if } 22 < y < 26, \\ 0, & \text{if } y \ge 26. \end{cases} \qquad (3\text{-}33)$$

(see Fig. 3.5)

Then,

$\text{CORE}(Z) = \{y \mid 20 \le y \le 22\}$;

$\text{BOUNDARY}(Z) = \{y \mid 16 < y < 20 \text{ and } 22 < y < 26\}$;

$0.5\text{-BOUNDARY}(Z) = \{18\} \cup \{24\}$, and $\underline{Z} = \{y \mid 18 \le y \le 24\}$;

$\text{EDGE}(Z) = \{16\} \cup \{26\}$.

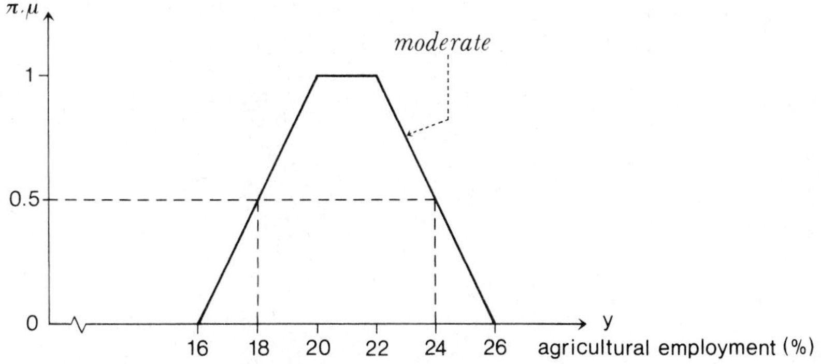

Fig. 3.5 Possibility distribution function of "*moderate*"

Example 3.4. Let $\underline{\mathcal{Y}}_1$ = "annual per capita income" and $\underline{\mathcal{Y}}_2$ = "years of education" be two differentiating characteristics. Let $Y_1 \triangleq$ "income" and $Y_2 \triangleq$ "education". Let Z be characterized as a *medium* income and *moderate* education region. Let the joint linguistic proposition be:

$$P: Y_1(Z) \text{ is } \textit{medium} \text{ and } Y_2(Z) \text{ is } \textit{moderate}, \tag{3-34}$$

which induces a joint possibility distribution

$$\Pi_{[Y_1(Z),\ Y_2(Z)]} = \textit{medium} \times \textit{moderate}, \tag{3-35}$$

such that

$$\text{Poss}(Y_1(Z)=y_1,\ Y_2(Z)=y_2) = \pi_{[Y_1(Z),\ Y_2(Z)]}(y_1, y_2)$$

$$= \mu_{\textit{medium} \times \textit{moderate}}(y_1, y_2) = \min[\mu_{\textit{medium}}(y_1),\ \mu_{\textit{moderate}}(y_2)]. \tag{3-36}$$

Let

$\underline{Y}_1 = \{7{,}000,\ 10{,}000,\ 16{,}000,\ 20{,}000,\ 24{,}000,\ 28{,}000,\ 32{,}000,\ 40{,}000,\ 50{,}000\}$,

$\underline{Y}_2 = \{0,\ 1,\ 4,\ 6,\ 12,\ 16,\ 18,\ 21,\ >21\}$.

Let

"*medium* income" = (0, 0, 0.3, 0.5, 1, 0.5, 0.3, 0, 0),

"*moderate* education" = (0, 0, 0.1, 0.4, 1, 0.6, 0.4, 0, 0).

Then *medium* × *moderate* is a fuzzy binary relation.

(see Fig. 3.6).

CORE(Z) = $\{(24{,}000,\ 12)\}$;

BOUNDARY(Z) = $\{16{,}000,\ 20{,}000,\ 24{,}000,\ 28{,}000,\ 32{,}000\} \times \{4,\ 6,\ 12,\ 16,\ 18\}$ [excluding (24000, 12)];

0.5-BOUNDARY(Z) = $\{(20{,}000,\ 12),\ (20{,}000,\ 16),\ (28{,}000,\ 12),\ (28{,}000,\ 16)\}$, and
$\underline{Z} = \{20{,}000,\ 24{,}000,\ 28{,}000\} \times \{12,\ 16\}$;

$$\text{EDGE}(Z) = [\{10{,}000,\ 40{,}000\} \times \{1,\ 4,\ 6,\ 12,\ 16,\ 18,\ 21\}] \cup [\{10{,}000,\ 16{,}000,\ 20{,}000,\ 24{,}000,\ 28{,}000,\ 32{,}000,\ 40{,}000\} \times \{1,\ 21\}].$$

(y_1) Annual per capita income \ Years of education (y_2)	0	1	4	6	12	16	18	21	>21
7,000	0	0	0	0	0	0	0	0	0
10,000	0	0	0	0	0	0	0	0	0
16,000	0	0	0.1	0.3	0.3	0.3	0.3	0	0
20,000	0	0	0.1	0.4	0.5	0.5	0.4	0	0
24,000	0	0	0.1	0.4	1	0.6	0.4	0	0
28,000	0	0	0.1	0.4	0.5	0.5	0.4	0	0
32,000	0	0	0.1	0.3	0.3	0.3	0.3	0	0
40,000	0	0	0	0	0	0	0	0	0
50,000	0	0	0	0	0	0	0	0	0

Fig. 3.6 Possibility distribution function of *medium income* and *moderate education*

3. 2. 3. Boundary as an Interface of Regions

In subsection 3.2.2, the boundary of a single region is defined. In a space with more than one region, however, boundary should be treated as the interface of any two regions. Based on the two-valued logic, the boundary between any two regions is precise and can be unambiguously delimited. In this situation, principle of the excluded-middle can be enforced.

As discussed previously, regions are fuzzy concepts with imprecise spatial connotations. Therefore, the gradual transition from one region to the other cannot guarantee a clear split of any two regions. Except for special cases, it is natural and reasonable to assume that interfaces are fuzzy. In general, regions are mutually exclusive only when precise boundaries are imposed. The implication then is that in reality two regions can only be separated to a certain degree. Overlapping of regions should be expected. If regional characterizations are linguistically-based, the separating boundaries would tend to appear as a zone rather than a line. In what follows, a two-region situation is first formalized. The argument is then extended to the multiple-region context.

According to the format of characterization in (3-1) to (3-3), let Z and W be two regions characterized respectively by:

$$\begin{cases} P\colon Y(Z) \text{ is } F, \\ \Pi_{Y(Z)} = F, \\ \mathrm{Poss}(Y(Z)=y) = \pi_{Y(Z)}(y) = \mu_F(y); \end{cases} \tag{3-37}$$

and,

$$\begin{cases} Q\colon Y(W) \text{ is } G, \\ \Pi_{Y(W)} = G, \\ \mathrm{Poss}(Y(W)=y) = \pi_{Y(W)}(y) = \mu_G(y). \end{cases} \tag{3-38}$$

Depending on how similar F and G are, Z and W are separated to a certain degree. Such a phenomenon can be formalized by the separation theorem of two bounded and convex fuzzy subsets.

Theorem 3.1. Let F and G be bounded and convex in $\underline{E}^n$ with maximal grades M_F and M_G such that

$$M_F = \sup_y \mu_F(y),$$

$$M_G = \sup_y \mu_G(y).$$

Let M be the maximal grade of $F \cap G$ such that

$$M = \sup_y \min[\mu_F(y), \mu_G(y)].$$

Then, the highest degree of separation of F and G is

$$S = 1 - M.$$

Theorem 3.1 is in fact an extension of the separation theorem of two ordinary convex sets in $\underline{E}^n$. It is well-known that if two convex sets $\underline{A}$ and $\underline{B}$ have no common inner points, then there exists a hyperplane H which separates $\underline{A}$ and $\underline{B}$ into two separate sides of H.

Evidently, such a situation also exists when bounded and convex fuzzy subsets are involved. The theorem states that a hyperplane H exists and passes through the point M such that $\mu_F(y) \leq m$ for all y on one side of H and $\mu_G(y) \leq m$ for all y on the other side of H. This is the point at which $\underline{\Omega}_F = \{y \mid \mu_F(y) > M\}$ and $\underline{\Omega}_G = \{y \mid \mu_G(y) > M\}$ are separated into two disjoint sets. Since M is the maximal grade of $F \cap G$, i.e. the highest degree of similarity of F and G, then the highest degree of separation of F and G is naturally $S = 1 - M$ and the hyperplane H is the only one possible for such a situation (see Fig. 3.7 for the case in $\underline{E}^1$, Fig. 3.8 for the case in $\underline{E}^2$, and the proof in appendix 2).

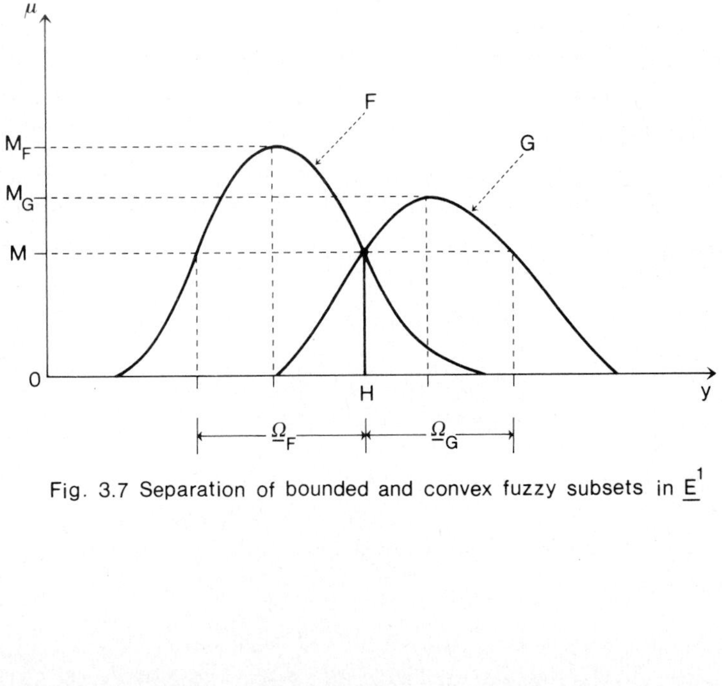

Fig. 3.7 Separation of bounded and convex fuzzy subsets in $\underline{E}^1$

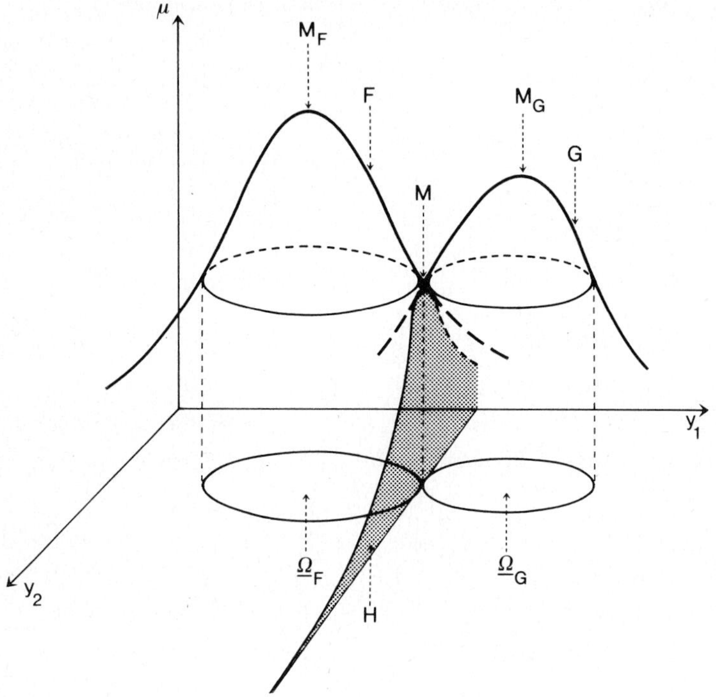

Fig. 3.8 Separation of bounded and convex fuzzy subsets in $\underline{E}^2$

Thus, if regions Z and W are characterized by bounded and convex fuzzy subsets F and G respectively, S then measures the highest degree of separation of F and G, i.e. of Z and W. Since F and G are both fuzzy relations in $\mathbb{R}^{\ell}$, then their being bounded and convex is guaranteed if each of the individual fuzzy subsets constituting F and G are bounded and convex.

Proposition 3. 1. Let $F,\ G \subseteq \underline{\mathscr{Y}}_1 \times \ldots \times \underline{\mathscr{Y}}_k \times \ldots \times \underline{\mathscr{Y}}_{\ell}$. Let $F = F_1 \times \ldots \times F_k \times \ldots \times F_m$ and $G = G_1 \times \ldots \times G_k \times \ldots \times G_{\ell}$. Then F and G are bounded and convex if F_k and G_k, $k = 1, \ldots, \ell$, are bounded and convex.

Having examined the situations under which two regions may

be separated, the boundary between any two regions can now be defined.

Definition 3.5. The boundary between regions Z and W, denoted as BOUNDARY(Z, W), is defined by

$$\text{BOUNDARY}(Z, W) = \left\{ y \,\middle|\, 0 < \min[\mu_F(y), \mu_G(y)] \leq M \right\}. \tag{3-39}$$

Depending on $F \cap G$, we have the following three cases:

Case 1. $F \cap G = \phi \Longrightarrow M = 0,\ S = 1$
$\Longrightarrow$ Z and W are completely different and can be unambiguously separated.

Case 2. $F = G \Longrightarrow F \cap G = F\ (= G) \Longrightarrow M = 1,\ S = 0$
$\Longrightarrow \text{BOUNDARY}(Z, W) = \left\{ y \,\middle|\, 0 < \mu_{F(\text{or } G)}(y) < 1 \right\}$,
(the single-region case).

Case 3. $F \cap G \neq \phi,\ F \not\subseteq G$ and $G \not\subseteq F \Longrightarrow$
$M \in (0, 1),\ S \in (0, 1) \Longrightarrow$
$\text{BOUNDARY}(Z, W) = \left\{ y \,\middle|\, 0 < \min[\mu_F(y), \mu_G(y)] \leq M \right\}$,
(Definition 3.5).

Thus, cases 1 and 2 are special cases of case 3. The three situations are illustrated respectively in Fig. 3.9a, b, and c.

By definition 3.5, suppose F and G are convex with $\mu_F(a) = \sup_y \mu_F(y) > \mu_G(a)$ and $\mu_G(b) = \sup_y \mu_G(y) > \mu_F(b)$, (see Fig. 3.9c), then the M-BOUNDARY(Z, W) $\cong$ y* is a precise "line" which separates Z and W into two disjoint regions:

$$\underline{Z} = \left\{ y \,\middle|\, \mu_F(y) > M \right\}, \text{ and } \underline{W} = \left\{ y \,\middle|\, \mu_G(y) > M \right\}, \tag{3-40}$$

with the degree of separation $S = 1 - M$.

Though such a precise boundary is possible, it does not tell the whole story. If $F \cap G \neq \phi$, the boundary should include all points in space whose characteristics are more or less compatible to the characterizations of both region Z and W.

To have a controlled overlapping of regions, a separation threshold

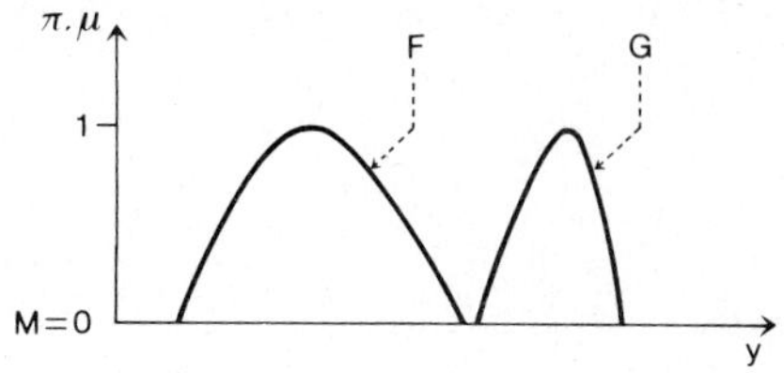

Fig. 3.9a A situation of $F \cap G = \Phi$

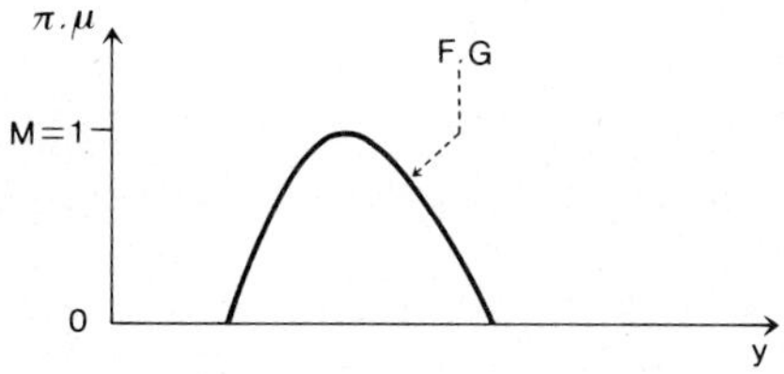

Fig. 3.9b A situation of $F \cap G = F$ $(=G)$

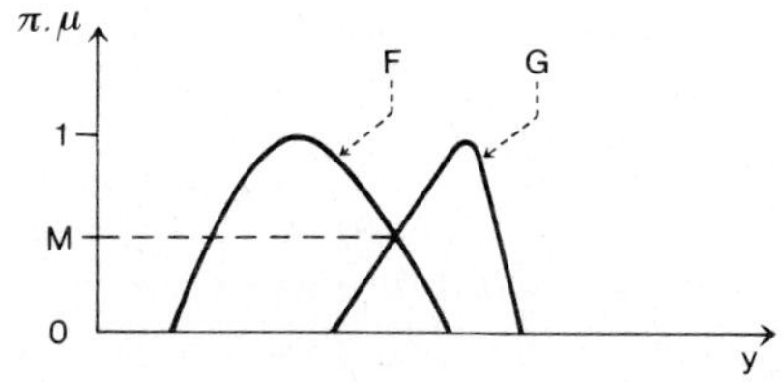

Fig. 3.9c A situation of $F \cap G \neq \Phi$, $F \nsubseteq G$ and $G \nsubseteq F$

$$\ell < \sup_{y} \min[\mu_F(y), \mu_G(y)] = M, \tag{3-41}$$

can be employed to delimit the boundary between Z and W. The spread of the boundary is then controlled by ℓ and the membership function $\mu_{F \cap G}$.

Definition 3. 6. The α-boundary between regions Z and W, denoted as α-BOUNDARY(Z, W), is defined by

$$\alpha\text{-BOUNDARY}(Z, W) = \left\{y \mid \min[\mu_F(y), \mu_G(y)] \geq \alpha\right\}, \quad 0 < \alpha \leq \ell. \tag{3-42}$$

Therefore, the α-boundary between two regions is generally a zone rather than a line. The coverage of Z and W depends on the value of α. It also determines the extent to which they overlap. Individually, Z and W are respectively delimited as

$$\underline{Z} = \left\{y \mid \mu_F(y) \geq \alpha\right\}, \text{ and } \underline{W} = \left\{y \mid \mu_G(y) \geq \alpha\right\}. \tag{3-43}$$

Example 3.5. Let Z be a region of *moderate* agricultural employment and is defined by (3-33). Let W be a region of *rather high* agricultural employment and is defined by

$$\mu_{rather\ high}(y) = \begin{cases} 0, & \text{if } y \leq 24, \\ \frac{y - 24}{4}, & \text{if } 24 < y < 28, \\ 1, & \text{if } y = 28, \\ \frac{32 - y}{4}, & \text{if } 28 < y < 32, \\ 0, & \text{if } y \geq 32. \end{cases} \tag{3-44}$$

(see Fig. 3.10)

Then, M = 0.25 and S = 0.75. That is, $\underline{Z} = \{y \mid 17 < y < 25\}$ and $\underline{W} = \{y \mid 25 < y < 31\}$ are 0.75-separable if $y^* = 25$ is selected as the boundary. In general, the boundary between Z and W, by (3-39), is

$$\text{BOUNDARY}(Z, W) = \left\{y \mid 24 < y < 26\right\}.$$

Let $\ell < 0.25$ be the separation threshold. Let $\alpha = 0.2$, then the 0.2-boundary is

$$0.2\text{-BOUNDARY}(Z, W) = \left\{y \mid 24.8 \leq y \leq 25.2\right\}, \text{ and}$$

$$\underline{Z} = \left\{y \mid 16.8 \leq y \leq 25.2\right\},$$

$$\underline{W} = \left\{y \mid 24.8 \leq y \leq 31.2\right\}.$$

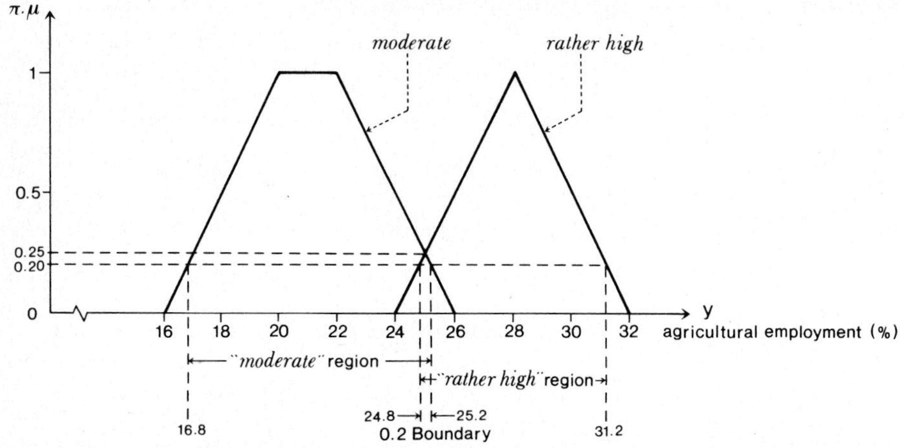

Fig. 3.10 Possibility distribution functions of *moderate* and *rather high*

Though the above discussion only focuses on the two-region case, the arguments can be extended to the multiple-region situation.

Let $\{Z_i\}$, $i = 1, \ldots, m$, be a system of m regions, $m > 2$. Let each region Z_i be characterized by

$$\begin{cases} P_i\colon\ Y(Z_i) \text{ is } F_i, \\ \Pi_{Y(Z_i)} = F_i, \\ \text{Poss}(Y(Z_i)=y) = \pi_{Y(Z_i)}(y) = \mu_{F_i}(y). \end{cases} \tag{3-45}$$

Let F_i, $i = 1, \ldots, m$, be bounded and convex. By proposition 3.1, $F_i \cap F_j$, $\forall\, i, j, i \neq j$, are bounded and convex also. By theorem 3.1, they all possess maximal grades. Let M_{ij} denotes the maximal grade of $F_i \cap F_j$. Then

$$M_{ij} = \sup_y \min[\mu_{F_i}(y), \mu_{F_j}(y)], \quad \forall\, i, j, i \neq j. \tag{3-46}$$

The pair-wise boundary problem follows the argument of the two-region case.

Definition 3.7. The boundary between regions Z_i and Z_j, for all i, j with $i \neq j$, denoted as BOUNDARY(Z_i, Z_j), is defined by

$$\text{BOUNDARY}(Z_i, Z_j) = \left\{ y \mid 0 < \min[\mu_{F_i}(y), \mu_{F_j}(y)] \le M_{ij} \right\}. \quad (3\text{-}47)$$

Taking all regions into consideration, however, a separation threshold

$$\ell < M = \min_{\substack{ij \\ i \neq j}} M_{ij} = \min_{\substack{ij \\ i \neq j}} \sup_y \min[\mu_{F_i}(y), \mu_{F_j}(y)], \quad (3\text{-}48)$$

can be imposed so that overlapping of all regions can be controlled. That is, the separation threshold is less than the minimum of the maximal grades of $F_i \cap F_j$, $\forall$ i, j, $i \neq j$. It guarantees that to-be-classified regions belong to at least one prespecified region.

Definition 3.8. The α-boundary between regions Z_i and Z_j, for all i, j with $i \neq j$, denoted as α-BOUNDARY(Z_i, Z_j), is defined by

$$\alpha\text{-BOUNDARY}(Z_i, Z_j) = \left\{ y \mid \min[\mu_{F_i}(y), \mu_{F_j}(y)] \ge \alpha \right\}, \quad 0 < \alpha \le \ell. \quad (3\text{-}49)$$

Therefore, regions Z_i and Z_j become

$$\underline{Z}_i = \left\{ y \mid \mu_{F_i}(y) \ge \alpha \right\}, \text{ and } \underline{Z}_j = \left\{ y \mid \mu_{F_j}(y) \ge \alpha \right\}. \quad (3\text{-}50)$$

They generally overlap to a certain degree.

Example 3.6. Let Z_1 be a region of *moderate* agricultural employment and is defined by (3-33). Let Z_2 be a region of *rather high* agricultural employment and is defined by (3-44). Let Z_3 be a region of *high* agricultural employment and is defined by (3-30). Then $M_{12} = 0.25$, $M_{13} = 0.168$, and $M_{23} = 0.667$. Thus, M = min(0.25, 0.168, 0.667) = 0.168.

Let $\alpha = 0.16$. Then, the three boundaries are respectively:

$$0.16\text{-BOUNDARY}(Z_1, Z_2) = \left\{ y \mid 24.64 \le y \le 25.36 \right\},$$

$$0.16\text{-BOUNDARY}(Z_1, Z_3) = \left\{ y \mid 25.28 \le y \le 25.36 \right\},$$

$$0.16\text{-BOUNDARY}(Z_2, Z_3) = \left\{ y \mid 25.28 \le y \le 31.36 \right\},$$

and,

$$\underline{Z}_1 = \left\{ y \mid 16.64 \le y \le 25.36 \right\},$$

$$\underline{Z}_2 = \left\{ y \mid 24.64 \le y \le 31.36 \right\},$$

$$\underline{Z}_3 = \left\{ y \mid y \ge 25.28 \right\}.$$

Therefore, the extent of transition from one region to another depends on how similar in characteristics the two regions are. The more similar they are, the more they overlap and the fuzzier their boundary becomes.

Remark. The min-operator in (3-48) is a very strict operator for determining the separation threshold. A very low value of may be obtained sometimes. Under this situation, a large extent of overlapping among regions is realized. Therefore, it should only be employed if controlled overlapping of all regions is desirable. Otherwise, pair-wise determination of boundary in (3-47) would be more appropriate.

3.3. ASSIGNMENT TO A REGION

In section 3.2, our discussion is centered on the characterization of regions when their differentiating characteristics are imprecise. Our task is to create a pattern which best describes the characteristics of a region.

The regional assignment problem to be discussed in this section involves the assignment of spatial units to a set of prespecified regions. It consists of a discrimination procedure by which the assignment can be determined. In practice, it is a pattern-matching process which matches a spatial unit with the most similar prespecified region.

Similar to the characterization of a region, a spatial

unit can be characterized with respect to a set of characteristics. The assignment procedure then attempts to make the best match of the two characterizations. The regional assignment procedure discussed here is mainly based on Leung (1985a).

Characterizations of a spatial unit is discussed in subsection 3.3.1. Precise and fuzzy characteristics are both considered. In subsection 3.3.2, assignment procedures which can handle both precise and fuzzy patterns are discussed.

3.3.1. Characterization of a Spatial Unit

A spatial unit can be characterized by a joint linguistic proposition with precise or fuzzy denotation. In general, we have two basic types of characterizations.

(a) Characterization with precise information.

Let $\{\underline{\mathscr{Y}}_k\}$, $k = 1, \ldots, \ell$, be a set of differentiating characteristics. Let X be a spatial unit characterized by a joint linguistic proposition

$$Q: Y(X) \text{ is } \beta, \tag{3-51}$$

which induces a possibility distribution function

$$\pi_{Y(X)}(y) = \begin{cases} 1, & \text{if } y = \beta, \\ 0, & \text{if } y \neq \beta, \end{cases} \tag{3-52}$$

where $\beta \in \mathbb{R}^\ell$ is a vector whose component β_k, $k = 1, \ldots, \ell$, on precise values in $\mathbb{R}$.

Thus, the joint linguistic proposition in (3-51) is in fact

$$Q: Y_1(X) \text{ is } \beta_1 \text{ and } \ldots \text{ and } Y_k(X) \text{ is } \beta_k \text{ and } \ldots \text{ and } Y_\ell(X) \text{ is } \beta_\ell, \tag{3-53}$$

with $\beta_k \in \mathbb{R}$, $k = 1, \ldots, \ell$.

The denotation of the proposition in (3-51) is precise since β is a vector with precise numerical values β_k's. Thus, there is no fuzziness involved.

(b) Characterization with imprecise information.

Let X be a spatial unit characterized by a joint linguistic proposition

$$Q: Y(X) \text{ is } H, \tag{3-54}$$

which induces a possibility distribution

$$\Pi_{Y(X)} = H, \tag{3-55}$$

such that

$$\begin{aligned} \text{Poss}(Y(X)=y) &= \pi_{Y(X)}(y) = \mu_H(y) \\ &= \min[\mu_{H_1}(y_1), \ldots, \mu_{H_k}(y_k), \ldots, \mu_{H_\ell}(y_\ell)]. \end{aligned} \tag{3-56}$$

This format of characterization is identical to that employed in regional characterization, (3-1) to (3-3). The information H_k, $k = 1, \ldots, \ell$, can take on fuzzy numbers, such as "*approximately* β_k", as its value, or it can just be any linguistic value, such as "*very low*", without any specific reference points. In either way, they are just fuzzy subsets.

Therefore, the assignment procedures should be able to assign spatial units with precise or fuzzy characteristics to prespecified regions with fuzzy characteristics.

3.3.2. Regional Assignment Procedures

Let P in (3-1) and Q in (3-54) be the characterizations of region Z and spatial unit X respectively. Then, the possibility of X being assigned to Z is obtained through the following assignment rule:

$$\begin{aligned} \text{Poss}(Q|P) &= \text{Poss}(Y(X) \text{ is } H | Y(Z) \text{ is } F) \\ &= \sup_y \min[\mu_H(y), \mu_F(y)] \\ &= \min\Big\{\sup_{y_1} \min[\mu_{H_1}(y_1), \mu_{F_1}(y_1)], \ldots, \\ &\qquad \sup_{y_k} \min[\mu_{H_k}(y_k), \mu_{F_k}(y_k)], \ldots, \\ &\qquad \sup_{y_\ell} \min[\mu_{H_\ell}(y_\ell), \mu_{F_\ell}(y_\ell)]\Big\}. \end{aligned} \tag{3-57}$$

The assignment rule says that the possibility of assigning spatial unit X with characteristic H to region Z with characteristic F is equal to the minimum of the heights of the intersections $H_k \cap F_k$, $k = 1, \ldots, \ell$. It is the minimum threshold applied to all characteristics and is the possibility of X being considered as Z. The possibility of assigning spatial unit X to region Z then depends on how well the pattern H characterizing X matches F characterizing Z.

Example 3.7. Let Z be a region of *moderate* income and *high* industrial employment and is characterized by (3-14) with the corresponding membership functions depicted in Fig. 3.1 and Fig. 3.2. Let spatial unit X be characterized by

Q: Income(X) is *approximately* 12,000 and
Industrial employment(X) is *approximately* 40%, (3-58)

which induces a joint possibility distribution

$$\Pi_{[\text{Income}(X),\ \text{Industrial employment}(X)]} = \textit{approximately}\ 12{,}000 \times \textit{approximately}\ 40\%$$

such that

$$\begin{aligned}&\text{Poss}(\text{income}(X)=a,\ \text{industrial employment}(X)=b)\\ &= \pi_{[\text{Income}(X),\ \text{industrial employment}(X)]}(a, b)\\ &= \min[\mu_{\textit{approximately}\ 12{,}000}(a),\ \mu_{\textit{approximately}\ 40\%}(b)]. \qquad (3\text{-}59)\end{aligned}$$

Let the membership functions $\mu_{\textit{approximately}\ 12{,}000}$ and $\mu_{\textit{approximately}\ 40\%}$ be depicted respectively in Fig. 3.11a and b. Then, the possibility of X being assigned to region Z is

$$\begin{aligned}&\text{Poss}(\text{Income}(X)\ \text{is}\ \textit{approximately}\ 12{,}000\ \text{and}\\ &\quad \text{industrial employment}(X)\ \text{is}\ \textit{approximately}\ 40\% \mid \text{Income}(Z)\\ &\quad \text{is}\ \textit{moderate}\ \text{and industrial employment}(Z)\ \text{is}\ \textit{high})\\ &= \sup_{y=(y_1,y_2)} \min[\mu_{\textit{approximately}\ 12{,}000 \times \textit{approximately}\ 40\%}(y_1, y_2),\\ &\qquad \mu_{\textit{moderate} \times \textit{high}}(y_1, y_2)]\end{aligned}$$

$$= \min\left\{\sup_{y_1} \min[\mu_{approximately\ 12,000}(y_1), \mu_{moderate}(y_1)],\right.$$

$$\left.\sup_{y_2} \min[\mu_{approximately\ 40\%}(y_2), \mu_{high}(y_2)]\right\}$$

$$= \min\{0.4, 0.2\} = 0.2. \qquad (3\text{-}60)$$

(see Fig. 3.11a and b)

Let $\{Z_1, \ldots, Z_j, \ldots, Z_n\}$ be a set of n regions to which a set of m spatial units $\{X_1, \ldots, X_i, \ldots, X_m\}$ is to be assigned. Let $\{\underline{\mathscr{Y}}_1, \ldots, \underline{\mathscr{Y}}_k, \ldots, \underline{\mathscr{Y}}_\ell\}$ be a set of characteristics by which regions and spatial units are characterized as F_j

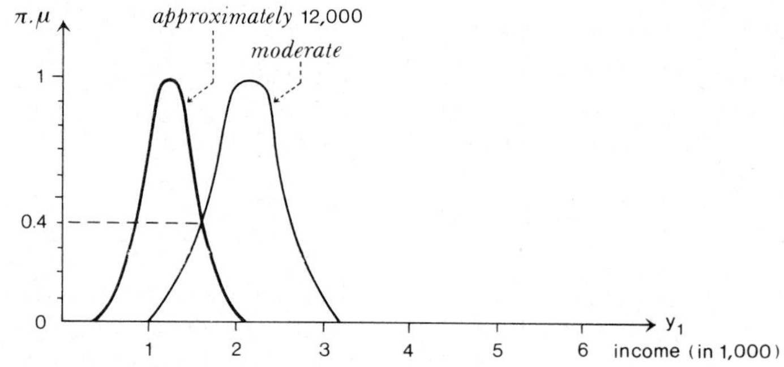

Fig. 3.11a Possibility of assigning *approximately* 12,000 to *moderate*

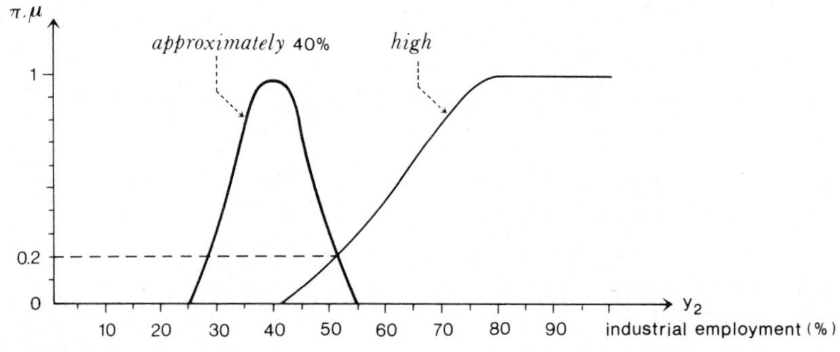

Fig. 3.11b Possibility of assigning *approximately* 40% to *high*

and H_i respectively. The regional assignment procedure then involves the determination of the possibility of assigning spatial unit X_i, i = 1, ..., m, to region Z_j, j = 1, ..., n, via:

$$\mathrm{Poss}(Q_i|P_j) = \mathrm{Poss}(Y(X_i) \text{ is } Q_i|Y(Z_j) \text{ is } P_j)$$

$$= \sup_y \min[\mu_{H_i}(y), \mu_{F_j}(y)]$$

$$= \min\Big\{\sup_{y_1} \min[\mu_{H_{i1}}(y_1), \mu_{F_{j1}}(y_1)], \ldots,$$

$$\sup_{y_k} \min[\mu_{H_{ik}}(y_k), \mu_{F_{jk}}(y_k)], \ldots,$$

$$\sup_{y_\ell} \min[\mu_{H_{i\ell}}(y_\ell), \mu_{F_{j\ell}}(y_\ell)]\Big\}. \qquad (3\text{-}61)$$

Example 3.8. Let Z_1, Z_2, and Z_3 be three prespecified regions characterized respectively by the following joint linguistic propositions:

P_1: Income(Z_1) is *high* and industrial employment(Z_1) is *rather high*; (3-62)

P_2: Income(Z_2) is *moderate* and industrial employment(Z_2) is *high*; (3-63)

P_3: Income(Z_3) is *low* and industrial employment(Z_3) is *sort of low*. (3-64)

(The graphs of the membership functions for income are depicted in Fig. 3.12a. Those in Fig. 3.13a and Fig. 3.14a are identical to that of Fig. 3.12a. Likewise, the graphs of the membership functions for industrial employment are depicted in Fig. 3.12b, Fig. 3.13b, and Fig. 3.14b).

Let X_1, X_2, and X_3 be three spatial units characterized respectively by the following joint linguistic propositions:

Q_1: Income(X_1) is *approximately* 36,000 and industrial employment(X_1) is *approximately* 75%; (3-65)

(see Fig. 3.12a for income and Fig. 3.12b for industrial employment)

Q_2: Income(X_2) is *approximately* 12,000 and
industrial employment(X_2) is *approximately* 40%; (3-66)

(see Fig. 3.13a for income and Fig. 3.13b for industrial employment)

Q_3: Income(X_3) is *approximately* 16,000 and
industrial employment(X_3) is *approximately* 50%. (3-67)

(see Fig. 3.14a for income and Fig. 3.14b for industrial employment)

Applying the assignment procedure in (3-61), the possibilities of assigning X_1 to Z_1, Z_2, and Z_3 are respectively:

$$\text{Poss}(Q_1|P_1) = 0.5; \ \text{Poss}(Q_1|P_2) = 0.1; \text{ and } \text{Poss}(Q_1|P_3) = 0. \tag{3-68}$$

(see Fig. 3.12a and b)

By the same token, the possibilities of assigning X_2 to regions Z_1, Z_2, and Z_3 are respectively:

$$\text{Poss}(Q_2|P_1) = 0.1; \ \text{Poss}(Q_2|P_2) = 0.2; \text{ and}$$
$$\text{Poss}(Q_2|P_3) = 0.8. \tag{3-69}$$

(see Fig. 3.13a and b)

Similarly, the possibilities of assigning X_3 to Z_1, Z_2, and Z_3 are respectively:

$$\text{Poss}(Q_3|P_1) = 0.1; \ \text{Poss}(Q_3|P_2) = 0.25; \text{ and}$$
$$\text{Poss}(Q_3|P_3) = 0.25. \tag{3-70}$$

(see Fig. 3.14a and b)

Examining the results summarized in the fuzzy assignment matrix W in Fig. 3.15, we can see that there is a certain degree of fuzziness in this regional assignment problem. The mutually exclusive property does not hold because a spatial unit is not assigned to one and only one region, but rather to all regions with varying possibilities. For example, spatial unit X_1 is more similar in characteristics to region Z_1 while it is also similar, to a lesser extent, to Z_2. The pattern characterizing X_2 is closer to that of Z_3 but is also similar

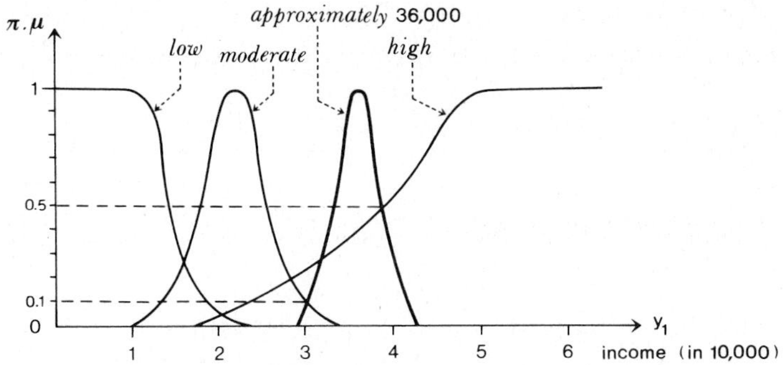

Fig. 3.12a Possibilities of assigning spatial unit X_1 to regions Z_1, Z_2 and Z_3 with respect to income

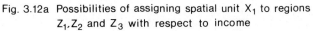

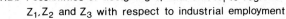

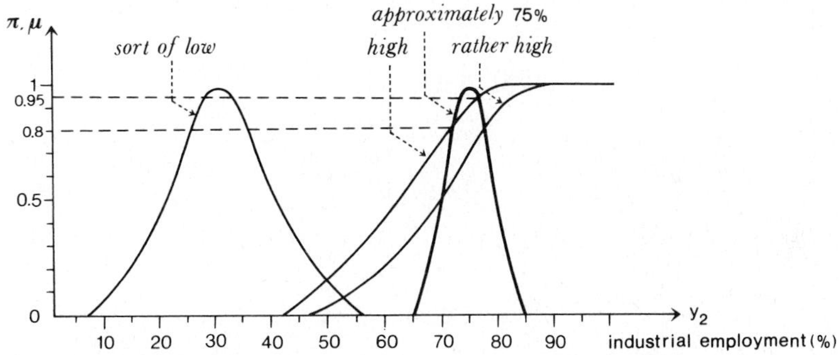

Fig. 3.12b Possibilities of assigning spatial unit X_1 to regions Z_1, Z_2 and Z_3 with respect to industrial employment

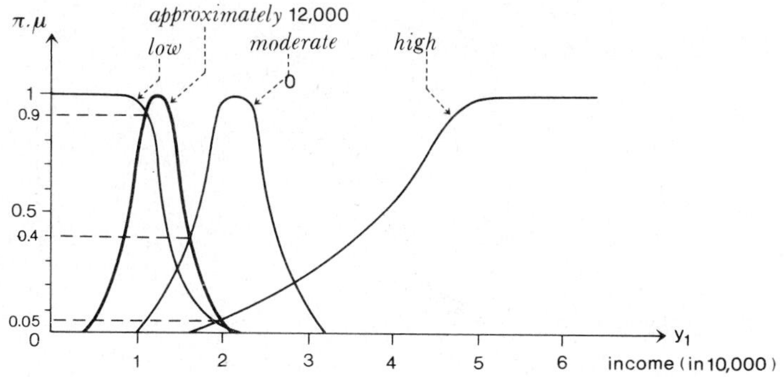

Fig. 3.13a Possibilities of assigning spatial unit X_2 to regions Z_1, Z_2 and Z_3 with respect to income

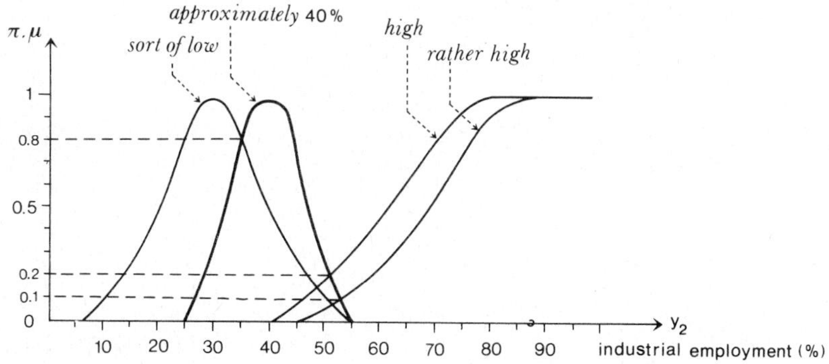

Fig. 3.13b Possibilities of assigning spatial unit X_2 to regions Z_1, Z_2 and Z_3 with respect to industrial employment

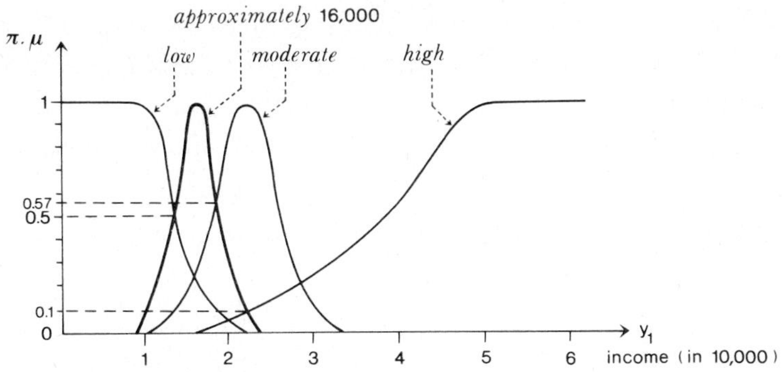

Fig. 3.14a Possibilities of assigning spatial unit X_3 to regions Z_1, Z_2 and Z_3 with respect to income

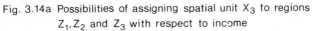

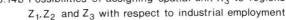

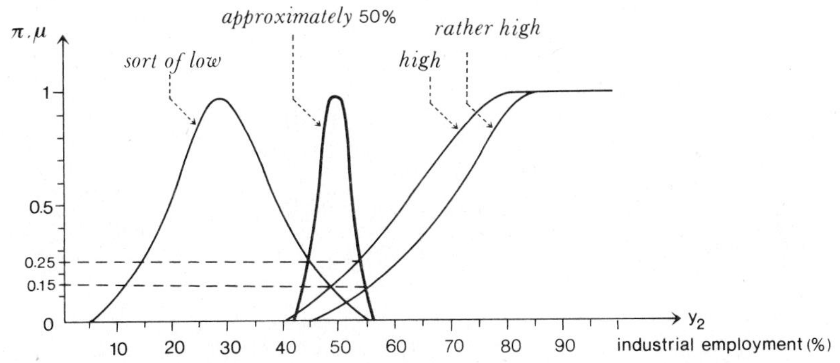

Fig. 3.14b Possibilities of assigning spatial unit X_3 to regions Z_1, Z_2 and Z_3 with respect to industrial employment

to that of Z_1 and Z_2 to some extent. The pattern of X_3 is fuzzier. It is almost equally possible to assign it to Z_1 or Z_2 or Z_3. Therefore, a clear-cut assignment is not realized. In general, unambiguous assignment is an exception when the regions and spatial units possess fuzzy characteristics. Nevertheless, clear-cut assignment can always be obtained from the fuzzy assignment results if necessary.

		regions		
		Z_1	Z_2	Z_3
W = spatial units	X_1	0.50	0.10	0
	X_2	0.10	0.20	0.80
	X_3	0.10	0.25	0.25

Fig. 3.15 Fuzzy assignment matrix for example 3.8

Let $W = \{w_{ij}\}$, $i = 1, \ldots, m$; $j = 1, \ldots, n$, be the fuzzy assignment matrix with $w_{ij} = \text{Poss}(Q_i | P_j)$. For example, $w_{23} = 0.8$ in Fig. 3.15 is the possibility of assigning X_2 to Z_3. To achieve a clear-cut assignment, an ordinary assignment matrix $\underline{W}$ can be obtained by selecting an assignment threshold α such that

$$f_{\underline{W}}(w_{ij}) = \begin{cases} 1, & \text{if } w_{ij} \geq \alpha, \\ 0, & \text{if } w_{ij} < \alpha. \end{cases} \tag{3-71}$$

That is, if the possibility of assigning X_i to Z_j is greater than or equal to α, then X_i is assigned to Z_j. If $\alpha = 0.5$ in (3-71), we obtain the ordinary assignment matrix $\underline{W}$ which is closest to the fuzzy assignment matrix W under the concept of generalized relative Hamming distance [see (3-94)]. For example, the ordinary matrix, $\underline{W}$, closest to the fuzzy assignment matrix in Fig. 3.15 is depicted in Fig. 3.16. Under this situation, X_1 is assigned to Z_1, X_2 is assigned to Z_2, and X_3 cannot be assigned to any of the three regions.

regions

$$\underline{W} = \text{spatial units} \begin{array}{c} X_1 \\ X_2 \\ X_3 \end{array} \overset{\begin{array}{ccc} Z_1 & Z_2 & Z_3 \end{array}}{\begin{bmatrix} 1 & 0 & 0 \\ 0 & 0 & 1 \\ 0 & 0 & 0 \end{bmatrix}}$$

Fig. 3.16 Precise assignment matrix derived from Fig. 3.15

Of course, if we are willing to accept the highest possibility as indicative of X being Z, then we would assign X_1 to Z_1, X_2 to Z_3, and X_3 to Z_2 and Z_3 in Fig. 3.15.

On the other hand, if a controlled overlapping of regions is desirable, the separation threshold defined in (3-48) can be applied to the fuzzy assignment matrix to determine the final assignment. Consider Fig. 3.15 of example 3.8, the separation threshold is $\ell < \min(0.1, 0.1, 0.25) = 0.1$ (see Fig. 3.17).

	$Z_1 \cap Z_2$	$Z_1 \cap Z_3$	$Z_2 \cap Z_3$
	0.10	0	0
	0.10	0.10	0.20
	0.10	0.10	0.25
max	0.10	0.10	0.25

Fig. 3.17 Matrix for the determination of the separation threshold for Fig. 3.15

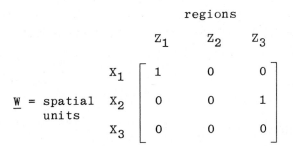

Then, in this particular case, X_1 can be assigned to Z_1 and Z_2, while X_2 and X_3 can be assigned to Z_1, Z_2 and Z_3. That is, X_1, X_2, and X_3 are in fact not distinctive enough to guarantee a complete assignment to a single region.

As stated in (3-51), the characteristics of a spatial unit can be precisely measured. If the joint linguistic proposition of X assumes the format in (3-53), then the possibility of assigning X to Z which is characterized by (3-1) becomes

$$\mathrm{Poss}(Q|P) = \min[\mu_{F_1}(\beta_1), \ldots, \mu_{F_k}(\beta_k), \ldots, \mu_{F_\ell}(\beta_\ell)]. \tag{3-72}$$

Thus, the assignment procedure in (3-57), i.e. (3-61), can handle regional assignment problems with precise and fuzzy information.

In fact, if both P and Q are nonfuzzy, (3-57) reduces to

$$\mathrm{Poss}(Q|P) = \begin{cases} 1, & \text{if } H \cap F \neq \phi, \\ 0, & \text{if } H \cap F = \phi. \end{cases} \tag{3-73}$$

Since what is possible about an assignment does not guarantee what is certain about an assignment, a measure of certainty is then necessary. With respect to the assignment rule in (3-57), the degree of certainty about equating the linguistic characterization Q in (3-54) with the linguistic characterization P in (3-1) is defined by

$$\mathrm{Cert}(Q|P) = \inf_y \max[\mu_H(y), 1 - \mu_F(y)]. \tag{3-74}$$

When P and Q are nonfuzzy statements, (3-74) reduces to

$$\mathrm{Cert}(Q|P) = \begin{cases} 1, & \text{if } F \subseteq H, \\ 0, & \text{otherwise.} \end{cases} \tag{3-75}$$

Based on (3-74), a possibility-certainty duality (Cayrol, Farreny, and Prade, 1980) can be defined by

$$\begin{aligned} &\mathrm{Cert}(Q|P) \\ &= 1 - \mathrm{Poss}(\bar{Q}|P) \\ &= 1 - \sup_y \min[1 - \mu_H(y), \mu_F(y)] \end{aligned}$$

$$= \inf_{y} \max[\mu_H(y), \mu_{1-F}(y)]. \qquad (3\text{-}76)$$

Its interpretation is that the degree of certainty of assigning spatial unit X to region Z is equal to the degree of impossibility of not assigning X to Z. If P and Q are linguistic propositions generating two fuzzy patterns F and H respectively, a perfect match of F and G would only give a certainty of 0.5 instead of 1 in (3-76), i.e. Cert(Q|P) = 0.5. The result is due to the fact that since both F and H are fuzzy, then $\bar{H} \cap F$ is nonempty. Therefore, requiring absolute certainty is to require $\bar{H} \cap F = \phi$, i.e. both F and H are nonfuzzy. Fig. 3.18 and Fig. 3.19 depict the relationships between the possibility and certainty of regional assignment for a one characteristic case.

Example 3.9. (A simple classification of climatic regions, an application)

To demonstrate the versatility of the regional classification framework discussed in sections 3.2 and 3.3, Leung (1987a) has applied it to a simple classification of climatic regions in Taiwan. The purpose was not to launch a full scale climatic classification of the island but to show, as an example, what one can get from the fuzzy set approach.

In the empirical study, annual precipitation, $\mathcal{Y}_1$ (measured in mm.) and annual temperature, $\mathcal{Y}_2$ (measured in °C) are base variables by which climatic regions are classified. With respect to the linguistic variable Y_1 = "precipitation", regions can be prespecified as *abundant* (A), *substantial* (S), and *adequate* (Ad) which are defined respectively by the following membership functions:

$$\mu_A(y_1) = \begin{cases} 1, & \text{if } y_1 \geq 3{,}000 \text{ mm}, \\ \dfrac{y_1 - 2{,}000 \text{ mm}}{1{,}000 \text{ mm}}, & \text{if } 2{,}000 \text{ mm} < y_1 < 3{,}000 \text{ mm}, \\ 0, & \text{if } y_1 \leq 2{,}000 \text{ mm}; \end{cases} \qquad (3\text{-}77)$$

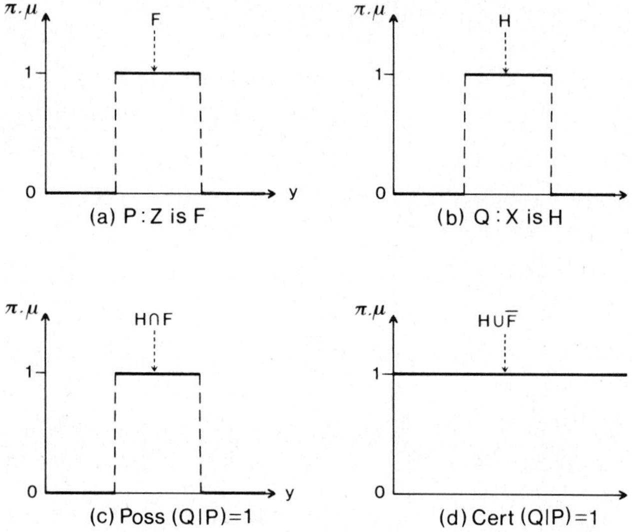

Fig. 3.18 Possibility–certainty duality when both F and H are non-fuzzy

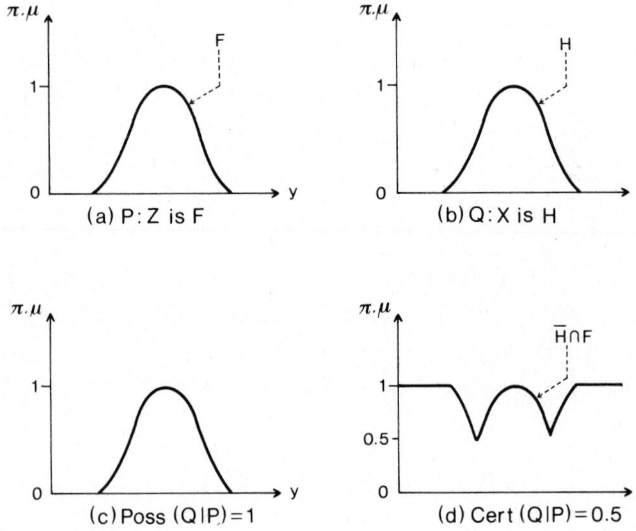

Fig. 3.19 Possibility–certainty duality when both F and H are fuzzy

$$\mu_S(y_1) = \begin{cases} 0, & \text{if } y_1 \le 1{,}300 \text{ mm}, \\ \dfrac{y_1 - 1{,}300 \text{ mm}}{400 \text{ mm}}, & \text{if } 1{,}300 \text{ mm} < y_1 < 1{,}700 \text{ mm}, \\ 1, & \text{if } 1{,}700 \text{ mm} \le y_1 \le 1{,}900 \text{ mm}, \\ \dfrac{2{,}300 \text{ mm} - y_1}{400 \text{ mm}}, & \text{if } 1{,}900 \text{ mm} < y_1 < 2{,}300 \text{ mm}, \\ 0, & \text{if } y_1 \ge 2{,}300 \text{ mm}; \end{cases} \tag{3-78}$$

$$\mu_{Ad}(y_1) = \begin{cases} 0, & \text{if } y_1 \le 500 \text{ mm}, \\ \dfrac{y_1 - 500 \text{ mm}}{400 \text{ mm}}, & \text{if } 500 \text{ mm} < y_1 < 900 \text{ mm}, \\ 1, & \text{if } 900 \text{ mm} \le y_1 \le 1{,}100 \text{ mm}, \\ \dfrac{1{,}500 \text{ mm} - y_1}{400 \text{ mm}}, & \text{if } 1{,}100 \text{ mm} < y_1 < 1{,}500 \text{ mm}, \\ 0, & \text{if } y_1 \ge 1{,}500 \text{ mm}. \end{cases} \tag{3-79}$$

With respect to the linguistic variable Y_2 = "temperature", regions can be prespecified as *hot* (H), *warm* (W), *cool* (Cl), and *cold* (Cd) which are defined respectively by the following membership functions:

$$\mu_H(y_2) = \begin{cases} 1, & \text{if } y_2 \ge 32°\text{C}, \\ \dfrac{y_2 - 24°\text{C}}{8°\text{C}}, & \text{if } 24°\text{C} < y_2 < 32°\text{C}, \\ 0, & \text{if } y_2 \le 24°\text{C}; \end{cases} \tag{3-80}$$

$$\mu_W(y_2) = \begin{cases} 0, & \text{if } y_2 \le 16°\text{C}, \\ \dfrac{y_2 - 16°\text{C}}{4°\text{C}}, & \text{if } 16°\text{C} < y_2 < 20°\text{C}, \\ 1, & \text{if } 20°\text{C} \le y_2 \le 22°\text{C}, \\ \dfrac{26°\text{C} - y_2}{4°\text{C}}, & \text{if } 22°\text{C} < y_2 < 26°\text{C}, \\ 0, & \text{if } y_2 \ge 26°\text{C}; \end{cases} \tag{3-81}$$

$$\mu_{Cl}(y_2) = \begin{cases} 0, & \text{if } y_2 \leq 8°C, \\ \dfrac{y_2 - 8°C}{5°C}, & \text{if } 8°C < y_2 < 13°C, \\ 1, & \text{if } y_2 = 13°C \\ \dfrac{18°C - y_2}{5°C}, & \text{if } 13°C < y_2 < 18°C, \\ 0, & \text{if } y_2 \geq 18°C; \end{cases} \tag{3-82}$$

$$\mu_{Cd}(y_2) = \begin{cases} 1, & \text{if } y_2 \leq 5°C, \\ \dfrac{10°C - y_2}{5°C}, & \text{if } 5°C < y_2 < 10°C, \\ 0, & \text{if } y_2 \geq 10°C. \end{cases} \tag{3-83}$$

To classify Taiwan into climatic regions, annual precipitation (y_1) and annual temperature (y_2) from 80 stations are extracted from the 1984 Climatological Data Annual Report (see Table 3.1). The data are only single-year data and the climatic stations are limited in number. However, they are sufficient to show how the classification framework works.

Taking annual precipitation and annual temperature into consideration, there should theoretically be 12 climatic types. Nevertheless, some of the types do not exist in Taiwan. Thus, only 9 are relevant. They are: Abundant Precipitation and Hot (AH), Abundant Precipitation and Warm (AW), Abundant Precipitation and Cool (ACl), Abundant Precipitation and Cold (ACd), Substantial Precipitation and Hot (SH), Substantial Precipitation and Warm (SW), Substantial Precipitation and Cool (SCl), Substantial Precipitation and Cold (SCd), and Adequate Precipitation and Warm (AdW). Theoretically, there are $\binom{9}{2} = 36$ boundaries to be determined. Practically, only those between adjacent regions need to be delimited.

The possibility (grade of membership) of assigning each climatic station to the climatic classes are first determined. Since the measurements y_1 and y_2 are precise, the assignment rule in (3-72) is employed and the results are tabulated in

Table 3.1. Applying the arguments in subsection 3.2.3, the climatic regions together with their separating boundaries (clear-cut) and fuzzy boundaries are obtained (see Table 3.2, Fig. 3.20, Fig. 3.21). For example, AW and SW are adjacent, then by (3-46) $M_{AW,\ SW} = \min(M_{AS}, M_W) = \min(0.214, 1) = 0.214$ and the highest degree of separation of AW and SW is 0.786. The precise boundary (2,214 mm; 16.856°C) separates $\underline{AW} = \{(y_1, y_2) | y_1 > 2{,}214 \text{ mm and } 16.856°C < y_2 < 25.144°C\}$, and $\underline{SW} = \{(y_1, y_2) | 1{,}385.6 \text{ mm} < y_1 < 2{,}214 \text{ mm and } 16.856°C < y_2 < 25.144°C\}$ into two disjoint regions to 0.786 degree. Here, the clear-cut boundary is 2,214 mm since both AW and SW are within a warm region. Similarly, for SW and ACl, $M_{SW,\ ACl} = \min(M_{SA}, M_{W,\ Cl}) = \min(0.214, 0.22) = 0.214$ and the highest degree of separation is 0.786. The precise boundary (2,214 mm; 16.856°C) separates $\underline{SW} = \{(y_1, y_2) | 1{,}385.6 \text{ mm} < y_1 < 2{,}214 \text{ mm and } 16.856°C < y_2 < 25.144°C\}$ and $\underline{ACl} = \{(y_1, y_2) | y_1 > 2{,}214 \text{ mm and } 9.07°C < y_2 < 16.856°C\}$ into two disjoint regions. Thus, when boundaries are set for the values obtained by (3-46), we obtain clear-cut boundaries and regions are completely disjoint at $1 - M_{ij}$ degree of separation (Fig. 3.20). However, they do not show the intrinsic fuzziness completely. It is only when definition 3.7 is made effective do we manage to depict the genuine fuzziness of the boundaries (Fig. 3.21).

The empirical study shows that regions are generally fuzzy constructs with fuzzy boundaries. The fuzzy set framework not only can delimit clear-cut boundaries but can also precisely delimit the zones of transition, the edges and cores of regions. Based on the grades of membership of the climatic stations, core, gray, and unusual areas can be identified. Obviously, such a task cannot be accomplished by the conventional logic of regional classification. Within the latter framework, valuable information would be artifically eliminated.

3.4. GROUPING FOR REGIONS

In section 3.3, I have discussed the characterization of a

Table 3.1 Climatic stations and their grades of membership in various climatic regions

Stations			Precipitation				Temperature					Climatic Regions (grades of membership)								
				grades of membership				grades of membership												
Province	Name	No.	(mm)	A	S	Ad	(°C)	H	W	Cl	Cd	AH	AW	ACl	ACd	SH	SW	SCl	AdH	AdW
Taibei	Anbu	1	6404	1.000			16.4		0.100	0.320			0.100	0.320						
Taibei	Zhuzi Hu	2	6081	1.000			18.1		0.525				0.525							
Taibei	Danshui	3	2621	0.621			22.0		1.000				0.621							
Taibei	Jilong	4	4836	1.000			22.0		1.000				1.000							
Taibei	Taibei	5	2711	0.711			22.0		1.000				0.711							
Taoyuan	Yangmei	6	2562	0.562			20.3		1.000				0.562							
Yilan	Yilan	7	3288	1.000			21.8		1.000				1.000							
Yilan	Sanxing	8	4784	1.000			22.6		0.850				0.850							
Yilan	Su'ao	9	4178	1.000			22.0		1.000				1.000							
Xinzhu	Xinfeng	10	2017	0.017	0.708		21.4		1.000				0.017				0.708			
Xinzhu	Xinzhu	11	1962		0.845		21.9		1.000								0.845			
Xinzhu	Guanxi	12	3106	1.000			21.5		1.000				1.000							
Xinzhu	Wufeng	13	3236	1.000			16.8		0.200	0.240			0.200	0.240						
Miaoli	Zhunan	14	2095	0.095	0.513		21.2		1.000				0.095				0.513			
Miaoli	Nanzhuang	15	2775	0.775			17.8		0.450	0.040			0.450	0.040						
Taizhong	Qingshui	16	1030			1.000	21.7		1.000											1.000
Taizhong	Wuqi	17	1092			1.000	21.9		1.000											1.000
Taizhong	Taizhong	18	1548		0.620		22.5		0.875								0.620			
Taizhong	Dali	19	1662		0.905		22.9		0.775								0.775			
Taizhong	Wufeng	20	1653		0.883		22.7		0.825								0.825			
Zhanghua	Dacun	21	599			0.248	23.5		0.625											0.248
Zhanghua	Tianzhong	22	1572		0.680		22.4		0.900								0.680			
Zhanghua	Xizhou	23	1000			1.000	22.2		0.950											0.950
Nantou	Caotun	24	1757		1.000		22.9		0.775								0.775			
Nantou	Nantou	25	1994		0.765		23.2		0.700								0.700			
Nantou	Puli	26	2016	0.016	0.710		22.6		0.850				0.016				0.710			
Nantou	Puli	27	2072	0.072	0.570		21.6		1.000				0.072				0.570			

Table 3.1 (cont'd)

Stations			Precipitation				Temperature					Climatic Regions (grades of membership)								
				grades of membership				grades of membership												
Province	Name	No.	(mm)	A	S	Ad	(ºC)	H	W	Cl	Cd	AH	AW	ACl	ACd	SH	SW	SCl	AdH	AdW
Nantou	Yuchi	28	2167	0.167	0.333		21.0		1.000				0.167				0.333			
Nantou	Riyue Tan	29	2198	0.198	0.255		19.1		0.775				0.198				0.255			
Nantou	Lugu	30	2551	0.551			16.8		0.200	0.240			0.200	0.240						
Nantou	Danda	31	1732		1.000		10.8			0.560								0.560		
Yunlin	Huwei	32	1117			0.958	22.1		0.975											9.958
Hualian	Xiulin	33	3114	1.000			13.8			0.840				0.840						
Hualian	Hualian	34	2246	0.246	0.135		22.6		0.850				0.246				0.135			
Hualian	Shoufeng	35	2163	0.163	0.343		22.5		0.875				0.163				0.343			
Hualian	Wanrong	36	1889		1.000		14.7			0.660								0.660		
Hualian	Ruisui	37	1943		0.893		22.3		0.925								0.893			
Hualian	Zhuo Xi	38	3175	1.000			18.7		0.675				0.675							
Jiayi	Ali Shan	39	3166	1.000			10.6			0.520				0.520						
Jiayi	Yu Shan	40	2605	0.605			4.0				1.000				0.605					
Jiayi	Wufeng	41	4039	1.000			10.7			0.540				0.540						
Jiayi	Fanlu	42	2188	0.188	0.280		17.2		0.300	0.160			0.188	0.160			0.280	0.160		
Jiayi	Jiayi	43	1379		0.198	0.303	22.5		0.875								0.198			0.303
Jiayi	Pozi	44	986			1.000	22.3		0.925											0.925
Jiayi	Budai	45	955			1.000	23.4		0.650											0.650
Jiayi	Yizhu	46	1062			1.000	24.5	0.063	0.375										0.063	0.375
Tainan	Dongji Yu	47	516			0.040	23.1		0.725											0.040
Tainan	Xinying	48	1215			0.713	22.7		0.825											0.713
Tainan	Beimen	49	1198			0.755	23.0		0.750											0.750
Tainan	Qigu	50	1245			0.638	22.9		0.775											0.638
Tainan	Shanhua	51	1754		1.000		23.5		0.625								0.625			
Tainan	Xinhua	52	1758		1.000		23.1		0.725								0.725			
Tainan	Tainan	53	1373		0.183	0.318	23.6		0.600								0.183			0.318
Tainan	Guanziling	54	2244	0.244	0.140		19.7		0.925				0.244				0.140			

Table 3.1 (cont'd)

Stations			Precipitation				Temperature					Climatic Regions (grades of membership)								
				grades of membership				grades of membership												
Province	Name	No.	(mm)	A	S	Ad	(°C)	H	W	Cl	Cd	AH	AW	ACl	ACd	SH	SW	SCl	AdH	AdW
Gaoxiong	Liugui	55	2058	0.058	0.605		24.3	0.038	0.425			0.038	0.058			0.038	0.425			
Gaoxiong	Shanlin	56	1727		1.000		23.9		0.525								0.525			
Gaoxiong	Maolin	57	2460	0.460			20.7		1.000				0.460							
Gaoxiong	Xinwei	58	2296	0.296	0.010		19.8		0.950				0.296				0.010			
Gaoxiong	Meinong	59	1769		1.000		23.5		0.625								0.625			
Gaoxiong	Meinong	60	1907		0.983		22.9		0.775								0.775			
Gaoxiong	Tianliao	61	1762		1.000		22.6		0.850								0.850			
Gaoxiong	Gaoxiong	62	1629		0.823		24.3	0.038	0.425							0.038	0.425			
Gaoxiong	Fengshan	63	1810		1.000		25.6	0.200	0.100							0.200	0.100			
Taidong	Xingang	64	2443	0.443			23.6		0.600				0.443							
Taidong	Yanping	65	2977	0.977			14.8			0.640				0.640						
Taidong	Beinan	66	4156	1.000			19.0		0.750				0.750							
Taidong	Taidong	67	2181	0.181	0.298		24.1	0.013	0.475			0.013	0.181			0.013	0.298			
Taidong	Taima	68	2024	0.024	0.690		24.1	0.013	0.475			0.013	0.024			0.013	0.475			
Taidong	Dawu	69	2015	0.015	0.713		24.6	0.075	0.350			0.015	0.015			0.075	0.350			
Pingdong	Sandi	70	1774		1.000		17.7		0.425	0.060							0.425	0.060		
Pingdong	Gaoshu	71	1551		0.628		24.6	0.075	0.350							0.075	0.350			
Pingdong	Ligang	72	1519		0.548		24.8	0.100	0.300							0.100	0.300			
Pingdong	Jiuru	73	1557		0.643		24.3	0.038	0.425							0.038	0.425			
Pingdong	Wutai	74	2005	0.005	0.738		19.2		0.800				0.005				0.738			
Pingdong	Pingdong	75	1701		1.000		24.8	0.100	0.300							0.100	0.300			
Pingdong	Neipu	76	2030	0.030	0.675		22.4		0.900				0.030				0.675			
Pingdong	Wandan	77	1544		0.610		25.8	0.225	0.050							0.225	0.050			
Pingdong	Chunri	78	1922		0.945		18.0		0.500								0.500			
Pingdong	Hengchun	79	2194	0.194	0.265		24.5	0.063	0.375			0.063	0.194			0.063	0.265			
Pingdong	Lan Yu	80	2716	0.716			22.7		0.825				0.716							

Table 3.2 Climatic regions and their separating and fuzzy boundaries

Climatic Regions		Maximal Grades (M)	Highest Degrees of Separation (S)	Separating Boundaries	Fuzzy Boundaries
AH	AW	0.166	0.834	25.33°C	(24°C, 26°C)
AW	ACl	0.220	0.780	16.88°C	(16°C, 18°C)
	SW	0.214	0.786	2,214 mm	(2,000 mm, 2,300 mm)
	SCl	0.214	0.786	2,214 mm; 16.856°C	(2,000 mm, 2,300 mm) ∩ (16°C, 18°C)
ACl	ACd	0.200	0.800	9°C	(8°C, 10°C)
	SW	0.214	0.786	2,214 mm; 16.856°C	(2,000 mm, 2,300 mm) ∩ (16°C, 18°C)
	SCl	0.214	0.786	2,214 mm	(2,000 mm, 2,300 mm)
	SCd	0.240	0.786	2,214 mm; 9°C	(2,000 mm, 2,300 mm) ∩ (8°C, 10°C)
SH	SW	0.166	0.834	25.33°C	(24°C, 26°C)
SW	SCl	0.220	0.780	16.88°C	(16°C, 18°C)
	AdW	0.250	0.750	1,400 mm	(1,300 mm, 1,500 mm)

Fig. 3.20 Disjoint climatic regions based on annual precipitation and annual temperature

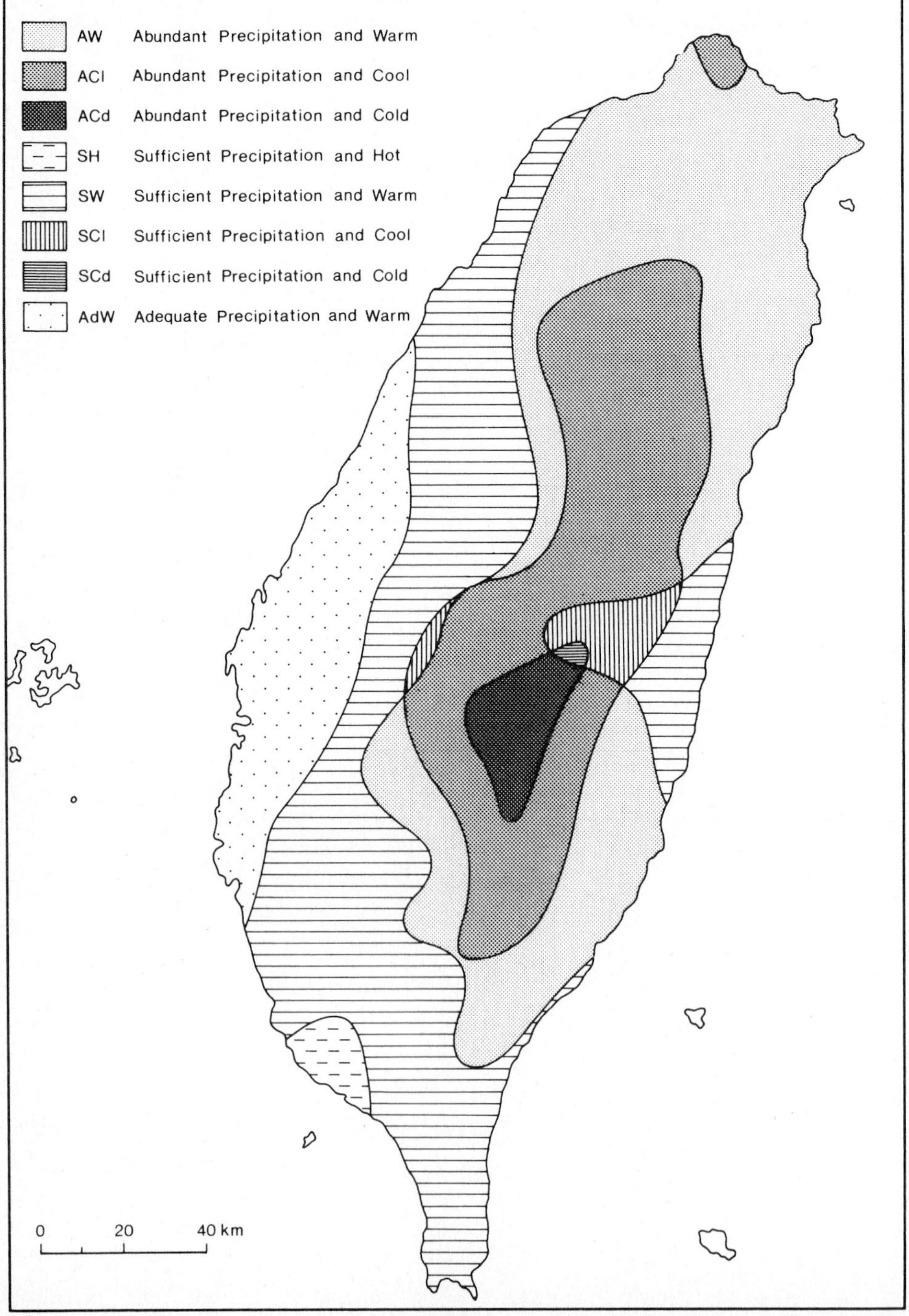

Fig. 3.21
Overlappings of climatic regions based on annual precipitation and annual temperature

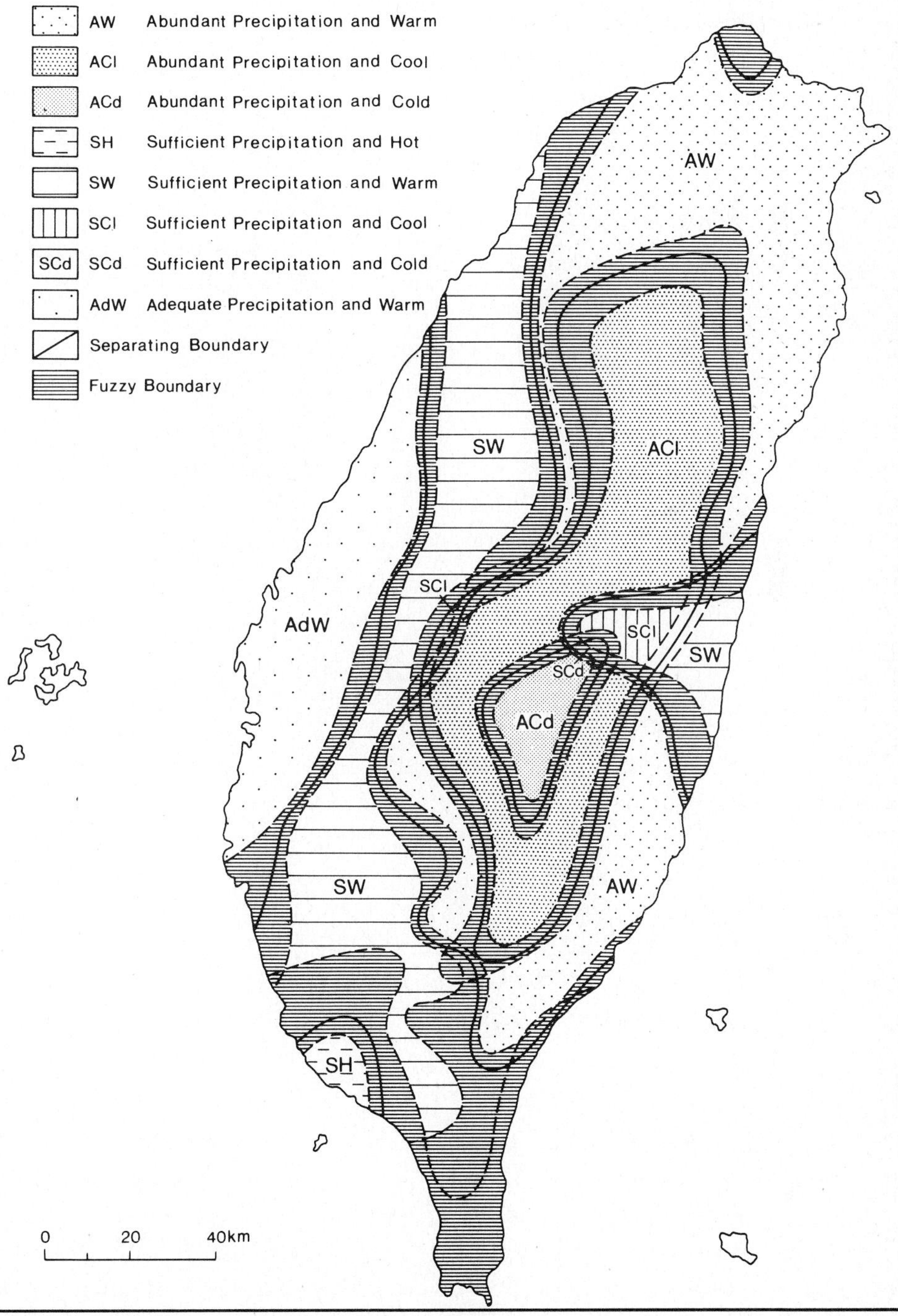

region and the assignment of spatial units to a set of prespecified regions. In this section, our discussion centers on regionalization problems which involve the grouping of spatial units into regions in accordance with their similarities, differences, or interrelationships. To accommodate qualitative and quantitative data with varying degrees of imprecision, a flexible framework for regionalization is necessary. The analysis in this section is partly based on Leung (1984a).

In subsection 3.4.1, a fuzzy set concept of smilarity is formulated to serve as a basis for regionalization. The construction of a similarity matrix with precise data is discussed. In subsection 3.4.2, it is demonstrated that similarity matrices can also be derived from imprecise data.

Based on the concept of similarity, a grouping procedure is formulated in subsection 3.4.3 to group spatial units into regions in accordance with their homogeneities in characteristics.

In subsection 3.4.4, fuzzy set concepts of distance or dissimilarity are formulated to group spatial units into regions. It is shown that grouping on the basis of distance is the counterpart of that of similarity.

Procedures in subsections 3.4.1, 3.4.2, 3.4.3, and 3.4.4 are all constructed to derive homogeneous regions. Functional regions derived from relationships among spatial units are examined in subsection 3.4.5.

In subsection 3.4.6, the issue of mutual exclusivity is discussed. Situations leading to disjoint and nondisjoint regions are scrutinized.

3.4.1. Construction of Similarity Matrices with Precise Data

Let $\underline{X} = \{x_i\}$, i = 1, ..., m, be a set of spatial units and Let $\underline{Y} = \{y_k\}$, k = 1, ..., ℓ, be a set of characteristics. Let $\underline{A} = \{a_{ik}\}$, i = 1, ..., m; k = 1, ..., ℓ, be a $m \times \ell$ data matrix whose element, a_{ik}, is the value of spatial unit i under characteristic k. To group spatial units into regions with respect to their similarities in characteristics, a similarity matrix needs to be constructed from the data matrix $\underline{A}$. In this section, methods for the construction of a similarity matrix are proposed.

Let the value of a_{ik}, i = 1, ..., m, k = 1, ..., ℓ, be precise. Let R be a m × m similarity matrix derived from the data matrix $\underline{A}$. Its element r_{ij}, i = j = 1, ..., m, is the degree of similarity of spatial units i and j. Then r_{ij}, $0 \leq r_{ij} \leq 1$, can be obtained by any of the following methods:

(a) Correlation method.

$$r_{ij} = \frac{\sum_{k=1}^{\ell} |a_{ik} - \bar{a}_i| \cdot |a_{jk} - \bar{a}_j|}{\sqrt{\sum_{k=1}^{\ell} (a_{ik} - \bar{a}_i)^2} \sqrt{\sum_{k=1}^{\ell} (a_{jk} - \bar{a}_j)^2}}, \tag{3-84}$$

where

$$\bar{a}_i = \frac{1}{\ell} \sum_{k=1}^{\ell} a_{ik} \text{ and } \bar{a}_j = \frac{1}{\ell} \sum_{k=1}^{\ell} a_{jk};$$

Thus, r_{ij} is the correlation coefficient between spatial units x_i and x_j. The larger the value of r_{ij} is, the higher the similarity of spatial units i and j becomes.

(b) Method of cosine.

$$r_{ij} = \frac{\left|\sum_{k=1}^{\ell} a_{ik} a_{jk}\right|}{\sqrt{\left(\sum_{k=1}^{\ell} a_{ik}^2\right)\left(\sum_{k=1}^{\ell} a_{jk}^2\right)}}, \tag{3-85}$$

where r_{ij} is the cosine of the angle between two vectors.

(c) Exponential similarity method.

$$r_{ij} = \frac{1}{\ell} \sum_{k=1}^{\ell} \exp\left[-\left(\frac{3}{4}\right)\frac{(a_{ik} - a_{jk})^2}{\beta_k^2}\right], \quad \beta_k > 0, \tag{3-86}$$

where r_{ij} is the exponential similarity coefficient, and $\exp \equiv e$.

(d) Absolute-valued exponent method.

$$r_{ij} = e^{-\beta \sum_{k=1}^{\ell} |a_{ik} - a_{jk}|}, \quad \beta > 0. \tag{3-87}$$

(e) Absolute-valued subtraction method.

$$r_{ij} = \begin{cases} 1, & \text{if } i = j, \\ 1 - \beta \sum_{k=1}^{\ell} |a_{ik} - a_{jk}|, & \text{if } i \neq j, \end{cases} \tag{3-88}$$

where β is selected so that $0 \leq r_{ij} \leq 1$.

(f) Absolute-valued reciprocal method.

$$r_{ij} = \begin{cases} 1, & \text{if } i = j, \\ 1 - \dfrac{\beta}{\sum_{k=1}^{\ell} |a_{ik} - a_{jk}|}, & \text{if } i \neq j, \end{cases} \tag{3-89}$$

where β is selected so that $0 \leq r_{ij} \leq 1$.

(g) Maximum-minimum method.

$$r_{ij} = \frac{\sum_{k=1}^{\ell} \min(a_{ik}, a_{jk})}{\sum_{k=1}^{\ell} \max(a_{ik}, a_{jk})}. \tag{3-90}$$

(h) Arithmetic-mean minimum method.

$$r_{ij} = \frac{\sum_{k=1}^{\ell} \min(a_{ik}, a_{jk})}{\left(\frac{1}{2}\right) \sum_{k=1}^{\ell} (a_{ik} + a_{jk})}. \tag{3-91}$$

(i) Geometric-mean minimum method.

$$r_{ij} = \frac{\sum_{k=1}^{\ell} \min(a_{ik}, a_{jk})}{\sum_{k=1}^{\ell} \sqrt{a_{ik}\, a_{jk}}} . \qquad (3\text{-}92)$$

(j) Subjective evaluation method.

The degree of similarity is subjectively assigned on the basis of belief, experience, or perception.

Of course, there are many other methods by which a similarity matrix can be constructed. Depending on the nature and objective of a regionalization problem, one method may be more appropriate than the other. The derived similarity matrix, which in appearance is no different from the ordinary similarity matrix, is then employed as a basis for regionalization.

3.4.2. Construction of Similarity Matrices with Fuzzy Data

Sometimes, regionalization problems may consist of characteristics whose denotations are fuzzy, or spatial units whose values under specific characteristics are imprecise. This is especially common when valuative criteria are involved. For example, in the grouping of months into seasons, variables such as *wet* and *hot* are major differentiating characteristics and can be defined as fuzzy subsets on some measurable base variables such as precipitation and temperature respectively. Furthermore, to-be-grouped spatial units may take on approximate values such as "*approximately* 1,100 mm" and "*approximately* 28°C" as records of precipitation and temperature respectively.

In the grouping for neighborhoods, characteristics such as *heavily industrial*, *polluted*, and *overcrowded* have fuzzy denotations. The values of a spatial unit with respect to these fuzzy characteristics may again be fuzzy or precise. The idea then is to determine the grade of membership of a precise or fuzzy measurement in the corresponding fuzzy characteristic.

The obtained grades in turn constitute the characteristics of a spatial unit.

Let A_k be a fuzzy characteristic defined by its membership function μ_{A_k}. Then the grade of membership of a spatial unit x with a precise measurement β in the base variable of A_k is $\mu_{A_k}(\beta)$. If the measurement is fuzzy, such as "*approximately* β", then its grade of membership in A_k is determined by:

$$\mu_{A_k}(\textit{approximately } \beta) = \sup_{y_k} \min[\mu_{\textit{approximately } \beta}(y_k),\ \mu_{A_k}(y_k)]. \tag{3-93}$$

If the characteristic is not measurable, i.e. no measurable base variable on which A_k is defined, then the grade of membership is subjectively determined.

Therefore, elements of the data matrix with fuzzy characteristics, regardless of the precision of their measurements, all take on values within the closed interval [0, 1]. With reference to all characteristics, each spatial unit can then be regarded as a fuzzy subset. The similarity of any two spatial units can be determined by how far apart in characteristics they are from one another.

Thus, the similarity of two spatial units x_i and x_j with respect to ℓ characteristics can be defined by

$$1 - d(x_i, x_j) = 1 - \frac{1}{\ell} \sum_{k=1}^{\ell} |\mu_{x_i}(x_k) - \mu_{x_j}(x_k)|, \tag{3-94}$$

where $d(x_i, x_j)$ is the generalized relative Hamming distance between spatial units x_i and x_j. It measures the distance between two fuzzy subsets and takes on its value in the closed interval [0, 1].

Thus, when $d(x_i, x_j) = 0$, the similarity of x_i and x_j is 1 (completely similar). When $d(x_i, x_j) = 1$, however, the similarity of the two spatial units is 0 (completely dissimilar). For any value in-between, similarity is only to a certain degree.

Example 3.10. Let $\{x_1, \ldots, x_7\}$ be a set of spatial units. Let $\{A_1, A_2, A_3, A_4\}$ be a set of fuzzy characteristics defined respectively as "*heavily industrial*", "*obnoxious smell*", "*overcrowded*", and "*tranquil*". Let Fig. 3.22 be the original data matrix in which the grades of membership of the spatial units in characteristics A_2, A_3, and A_4 are subjectively determined.

	A_1	A_2	A_3	A_4
x_1	95	0.9	0.9	0
x_2	90	0.9	1.0	0.1
x_3	40	0.3	0.4	0.6
x_4	0	0	0.1	1.0
x_5	~35*	0.4	0.3	0.5
x_6	~92	0.2	0.6	0.2
x_7	85	0.3	0.5	0.1

*: $\sim \triangleq$ approximately

Fig. 3.22 Original data matrix of example 3.10

Let A_1 be defined by

$$\mu_{A_1}(y_1) = \begin{cases} 0, & \text{if } y_1 \leq 10, \\ \dfrac{y_1 - 10}{90 - 10}, & \text{if } 10 < y_1 \leq 90, \\ 1, & \text{if } y_1 > 90, \end{cases} \qquad (3\text{-}95)$$

where $y_1 \triangleq$ percent industrial employment.

Let the first column of the data matrix be industrial employments of the spatial units measured in percentage. Except for x_5 and x_6, the measurements are precise. Then grades of membership can be obtained through (3-95).

If $\mu_{\sim 35}$ and $\mu_{\sim 92}$ are defined by the membership functions in Fig. 3.23, then the grades of membership of spatial units x_5 and x_6 are determined by (3-93).

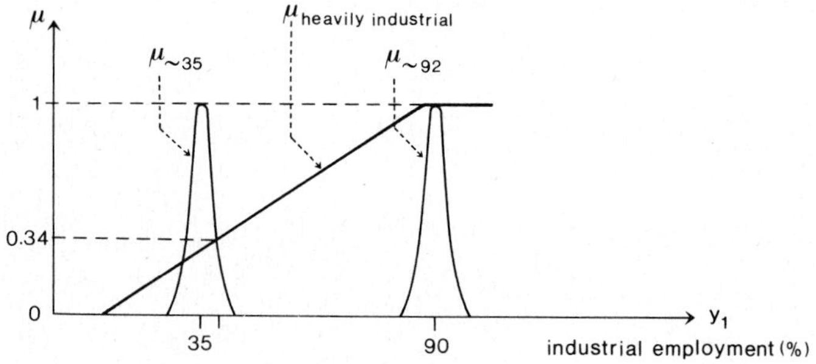

Fig. 3.23 Determination of grades of membership for example 3.10

The original data matrix then becomes a fuzzy binary matrix (Fig. 3.24). Calculating the generalized relative

	A_1	A_2	A_3	A_4
x_1	1.0	0.9	0.9	0
x_2	1.0	0.9	1.0	0.1
x_3	0.4	0.3	0.4	0.6
x_4	0	0	0.1	1.0
x_5	0.3	0.4	0.3	0.5
x_6	1.0	0.2	0.6	0.2
x_7	0.9	0.3	0.5	0.1

Fig. 3.24 Fuzzy binary matrix derived from the original data matrix in Fig. 3.22

Hamming distance between the spatial units, the distance matrix is obtained (Fig. 3.25). Applying (3-94), the similarity matrix (Fig. 3.26) is derived for the grouping procedure.

	x_1	x_2	x_3	x_4	x_5	x_6	x_7
x_1	0	0.05	0.58	0.93	0.58	0.30	0.30
x_2	0.05	0	0.58	0.93	0.58	0.30	0.30
x_3	0.58	0.58	0	0.35	0.10	0.33	0.28
x_4	0.93	0.93	0.35	0	0.35	0.63	0.63
x_5	0.58	0.58	0.10	0.35	0	0.38	0.33
x_6	0.30	0.30	0.33	0.63	0.38	0	0.10
x_7	0.30	0.30	0.28	0.63	0.33	0.10	0

Fig. 3.25 Distance matrix based on the fuzzy binary matrix in Fig. 3.24

	x_1	x_2	x_3	x_4	x_5	x_6	x_7
x_1	1	0.95	0.42	0.07	0.42	0.70	0.70
x_2	0.95	1	0.42	0.07	0.42	0.70	0.70
x_3	0.42	0.42	1	0.65	0.90	0.67	0.72
x_4	0.07	0.07	0.65	1	0.65	0.37	0.37
x_5	0.42	0.42	0.90	0.65	1	0.62	0.67
x_6	0.70	0.70	0.67	0.37	0.62	1	0.90
x_7	0.70	0.70	0.72	0.37	0.67	0.90	1

Fig. 3.26 Similarity matrix derived from the distance matrix in Fig. 3.25

Remark. Methods for deriving similarity matrices in subsection 3.4.1 are in fact conventional methods for defining similarity with precise data. Of course, when values of the variables are within [0, 1] (the fuzzy-set-like format), the methods still apply. The methods in subsection 3.4.2 is, however, constructed to derive similarity matrices from fuzzy or precise data or both. This is where the fuzzy set approach shows its versatility.

3.4.3. Grouping Based on the Concept of Similarity

Let R be the similarity matrix. Without loss of generality, let R be a fuzzy binary relation with $\mu_R(x_i, x_j) \in [0, 1]$ indicating the degree to which spatial units x_i and x_j are similar. The higher its value is, the higher the similarity of x_i and x_j becomes. The relation R is obviously reflexive and symmetric and is thus a resemblance relation. However, regionalization cannot be meaningfully performed unless R is transitive also. That is, R should be a similitude relation with the following properties:

(a) $\mu_R(x_i, x_i) = 1, \ \forall \ (x_i, x_i) \in \underline{X} \times \underline{X},$ (3-96)

(b) $\mu_R(x_i, x_j) = \mu_R(x_j, x_i), \ \forall \ (x_i, x_j) \in \underline{X} \times \underline{X},$ (3-97)

(c) $\mu_R(x_i, x_k) \geq \max_{x_j} \min[\mu_R(x_i, x_j), \mu_R(x_j, x_k)],$

$\forall \ (x_i, x_j), (x_j, x_k), (x_i, x_k) \in \underline{X} \times \underline{X}.$ (3-98)

Reflexivity and symmetry of R are respectively stated in (3-96) and (3-97), while max-min transitivity is imposed by (3-98). The physical meaning of (3-98) is that the similarity of spatial units x_i and x_k should be at least as high as the lowest of the similarities of x_i and x_j and of x_j and x_k, for all intermediate x_j's. Compared to the notion of transitivity in conventional analysis, the present one defines a weak transitivity of similarity.

Since not all similarity matrices are max-min transitive, we need a procedure to transform an intransitive similarity matrix R into a transitive one. A natural method is to derive the transitive closure, $\hat{R}$, of R.

Definition 3.9 The max-min transitive closure of a fuzzy binary relation R is

$$\hat{R} = R \cup R^2 \cup R^3 \cup \ldots, \tag{3-99}$$

where $R^2 = R \circ R$ is the max-min composition of R. The element $\mu_{\hat{R}}(x_i, x_j)$ indicates the max-min transitive similarity of x_i and x_j.

The following theorem shows how such a transitive closure can be obtained.

Theorem 3.2. Let R be any fuzzy binary relation. If for some k, the max-min composition $R^{k+1} = R^k$, then the max-min transitive closure is $\hat{R} = R \cup R^2 \cup \ldots \cup R^k$.

Definition 3.10. The strongest path from spatial unit x_i to x_j is defined by

$$\ell^*(x_i, x_j) = \max_{C(x_i, x_j)} \ell(x_{q_1} = x_i, \ldots, x_{q_r} = x_j), \tag{3-100}$$

where $C(x_i, x_j)$ is the ordinary set of all paths from x_i to x_j with the value of each path $(x_{q_1}, \ldots, x_{q_r})$ defined by

$$\ell(x_{q_1}, \ldots, x_{q_r}) = \min[\mu_R(x_{q_1}, x_{q_2}), \ldots, \mu_R(x_{q_{r-1}}, x_{q_r})]. \tag{3-101}$$

Thus, the strongest path from x_i to x_j is the strongest of the weakest link between x_i and x_j. Since

$$\mu_{\hat{R}}(x_i, x_j) = \ell^*(x_i, x_j), \quad \forall\, (x_i, x_j) \in \underline{X} \times \underline{X}, \tag{3-102}$$

then, the max-min transitive similarity of x_i and x_j is actually the strongest path from x_i to x_j. Thus, the similarity of x_i and x_j can be interpreted as the strongest of the weakest similarities of x_i and x_j in the path.

Based on the decomposition theorem (theorem 2.1), the similitude relation $\hat{R}$ (or R, if R is transitive) can be decomposed into

$$\hat{R} = \bigcup_{\alpha \in [0, 1]} \alpha \cdot \underline{\hat{R}}_{\alpha}, \qquad (3\text{-}103)$$

where

$$f_{\underline{\hat{R}}}(x_i, x_j) = \begin{cases} 1, & \text{if } \mu_{\hat{R}}(x_i, x_j) \geq \alpha, \\ 0, & \text{if } \mu_{\hat{R}}(x_i, x_j) < \alpha, \end{cases} \qquad (3\text{-}104)$$

and

$$\alpha_1 > \alpha_2 \Longrightarrow \underline{\hat{R}}_{\alpha_1} \subset \underline{\hat{R}}_{\alpha_2}. \qquad (3\text{-}105)$$

Since $\underline{\hat{R}}_{\alpha}$ is reflexive, symmetric, and transitive in the sense of ordinary sets, then it is an equivalence class (equivalence relation) of level α. Within each α-level equivalence class, the similarity of any two spatial units is no less than α. For example, if $\alpha = 0.6$, then $\mu_R(x_i, x_j) > 0.6$, $\forall (x_i, x_j) \in \underline{R}_{0.6}$.

Therefore, it is natural to equate each equivalence class with a region, i.e. $\underline{\hat{R}}_{\alpha} \triangleq \alpha$-region. All spatial units within a α-region is α-similar with degree of similarity $\geq \alpha$. For a specific value of α, we have a specific grouping of spatial units into regions. By varying the value of α, we could observe the variations of grouping for regions. Such a hierarchical regionalization procedure is natural and more informative. It can model the gradual and abrupt transitions from membership to non-membership of spatial units among regions.

Example 3.11. Consider the regionalization problem in example 3.10. The similarity matrix R (Fig. 3.26) is obviously reflexive and symmetric, but intransitive. To have a meaningful grouping for regions, the transitive closure $\hat{R}$ needs to be obtained first.

Since $R^4 = R^3$ in the series of max-min compositions in Fig. 3.27, then by theorem 3.2, the transitive closure $\hat{R} = R \cup R^2 \cup R^3 = R^3$. The element $\mu_{\hat{R}}(x_2, x_4) = 0.65$, for example, is the strongest path between spatial units x_2 and x_4, indicating the max-min transitive similarity of x_2 and x_4.

Since $\hat{R}$ is a similitude relation, then it can be decomposed into equivalence classes with respect to the degree of similarity α, $\alpha = 1, 0.95, 0.90, 0.72, 0.70$, and 0.65. Fig. 3.28 depicts the corresponding ordinary relation for $\alpha = 0.72$. The equivalence classes are then $\{x_4\}$, $\{x_1, x_2\}$, and $\{x_3, x_5, x_6, x_7\}$. That is, there are three 0.72-regions. The similarity of any two spatial units within any of the 0.72-regions is greater than or equal to 0.72. The similarity of any spatial units between two different 0.72-regions is less than 0.72.

Thus, for each degree of similarity α, a regionalization pattern can be derived. By taking the values of α in ascending or descending order, we obtain a hierarchical regionalization pattern depicted in Fig. 3.29. By scrutinizing the decomposition tree in Fig. 3.29, transition of memberships between regions can be determined.

Remark. To ensure that contiguous spatial units are grouped together in the grouping process, the following spatial contiguity constraint:

$$\mu_R(x_i, x_j) = 1, \text{ if } x_i \text{ and } x_j \text{ are contiguous}, \ \forall\, i, j, \qquad (3\text{-}106)$$

can be imposed on the similarity matrix R. However, I would not recommend the imposition of such a constraint unless it is a necessity. In general, if phenomena vary over space in a more or less continuous manner, spatial units would tend to fall into the same region without the contiguity constraint anyway. If, however, two adjacent regions are rather different in characteristics, forcing them together through the contiguity constraint would produce an unnatural pattern of regionalization.

$$
\begin{bmatrix}
1 & 0.95 & 0.42 & 0.07 & 0.42 & 0.70 & 0.70 \\
0.95 & 1 & 0.42 & 0.07 & 0.42 & 0.70 & 0.70 \\
0.42 & 0.42 & 1 & 0.65 & 0.90 & 0.67 & 0.72 \\
0.07 & 0.07 & 0.65 & 1 & 0.65 & 0.37 & 0.37 \\
0.42 & 0.42 & 0.90 & 0.65 & 1 & 0.62 & 0.67 \\
0.70 & 0.70 & 0.67 & 0.37 & 0.62 & 1 & 0.90 \\
0.70 & 0.70 & 0.72 & 0.37 & 0.67 & 0.90 & 1
\end{bmatrix},
$$

$$R$$

$$
\begin{bmatrix}
1 & 0.95 & 0.70 & 0.42 & 0.67 & 0.70 & 0.70 \\
0.95 & 1 & 0.70 & 0.42 & 0.67 & 0.70 & 0.70 \\
0.70 & 0.70 & 1 & 0.65 & 0.90 & 0.72 & 0.72 \\
0.42 & 0.42 & 0.65 & 1 & 0.65 & 0.65 & 0.65 \\
0.67 & 0.67 & 0.90 & 0.65 & 1 & 0.67 & 0.72 \\
0.70 & 0.70 & 0.72 & 0.65 & 0.67 & 1 & 0.90 \\
0.70 & 0.70 & 0.72 & 0.65 & 0.72 & 0.90 & 1
\end{bmatrix},
$$

$$R^2$$

$$
\begin{bmatrix}
1 & 0.95 & 0.70 & 0.65 & 0.70 & 0.70 & 0.70 \\
0.95 & 1 & 0.70 & 0.65 & 0.70 & 0.70 & 0.70 \\
0.70 & 0.70 & 1 & 0.65 & 0.90 & 0.72 & 0.72 \\
0.65 & 0.65 & 0.65 & 1 & 0.65 & 0.65 & 0.65 \\
0.70 & 0.70 & 0.90 & 0.65 & 1 & 0.72 & 0.72 \\
0.70 & 0.70 & 0.72 & 0.65 & 0.72 & 1 & 0.90 \\
0.70 & 0.70 & 0.72 & 0.65 & 0.72 & 0.90 & 1
\end{bmatrix},
$$

$$R^3$$

$$
\begin{bmatrix}
1 & 0.95 & 0.70 & 0.65 & 0.70 & 0.70 & 0.70 \\
0.95 & 1 & 0.70 & 0.65 & 0.70 & 0.70 & 0.70 \\
0.70 & 0.70 & 1 & 0.65 & 0.90 & 0.72 & 0.72 \\
0.65 & 0.65 & 0.65 & 1 & 0.65 & 0.65 & 0.65 \\
0.70 & 0.70 & 0.90 & 0.65 & 1 & 0.72 & 0.72 \\
0.70 & 0.70 & 0.72 & 0.65 & 0.72 & 1 & 0.90 \\
0.70 & 0.70 & 0.72 & 0.65 & 0.72 & 0.90 & 1
\end{bmatrix}.
$$

$$R^4$$

Fig. 3.27 A series of max-min compositions of the similarity matrix in Fig. 3.26

	x_1	x_2	x_4	x_3	x_5	x_6	x_7
x_1	1	1	0	0	0	0	0
x_2	1	1	0	0	0	0	0
x_4	0	0	1	0	0	0	0
x_3	0	0	0	1	1	1	1
x_5	0	0	0	1	1	1	1
x_6	0	0	0	1	1	1	1
x_7	0	0	0	1	1	1	1

Fig. 3. 28 Equivalence classes (regions) for 0.72-similarity

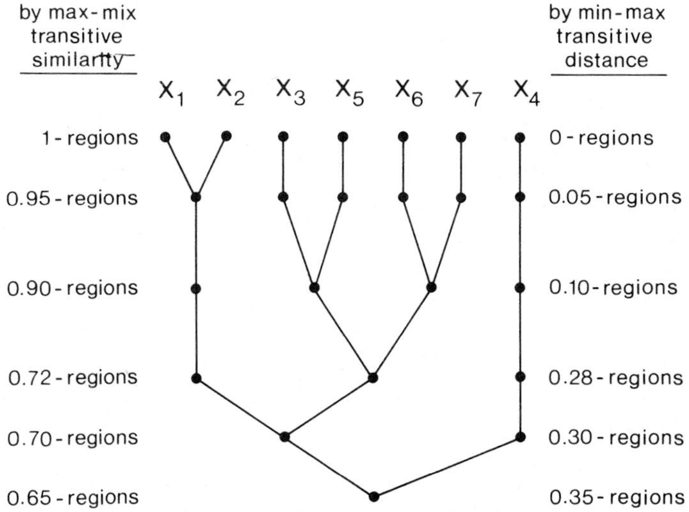

Fig. 3.29 Regionalization obtained through the decomposition of the max-min transitive closure in Fig. 3.27 and the min-max transitive closure in Fig. 3.30

3.4.4. Grouping Based on the Concept of Distance

Often, regionalization is based on the concept of distance. Conceptually, it is the opposite of the concept of similarity. In place of grouping together spatial units which are more similar in characteristics, spatial units which are of shorter distance are grouped together under the concept of distance. Thus, distance can be equated with dissimilarity. That is, the closer in characteristics the spatial units are, the shorter is the distance between them. Therefore, distance here does not mean physical distance in the ordinary sense.

Parallel to the concept of similarity, transitivity in distance is also essential for any grouping procedures. The basic fuzzy set concept which models closely such an idea is a dissimilitude relation R which satisfies the following properties:

(a) $\mu_R(x_i, x_i) = 0, \ \forall (x_i, x_i) \in \underline{X} \times \underline{X},$ (3-107)

(b) $\mu_R(x_i, x_j) = \mu_R(x_j, x_i), \ \forall (x_i, x_j) \in \underline{X} \times \underline{X},$ (3-108)

(c) $\mu_R(x_i, x_k) \leq \min_{x_j} \max[\mu_R(x_i, x_j), \mu_R(x_j, x_k)],$

$\forall (x_i, x_j), (x_j, x_k), (x_i, x_k) \in \underline{X} \times \underline{X}.$ (3-109)

Thus, a dissimilitude relation is antireflexive, (3-107), and symmetric, (3-108). Moreover, it is min-max transitive, (3-109). It is then a topological distance. The physical meaning of (3-109) is that the distance (dissimilarity) between spatial units x_i and x_k should be at most as long as the longest of the distances between x_i and x_j or that between x_j and x_k, for all intermediate x_j's.

If a fuzzy binary relation only satisfies antireflexivity and symmetry, then it is a dissemblance relation. Matrices whose elements are generalized relative Hamming distances are dissemblance relations (see for example Fig. 3.25). Since not all dissemblance relations are transitive, we need a procedure to obtain transitive closures from intransitive relations.

Definition 3.11. The min-max transitive closure of a fuzzy binary relation R is

$$\breve{R} = R \cap R^2 \cap R^3 \cap \ldots, \tag{3-110}$$

where $R^2 = R * R$ is the min-max composition of R. The element $\mu_{\breve{R}}(x_i, x_j)$ denotes the min-max transitive distance between x_i and x_j. The higher its value is, the higher the dissimilarity of x_i and x_j becomes.

The following theorem can be employed to obtain $\breve{R}$ from R.

Theorem 3.3. Let R be any fuzzy binary relation. If for some k, the min-max composition $R^{k+1} = R^k$, then the min-max transitive closure is $\breve{R} = R \cap R^2 \cap \ldots \cap R^k$.

The element $\mu_{\breve{R}}(x_i, x_j)$ is the min-max transitive distance between x_i and x_j. It is actually the weakest path from x_i to x_j.

Definition 3.12. The weakest path from spatial unit x_i to x_j is defined by

$$\ell_*(x_i, x_j) = \min_{C(x_i, x_j)} \ell(x_{q_1} = x_i, \ldots, x_{q_r} = x_j), \tag{3-111}$$

where $C(x_i, x_j)$ is the ordinary set of all paths from x_i to x_j with the value of each path $(x_{q_1} = x_i, \ldots, x_{q_r} = x_j)$ defined by

$$\ell(x_{q_1}, \ldots, x_{q_r}) = \max[\mu_R(x_{q_1}, x_{q_2}), \ldots, \mu_R(x_{q_{r-1}}, x_{q_\ell})]. \tag{3-112}$$

Thus, the weakest path from x_i to x_j is the shortest of the longest link between x_i and x_j. Since

$$\mu_{\breve{R}}(x_i, x_j) = \ell_*(x_i, x_j), \ \forall\ (x_i, x_j) \in \underline{X} \times \underline{X}, \tag{3-113}$$

then the distance between spatial units x_i and x_j is the weakest path from x_i to x_j. Therefore, the dissimilarity of x_i and x_j can be interpreted as the weakest of the strongest dissimilarities of x_i and x_j in the path.

Again, $\breve{R}$ can be decomposed into equivalence classes with respect to the degree of dissimilarity $\alpha \in [0, 1]$. Within each equivalence class, $\underline{\breve{R}}_\alpha$, the dissimilarity between any two spatial units is no larger than α. Therefore, the α-level sets can be treated as α-regions within which any two elements are α-distance from one another. For a specific value of α, a specific regionalization pattern is obtained. By considering all values of α in $\breve{R}$, a hierarchical pattern of regionalization is derived.

Since antireflexivity in (3-107) and reflexivity in (3-96) are complementary, and min-max transitivity in (3-109) is the complement of max-min transitivity in (3-98), then it is intuitive to expect that regionalization patterns based on the similitude relation and dissimilitude relation are identical. This relationship can easily be demonstrated.

Example 3.12. Consider the regionalization problem in example 3.10. The matrix R in Fig. 3.25 is the generalized relative Hamming distance matrix. It is the complement of the similarity matrix in Fig. 3.26. Since R is not min-max transitive, we need to obtain its transitive closure $\breve{R}$.

Since $R^4 = R^3$ in the series of min-max compositions in Fig. 3.30, then by theorem 3.3, the transitive closure $\breve{R} = R \cap R^2 \cap R^3 = R^3$. The element $\mu_{\breve{R}}(x_i, x_j)$ is the transitive distance (dissimilarity) between x_i and x_j. Observe also that $\breve{R}$ is the complement of $\hat{R}$ obtained in Fig. 3.27.

Select $\alpha = 0.28$ and decompose $\breve{R}$, the corresponding ordinary relation is depicted in Fig. 3.31. The equivalence classes are $\{x_4\}, \{x_1, x_2\}, \{x_3, x_5, x_6, x_7\}$. The distances between any two spatial units within these 0.28-regions is no longer than 0.28. The distance between any spatial units from different 0.28-regions is greater than 0.28. Observe also that this regionalization pattern is identical to the one obtained

$$
\begin{bmatrix}
0 & 0.05 & 0.58 & 0.93 & 0.58 & 0.30 & 0.30 \\
0.05 & 0 & 0.58 & 0.93 & 0.58 & 0.30 & 0.30 \\
0.58 & 0.58 & 0 & 0.35 & 0.10 & 0.33 & 0.28 \\
0.93 & 0.93 & 0.35 & 0 & 0.35 & 0.63 & 0.63 \\
0.58 & 0.58 & 0.10 & 0.35 & 0 & 0.38 & 0.33 \\
0.30 & 0.30 & 0.33 & 0.63 & 0.38 & 0 & 0.10 \\
0.30 & 0.30 & 0.28 & 0.63 & 0.33 & 0.10 & 0
\end{bmatrix},
$$

R

$$
\begin{bmatrix}
0 & 0.05 & 0.30 & 0.58 & 0.33 & 0.30 & 0.30 \\
0.05 & 0 & 0.30 & 0.58 & 0.33 & 0.30 & 0.30 \\
0.30 & 0.30 & 0 & 0.35 & 0.10 & 0.28 & 0.28 \\
0.58 & 0.58 & 0.35 & 0 & 0.35 & 0.35 & 0.35 \\
0.33 & 0.33 & 0.10 & 0.35 & 0 & 0.33 & 0.28 \\
0.30 & 0.30 & 0.28 & 0.35 & 0.33 & 0 & 0.10 \\
0.30 & 0.30 & 0.28 & 0.35 & 0.38 & 0.10 & 0
\end{bmatrix},
$$

R^2

$$
\begin{bmatrix}
0 & 0.05 & 0.30 & 0.35 & 0.30 & 0.30 & 0.30 \\
0.05 & 0 & 0.30 & 0.35 & 0.30 & 0.30 & 0.30 \\
0.30 & 0.30 & 0 & 0.35 & 0.10 & 0.28 & 0.28 \\
0.35 & 0.35 & 0.35 & 0 & 0.35 & 0.35 & 0.35 \\
0.30 & 0.30 & 0.10 & 0.35 & 0 & 0.28 & 0.28 \\
0.30 & 0.30 & 0.28 & 0.35 & 0.28 & 0 & 0.10 \\
0.30 & 0.30 & 0.28 & 0.35 & 0.28 & 0.10 & 0
\end{bmatrix},
$$

R^3

$$
\begin{bmatrix}
0 & 0.05 & 0.30 & 0.35 & 0.30 & 0.30 & 0.30 \\
0.05 & 0 & 0.30 & 0.35 & 0.30 & 0.30 & 0.30 \\
0.30 & 0.30 & 0 & 0.35 & 0.10 & 0.28 & 0.28 \\
0.35 & 0.35 & 0.35 & 0 & 0.35 & 0.35 & 0.35 \\
0.30 & 0.38 & 0.10 & 0.35 & 0 & 0.28 & 0.28 \\
0.30 & 0.30 & 0.28 & 0.25 & 0.28 & 0 & 0.10 \\
0.30 & 0.30 & 0.28 & 0.35 & 0.28 & 0.10 & 0
\end{bmatrix}.
$$

R^4

Fig. 3.30 A series of min-max compositions of the distance matrix in Fig. 3.25

by decomposing $\hat{R}$ with respect to the degree of similarity $\alpha = 0.72$ in Fig. 3.28.

Considering all values of α in $\breve{R}$, the hierarchical pattern of regionalization is derived (Fig. 3.29). Thus, patterns derived from decomposing the similitude and dissimilitude relations are identical.

Therefore, if distance is defined as the complement of similarity and vice versa, the regionalization patterns derived from both concepts are identical. However, distance needs not be defined as the complement of similarity. Regionalization patterns thus obtained may not be the same.

	x_1	x_2	x_4	x_3	x_5	x_6	x_7
x_1	1	1	0	0	0	0	0
x_2	1	1	0	0	0	0	0
x_4	0	0	1	0	0	0	0
x_3	0	0	0	1	1	1	1
x_5	0	0	0	1	1	1	1
x_6	0	0	0	1	1	1	1
x_7	0	0	0	1	1	1	1

Fig. 3.31 Equivalence classes (regions) for 0.28-distance

Remark. (a) The procedure discussed in this subsection does not guarantee the grouping together of contiguous spatial units. Should this become a necessity, the following spatial contiguity constraint can be imposed on the distance matrix:

$$\mu_R(x_i, x_j) = 0, \quad \text{if } x_i \text{ and } x_j \text{ are contiguous}, \ \forall\, i, j. \qquad (3\text{-}114)$$

(b) The min-max transitive closure of a dissemblence relation may yield transitively equidistant pairs of spatial units (see Example 3.13). Under this situation, the min-max transitive distance is not an appropriate basis for regionalization. Different notions of distance need to be employed. (See subsection 3.4.6 for other definitions of distance).

Example 3.13. Consider the data matrix of four spatial units x_1, x_2, x_3, x_4 with two characteristics y_1 and y_2 in Fig. 3.32. Its generalized relative Hamming distance matrix is depicted in Fig. 3.33. The corresponding min-max transitive closure is depicted in Fig. 3.34. Observe that all spatial units are equidistant though they are quite different in characteristics.

3.4.5. Grouping Based on Asymmetric Spatial Relations

Regionalization procedures discussed so far are for symmetric spatial relations. Derivation of homogeneous regions usually belongs to this category. Grouping for functional regions, however, is based on relationships which is ordinarily asymmetric in structure. Spatial interactions such as population movement, commodity flow, traffic volume, and information flow are typical examples.

If an aysmmetric spatial relation is decomposable, it can be decomposed into equivalence classes whose relations are antisymmetric. The idea for grouping is then to seek for a decomposable asymmetric fuzzy binary relation and to equate the derived equivalence classes with regions. The basic concept for such a grouping procedure is the notion of a fuzzy preorder.

Definition 3.13. A fuzzy binary relation R is a relation of fuzzy preorder if

$$\mu_R(x_i, x_i) = 1, \ \forall (x_i, x_i) \in \underline{X} \times \underline{X}, \tag{3-115}$$

	y_1	y_2
x_1	0	0
x_2	0	0.5
x_3	0.5	0
x_4	0	1

Fig. 3. 32 A data matrix

	x_1	x_2	x_3	x_4
x_1	0	0.25	0.25	0.50
x_2	0.25	0	0.5	0.25
x_3	0.25	0.50	0	0.75
x_4	0.50	0.25	0.75	0

Fig. 3. 33 The corresponding generalized relative Hamming distance matrix of Fig.3.32

	x_1	x_2	x_3	x_4
x_1	0	0.25	0.25	0.25
x_2	0.25	0	0.25	0.25
x_3	0.25	0.25	0	0.25
x_4	0.25	0.25	0.25	0

Fig. 3. 34 The derived min-max transitive closure of Fig. 3. 33

$$\mu_R(x_i, x_k) \geq \max_{x_j} \min[\mu_R(x_i, x_j), \mu_R(x_j, x_k)],$$

$$\forall (x_i, x_j), (x_j, x_k), (x_i, x_k) \in \underline{X} \times \underline{X}. \qquad (3\text{-}116)$$

Therefore, a fuzzy preorder is reflexive and max-min transitive. If it can be reduced into a set of similitude relations, then it is a reducible fuzzy preorder. The relation formed by the equivalence classes of a reducible fuzzy preorder via the concept of strongest path is necessarily antisymmetric and is hierarchic in structure. The result can be formally stated in the following theorem:

Theorem 3.4. Consider the strongest path from one class to another, the similitude classes in a reducible fuzzy preorder form among themselves a fuzzy order relation which satisfies the following conditions:

(a) $\mu_R(x_i, x_i) = 1, \ \forall (x_i, x_i) \in \underline{X} \times \underline{X};$ (3-117)

(b) $\mu_R(x_i, x_j) \neq \mu_R(x_j, x_i)$ or $\mu_R(x_i, x_j) = \mu_R(x_j, x_i) = 0,$

$\forall (x_i, x_j) \in \underline{X} \times \underline{X}$ and $x_i \neq x_j;$ (3-118)

(c) $\mu_R(x_i, x_k) \geq \max_{x_j} \min[\mu_R(x_i, x_j), \mu_R(x_j, x_k)],$

$\forall (x_i, x_j), (x_j, x_k), (x_i, x_k) \in \underline{X} \times \underline{X}.$ (3-119)

Conditions (3-117) and (3-119) are respectively reflexivity and max-min transitivity, and condition (3-118) is antisymmetry.

Therefore, regionalization pattern based on decomposable asymmetric relation is hierarchic in structure. Hierarchical functional regions can thus be derived.

Example 3.14. Consider the asymmetric spatial relation in Fig. 3.35. To group the nine spatial units into regions through the concept of a reducible fuzzy preorder, we first obtain the complement of the spatial relation (Fig. 3.36). Obviously, the

	x_1	x_2	x_3	x_4	x_5	x_6	x_7	x_8	x_9
x_1	0	0.4	1	1	1	1	1	1	1
x_2	0.4	0	1	1	1	1	1	1	1
x_3	0.4	0.3	0	0.3	0.5	0.9	1	1	1
x_4	0.4	0.3	0.3	0	0.5	0.9	1	1	1
x_5	0.4	0.2	0.5	0.5	0	0.9	1	1	1
x_6	0.7	0.7	0.8	0.8	0.8	0	1	1	1
x_7	0.4	0.4	1	1	1	0.8	0	0.8	1
x_8	0.4	0.3	1	1	1	0.8	0.8	0	1
x_9	0.4	0.4	1	1	1	1	0.9	0.9	0

Fig. 3.35 An asymmetric spatial relation

		R_1		R_2			R_3	R_4		R_5
		x_1	x_2	x_3	x_4	x_5	x_6	x_7	x_8	x_9
R_1	x_1	1	0.6	0	0	0	0	0	0	0
	x_2	0.6	1	0	0	0	0	0	0	0
R_2	x_3	0.6	0.7	1	0.7	0.5	0.1	0	0	0
	x_4	0.6	0.7	0.7	1	0.5	0.1	0	0	0
	x_5	0.6	0.8	0.5	0.5	1	0.1	0	0	0
R_3	x_6	0.3	0.3	0.2	0.2	0.2	1	0	0	0
R_4	x_7	0.6	0.6	0	0	0	0.2	1	0.2	0
	x_8	0.6	0.7	0	0	0	0.2	0.2	1	0
R_5	x_9	0.6	0.6	0	0	0	0	0.1	0.1	1

Fig. 3.36 The complement of the asymmetric spatial relation in Fig. 3.35

complement is a reducible fuzzy preorder which can be reduced to five similitude relations (regions) R_1, R_2, R_3, R_4, and R_5 (Fig. 3.36). They form among themselves a hierarchical relation (Fig. 3.37). The strength of the relationship between any two regions is the strongest path between them. For example, the relation between R_2 and R_1 is obtained as the global projection of the subrelation $R_{R_2R_1}$ as

$$g(R_{R_2R_1}) = \max_{x_i} \max_{x_j} \mu_{R_{R_2R_1}}(x_i, x_j) = 0.8, \; x_i \in R_1, \; x_j \in R_2. \tag{3-120}$$

The associated ordinary relation of the fuzzy binary relation in Fig. 3.37 is depicted in Fig. 3.38. The regions then form among themselves a hierarchical structure (Fig. 3.39).

3. 4. 6. Disjoint and Nondisjoint Regions

In the discussion of regional characterization in section 3.2, I argue that regions may be mutually exclusive, i.e. disjoint, or they may overlap to a certain degree, i.e. non-disjoint. In the analysis of grouping for regions in subsections 3.4.2, 3.4.3, and 3.4.4, procedures presented all produce disjoint regions.

In general, regions derived from the decomposition of a similitude relation R are disjoint. Similarly, the decomposition of a dissimilitude relation generally yields disjoint regions.

As discussed in subsection 3.4.4, min-max transitivity is not the only way to define distance. That is, regionalization in terms of distance does not have to rely on the decomposition of the min-max transitive closure of the generalized relative Hamming distance matrix. Instead of min-max transtivity, we can employ min-addition transitivity defined by:

$$\mu_R(x_i, x_k) = \min_{x_j}[\mu_R(x_i, x_j) + \mu_R(x_j, x_k)],$$

$$\forall (x_i, x_j), (x_j, x_k), (x_i, x_k) \in \underline{X} \times \underline{X}. \tag{3-121}$$

	R_1	R_2	R_3	R_4	R_5
R_1	1	0	0	0	0
R_2	0.8	1	0.1	0	0
R_3	0.3	0.2	1	0	0
R_4	0.7	0	0.2	1	0
R_5	0.6	0	0	0.1	1

Fig. 3. 37 Strongest paths between the similitude relations (regions)

	R_1	R_2	R_3	R_4	R_5
R_1	1	0	0	0	0
R_2	1	1	0	0	0
R_3	1	1	1	0	0
R_4	1	0	1	1	0
R_5	1	0	0	1	1

Fig. 3.38 The associated ordinary relation of Fig. 3. 37

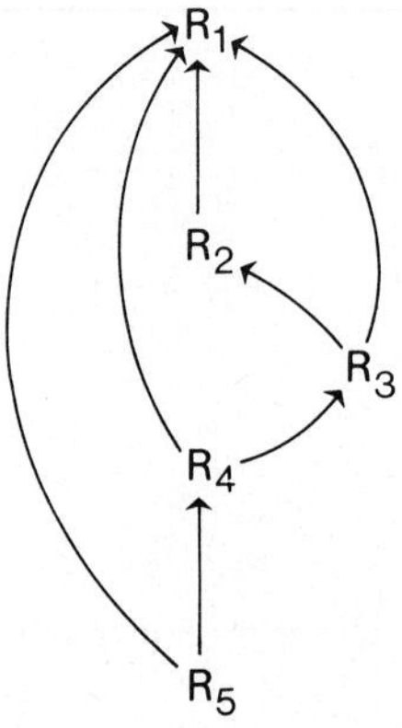

Fig. 3.39 The hierarchical regionalization pattern based on Fig. 3.38

Since any fuzzy binary relation which gives the generalized relative Hamming distance is its own min-addition transitive closure, it can then be decomposed directly. However, the decomposition of a min-addition transitive distance matrix may not give disjoint equivalence classes but nondisjoint maximal subrelations. (A subrelation of a relation is maximal if it is not contained in any other complete submatrix whose entries are all 1 in the sense used in ordinary set theory). By equating regions with maximal subrelations, mutually exclusive regions may not exist at all levels of decomposition. Overlapping of regions is then a common phenomenon.

Example 3.15. Consider the generalized relative Hamming distance matrix in Fig. 3.25. Since it is its own min-addition transitive closure, we can decompose it directly with respect to the min-addition transitive distance $\alpha \in [0, 1]$.

For $\alpha \leq 0.28$, the set of maximal subrelations can be derived through the application of Pichat's algorithm (see Kaufmann, 1975, pp. 398-400). Depicted in Fig. 3.40 is the upper triangular matrix of the ordinary relation associated with $\alpha \leq 0.28$. The Boolean expressions for all the lines are

$$\begin{aligned}
&\text{line } x_1: && x_1 + x_3 x_4 x_5 x_6 x_7, \\
&\text{line } x_2: && x_2 + x_3 x_4 x_5 x_6 x_7, \\
&\text{line } x_3: && x_3 + x_4 x_6, \\
&\text{line } x_4: && x_4 + x_5 x_6 x_7, \\
&\text{line } x_5: && x_5 + x_6 x_7, \\
&\text{line } x_6: && 1, \\
&\text{line } x_7: && 1.
\end{aligned} \tag{3-122}$$

Employing the following properties of Boolean sum $\dot{+}$ and Boolean product $\bullet$:

$$x \dot{+} x = x, \quad x \bullet x = x, \quad x \dot{+} xy = x, \tag{3-123}$$

the product of the expressions in (3-122) is

	x_1	x_2	x_3	x_4	x_5	x_6	x_7
x_1		1	0	0	0	0	0
x_2			0	0	0	0	0
x_3				0	1	0	1
x_4					0	0	0
x_5						0	0
x_6							1
x_7							

Fig. 3.40 Upper triangular matrix associated with 0.28 min-addition transitive distance

$$x = (x_1 \dot{+} x_3x_4x_5x_6x_7) \cdot (x_2 \dot{+} x_3x_4x_5x_6x_7) \cdot (x_3 \dot{+} x_4x_6) \cdot (x_4 \dot{+} x_5x_6x_7) \cdot (x_5 \dot{+} x_6x_7) \cdot 1 \cdot 1$$

$$= x_1x_2x_3x_4x_5 \dot{+} x_1x_2x_3x_5x_6x_7 \dot{+} x_1x_2x_4x_5x_7 \dot{+} x_1x_2x_4x_6x_7 \dot{+} x_3x_4x_5x_6x_7. \qquad (3\text{-}124)$$

The complement, $\bar{x}$, of x is

$$\bar{x} = x_6x_7 \dot{+} x_4 \dot{+} x_3x_7 \dot{+} x_3x_5 \dot{+} x_1x_2. \qquad (3\text{-}125)$$

Thus, the set of maximal subrelations (regions) is

$$\{(x_1, x_2), (x_3, x_5), (x_3, x_7), (x_4), (x_6, x_7)\}. \qquad (3\text{-}126)$$

The 0.28-regions with common elements are

$$\{(x_3, x_5)\} \cap \{(x_3, x_7)\} = x_3, \text{ and } \{(x_3, x_7)\} \cap \{(x_6, x_7)\} = x_7. \qquad (3\text{-}127)$$

Considering all values of α in the min-addition transitive closure, the decomposition tree is obtained in Fig. 3.41. The

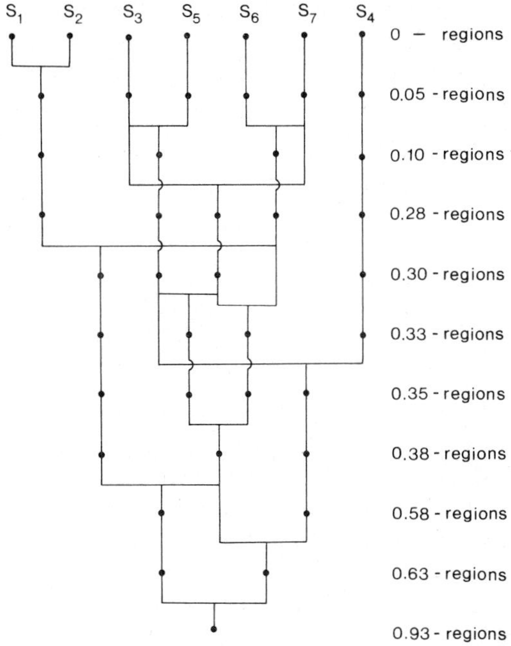

Fig. 3.41 Regionalization pattern obtained through the decomposition of the min-addition transitive distance matrix

overlapping of regions at all levels of decomposition is summarized in Table 3.3.

Example 3.16. (Economic regions of the European Economic Community, an application)

To unravel the regional structure of the European economy, Ponsard and Tranqui (1985) attempted to group 52 European Community Regions (ECR) with respect to 40 demographic and economic variables. The ECRs were grouped into regions in accordance with their dissimilarities (distances) in occupations, weekly work durations, education, health, and living standards.

The original data matrix $A = \{a_{ij}\}$, $i = 1, \ldots, 52$; $j = 1$,

Table 3.3 Overlapping of regions through the decomposition of the min-addition transitive distance matrix

Regions		Common Elements
0-regions:	$\{S_1\}, \{S_2\}, \{S_3\}, \{S_4\},$	none
	$\{S_5\}, \{S_6\}, \{S_7\}$	
0.05-regions:	$\{S_1, S_2\}, \{S_3\}, \{S_4\},$	none
	$\{S_5\}, \{S_6 , \{S_7\}$	
0.10-regions:	$\{S_1, S_2\}, \{S_3, S_5\}, \{S_4\},$	none
	$\{S_6, S_7\}$	
0.28-regions:	$\{S_1, S_2\}, \{S_3, S_5\}, \{S_3, S_7\}$	$\{S_3, S_5\} \cap \{S_3, S_7\} = \{S_3\}$
	$\{S_4\}, \{S_6, S_7\}$	$\{S_3, S_7\} \cap \{S_6, S_7\} = \{S_7\}$
0.30-regions:	$\{S_1, S_2, S_6, S_7\}, \{S_3, S_5\},$	$\{S_1, S_2, S_6, S_7\} \cap \{S_3, S_7\} = \{S_7\}$
	$\{S_3, S_7\}, \{S_4\}$	$\{S_3, S_5\} \cap \{S_3, S_7\} = \{S_3\}$
0.33-regions:	$\{S_1, S_2, S_6, S_7\}, \{S_3, S_5, S_7\}$	$\{S_1, S_2, S_6, S_7\} \cap \{S_3, S_5, S_7\} = \{S_7\}$
	$\{S_3, S_6, S_7\}, \{S_4\}$	$\{S_1, S_2, S_6, S_7\} \cap \{S_3, S_6, S_7\} = \{S_6, S_7\}$
		$\{S_3, S_5, S_7\} \cap \{S_3, S_6, S_7\} = \{S_3, S_7\}$

Table 3.3 (cont'd)

Regions	Common Elements
0.35-regions: $\{S_1, S_2, S_6, S_7\}$, $\{S_3, S_5, S_7\}$	$\{S_1, S_2, S_6, S_7\} \cap \{S_3, S_5, S_7\} = \{S_7\}$
$\{S_3, S_6, S_7\}$, $\{S_3, S_4, S_5\}$	$\{S_1, S_2, S_6, S_7\} \cap \{S_3, S_6, S_7\} = \{S_6, S_7\}$
	$\{S_3, S_5, S_7\} \cap \{S_3, S_6, S_7\} = \{S_3, S_7\}$
	$\{S_3, S_5, S_7\} \cap \{S_3, S_4, S_5\} = \{S_3, S_5\}$
	$\{S_3, S_6, S_7\} \cap \{S_3, S_4, S_5\} = \{S_3\}$
0.38-regions: $\{S_1, S_2, S_6, S_7\}$,	$\{S_1, S_2, S_6, S_7\} \cap \{S_3, S_5, S_6, S_7\} = \{S_6, S_7\}$
$\{S_3, S_5, S_6, S_7\}$, $\{S_3, S_4, S_5\}$	$\{S_3, S_5, S_6, S_7\} \cap \{S_3, S_4, S_5\} = \{S_3, S_5\}$
0.58-regions: $\{S_1, S_2, S_3, S_5, S_6, S_7\}$,	$\{S_1, S_2, S_3, S_5, S_6, S_7\} \cap \{S_3, S_4, S_5\} = \{S_3, S_5\}$
$\{S_3, S_4, S_5\}$	
0.63-regions: $\{S_1, S_2, S_3, S_5, S_6, S_7\}$,	$\{S_1, S_2, S_3, S_5, S_6, S_7\} \cap \{S_3, S_4, S_5, S_6, S_7\} = \{S_3, S_5, S_6, S_7\}$
$\{S_3, S_4, S_5, S_6, S_7\}$	
0.93-regions: $\{S_1, S_2, S_3, S_4, S_5, S_6, S_7\}$	all

..., 40, is of dimension 52 × 40. A new matrix $A' = \{a'_{ij}\}$, i = 1, ..., 52; j = 1, ..., 40, is obtained by dividing each element in the corresponding column of A by the highest value of that column. Then, each row of A' (i.e. each ECR) can be treated as a fuzzy subset with its element indicating the grade of membership in characteristic j. The generalized relative Hamming distances between all pairs of ECRs are then obtained to measure their dissimilarities. The resulting distance matrix R, with dimension 52 × 52, is antireflexive, symmetric, and min-addition transitive. Based on the discussion above, it can be decomposed with respective to the distance $\alpha \in [0, 1]$.

The analysis results are summarized as follows:

(a) The European Community is economically well-separated because there are no transnational regions. The only exception is for $\alpha = 0.04164$ (ranked 4th in terms of being least similar in the decomposition hierarchy), Luxembourg and Wales of United Kingdom are similar.

(b) ECRs not showing significant dissimilarities cannot be unambiguously separated, a natural phenomenon successfully captured by the fuzzy set approach.

(c) Coherent regions appear clearly within individual nations (Table 3.4). ECRs are grouped together rather quickly. In addition, the regions remain very stable once they are formed. That is, other ECRs can only be grouped into these regions after a few ranks of α values. The higher overlapping of regions in United Kingdom and the Netherlands show a more homogenous economic structure. A higher degree of heterogeneity is however found in Italy, France, and West Germany. This again demonstrates the robustness of the fuzzy set approach.

(d) Based on the ECRs grouping sequence with respect to the distance α, the procedure can differentiate the less distinctive from the more distinctive ECRs (Table 3.5 and 3.6), and can demonstrate the extreme heterogeneity of over-urbanized ECRs and large cities such as the Brussels region and the Flemish region.

Table 3.4 Rank and value of α for each region

Country	Value of α	Rank of α	Regions
United Kingdom[a]	0.03980	3	{West Midlands, North}
	0.04277	5	{Yorkshire and Humberside, East Midlands, South West}
	0.04353	7	{North, East Midlands}, {Yorkshire and Humberside, West Midlands}
	0.05062	10	{East Anglia, Wales}, {East Midlands, Wales}
	0.06078	25	{East Midlands, South West, Wales}
	0.06184	30	{West Midlands, North West}
	0.07160	45	{East Midlands, East Anglia, Wales}, {South West, West Midlands, Scotland}, {East Midlands, South West, West Midlands}
	0.07324	50	{East Midlands, South West, Wales}
Netherlands[b]	0.03255	2	{Oost Nederland, Zuid Nederland}
	0.04303	6	{Noord Nederland, Oost Nederland}, {Oost Nederland, Zuid Nederland}
	0.05679	15	{Noord Nederland, Zuidwest Nederland}
	0.06078	25	{Noord Nederland, Oost Nederland, Zuid Nederland}
Italy[c]	0.04734	8	{Nord Ovest, Centro}
	0.05679	15	{Campania, Sicilia}
	0.05930	20	{Nord Est, Centro}
	0.06184	30	{Emilia Romagna, Centro}

Table 3.4 (cont'd)

Country	Value of α	Rank of α	Regions
France[d]	0.05679	15	{Ouest, Centre Est}
	0.06078	25	{Sud Quest, Centre Est}, {Ouest, Centre Est}
	0.06644	35	{Sud Ouest, Mediterranee}
	0.07019	40	{Ouest, Sud Ouest, Centre Est}
	0.07160	45	{Nord Pas-de-Calais, Est}
	0.07324	50	{Bassin Parisien, Ouest}
Federal Republic of Germany[e]	0.05930	20	{Bremen, Saarland}, {Hamburg, Berlin (West)}
	0.06078	25	{Hamburg, Bremen}
	0.06184	30	{Schleswig-Holstein, Rheinland-Pfalz}
	0.06644	35	{Hessen, Niedersachsen}
	0.07160	45	{Hessen, Rheinland-Pfalz}

[a] All the ECRs aggregate to form regions except two: Northern Ireland and South East including London.

[b] The same remark can be repeated. Only West Nederland ECR including Amsterdam does not join a region.

[c] Three ECRs, Lombardy (with Milan), Latium (with Rome), and South do not associate with a region. We can note that the North ECRs present affinities relatively greater than those of the South.

[d] The aggregation speed of ECRs is relatively weak but all associate into regions except the Ile de France (with Paris).

[e] Three ECRs out of eleven do not appear in regions. They are Nordrhein-Westfalen, Bayern, and Baden-Wurttemberg.

Table 3.5 The less distinctive European Community Regions

Aggregation Speed (rank of α)	The Less Distinctive ECRs	Regions
6	Oost Nederland	{Oost Nederland, Noord Nederland} {Oost Nederland, Zuid Nederland}
7	Yorkshire and Humberside	{Yorkshire and Humberside, East Midlands, South West} {Yorkshire and Humberside, West Midlands}
	East Midlands	{East Midlands, Yorkshire and Humberside, South West} {East Midlands, North}
10	Wales	{Wales, East Anglia} {Wales, East Midlands} {Wales, Grand-Duche de Luxembourg}
15	Oost Nederland	{Oost Nederland, Zuid Nederland, Noord Nederland} {Oost Nederland, Noord Nederland, Zuidwest Nederland}

Table 3.6 The more distinctive European Community Regions

Aggregation Speed (rank of α)	The Less Distinctive ECRs	Regions
85	Region Bruxelloise	{Region Bruxelloise}
90	Region Bruxelloise	{Region Bruxelloise, Hamburg, Bremen}
155	Region Bruxelloise	{Region Bruxelloise, Hamburg, Bremen, Berlin (West)}
80	Danmark	{Danmark}
85	Danmark	{Danmark, Scotland}
90	Lazio (Roma)	{Lazio}
95	Lazio (Roma)	{Lazio, Sicilia}
125	Ireland	{Ireland}
130	Ireland	{Ireland, Northern Ireland}
165	Region Flamande	{Region Flamande}
170	Region Flamande	{Region Flamande, Est}
270	Ile de France (Paris)	{Ile de France}
300	Ile de France (Paris)	{Ile de France, Bassin Parisien}
900	Nordrhein-Westfalen	{Nordrhein-Westfalen}
940	Nordrhein-Westfalen	{Nordrhein-Westfalen, Bayern}
1010	South East (London)	{South East}
1020	South East (London)	{South East, Ile de France}

Remark. (a) In subsections 3.4.3 and 3.4.5, regionalization is based on the concept of max-min transitive similarity. In subsections 3.4.4 and 3.4.6, regionalization is based on the concept of min-max transitive distance and the min-addition transitive distance respectively. In addition to the min-max transitive distance and equivalently the max-min transitive similarity which give disjoint regions, and min-addition transitive distance which gives nondisjoint regions, other notions of distance or similarity can also be defined. In general, in defining max-min transitivity, the min-operator can be replaced by any associative and monotonic operator *. For example, the algebraic product, •, can replace the min-operator and gives max-product transitivity in terms of similarity. Its complement in turn gives min-sum transitive distance. The decomposition of their transitive closures yields again disjoint regions. Fig. 3.42 summarizes the fuzzy set procedures of grouping for regions. Depending on the regionalization problem, one procedure may prove to be more appropriate than the others. The system is thus very flexible, natural, and informative.

(b) Similar in spirit but different in approach, fuzzy clustering techniques (for a general review see Bezdek, 1981) can also be employed to group spatial units into regions. These techniques give more statistical arguments to the grouping procedures.

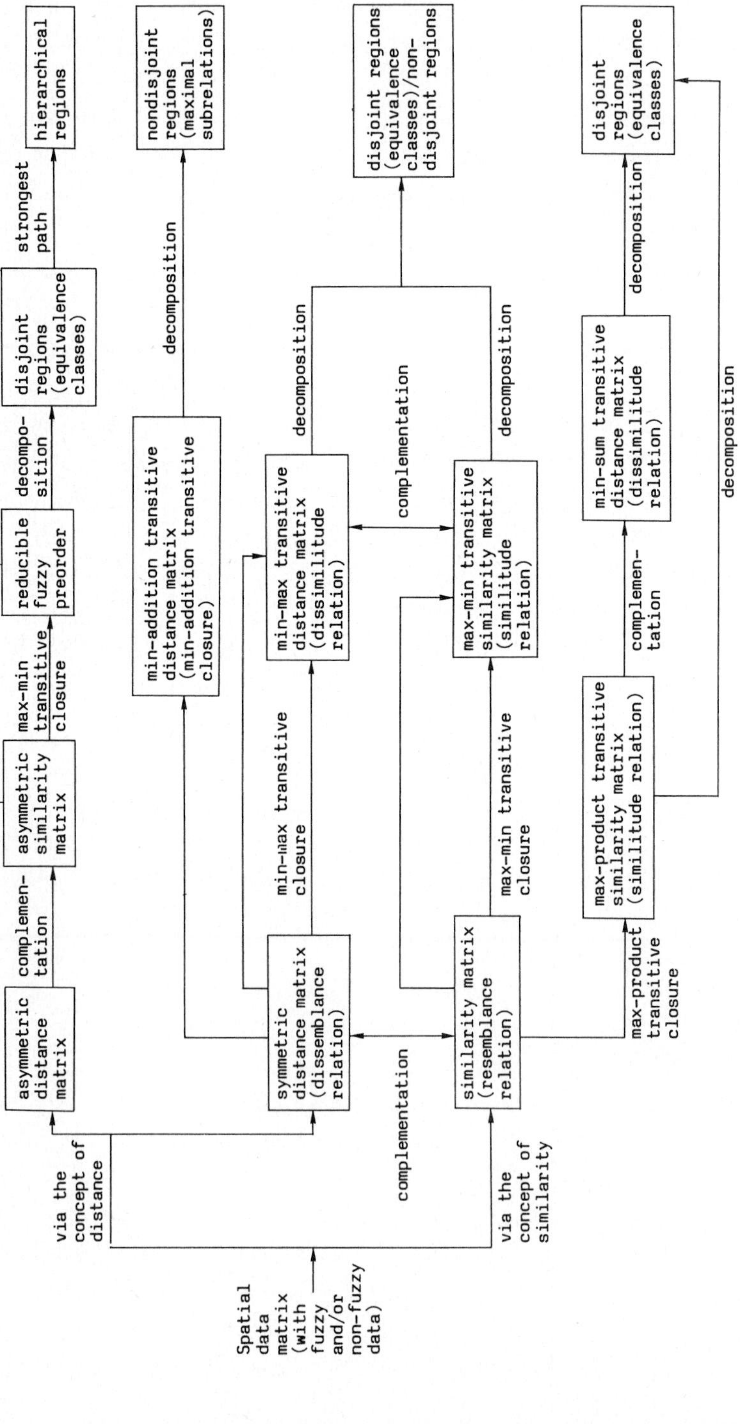

Fig. 3. 42 Schematic representation of various regionalization procedures

CHAPTER 4

IMPRECISIONS OF SPATIAL PREFERENCE, UTILITY, AND CHOICE

4.1. A NOTE ON THE IMPRECISION OF CHOICE

Most issues in spatial analysis involve the problem of choice. On the individual level, a person needs to make decisions with varying frequencies on matters such as the place of residence, place to shop, place to work, goods to consume, and transportation modes to take. On the organization level, basic decisions facing a private firm are locations of production, input-output combinations, market shares, logistics of distribution, pricing policies, and development or growth strategies. Likewise, agents in the public sector often need to decide on the optimal location-allocation schemes, efficient and equitable distributions of services, and the best public policies. Some of the choice problems may be simple, but most of them are complex. All decisionmaking processes, spatial or aspatial, are explicitly or implicitly governed by a preference-utility-choice structure which has been the backbone of theories or models of spatial choice.

In conventional spatial analysis, the preference-utility-choice structure is formulated on the basis of a two-valued logic. Information in the decisionmaking processes is assumed to be precise. Decisionmakers are aware of alternatives available to them and are capable of evaluating them in a clear-cut manner. Furthermore, they can make unambiguous discrimination of alternatives. Their preference structures are thus precise and totally ordered. Consequently, the associated utility functions can be precisely determined and are ordinal or cardinal in structure. Thus, imprecision of spatial choice is nonexistent in the conventional mode of analysis.

However, information is often imperfect and our decision-making processes are more or less imprecise. Individuals' preferences are determined by subjective factors and utility is a perceived but not an inherent characteristic of an alternative. Therefore, preference and utility usually involve a certain degree of imprecision. We may not be able to discriminate with precision different alternatives available to us but are capable of discriminating them to a certain extent. In place of asking whether or not one alternative is preferred to another, it makes more sense to ask how strongly one alternative is preferred to another one. As a result, we could only have a fuzzy ordering of alternatives.

Over the years, researchers such as Pipkin (1978), Orlovsky (1978, 1980), Leung (1979, 1982b), Nijkamp (1979), and Ponsard (1979, 1980, 1981) have argued for the formal treatment of imprecision in our preference-utility-choice structures. The theory of fuzzy subsets is regarded as one of the viable methods for a precise representation of these imprecise concepts. It enables us to formulate a more appropriate approximation to our imprecise preference-utility-choice structures. The fuzzy set approach is thus employed as a basis of formalization in this chapter. It focuses on the microeconomic foundations of the macroeconomic spaces, such as regions and nations, under imprecision. Without loss of generality, I restrict our discussion to the analysis of the choice problems in the theories of consumers' and producers' spatial behavior. The framework, however, is applicable to many other decisionmaking situations.

Preference, utility, and choice without constraints are analyzed in section 4.2. Construction of the imprecise preference-utility-choice structures in an unconstrained spatial context is discussed. Emphasis is placed on their structural properties.

In section 4.3, spatial choice in a constrained context is examined. The general framework of optimal choice with a fuzzy objective and a set of fuzzy constraints is presented.

In section 4.4, consumer's and producer's spatial partial equilibrium is investigated. Situations involving precise and fuzzy utilities are analyzed.

In section 4.5, general equilibrium in a fuzzy environment is scrutinized. Properties of the equilibrium solution are detailed.

4.2. UNCONSTRAINED SPATIAL CHOICE WITH IMPRECISE PREFERENCE STRUCTURES

In the classical theories of consumers' spatial behavior, an individual's decision is determined on the basis of his preference structure. Given a preference structure, a utility function is sought so that alternatives can be ranked in accordance with their assigned utility indices. The higher the utility index is, the more preferrable the alternative becomes. Here, an alternative can be a set of located commodities or any course of actions.

Conventionally, an individual is assumed to have a precise preference structure so that he can express his preference in one of the following statements:

(a) I prefer alternative A to alternative B;
(b) I prefer alternative B to alternative A;
(c) I am indifferent to alternative A and B.

As argued above, preference is usually imprecise. Thus, an individual may only be able to express his preference in one of the following statements:

(a) I prefer alternative A to alternative B to a certain extent;
(b) I prefer alternative B to alternative A to a certain extent;
(c) Alternatives A and B are indifferent to me to a certain extent.

Preference structure thus obtained is imprecise. The associated utilities, if they exist, are in turn fuzzy and alternatives may only be chosen with a certain degree of preference.

In subsection 4.2.1, a fuzzy set construction of the

preference relation is presented. Basic properties of the imprecise preference structure are examined.

Spatial choice without an explicit utility function is discussed in subsection 4.2.2. It is focused on the selection of optimal alternatives with respect to the fuzzy preference relation.

In subsection 4.2.3, spatial choice through an explicitly formulated function of imprecise utilities is analyzed. Conditions which guarantee the existence of a utility function are stated.

Throughout this section, only the unconstrained spatial choice problems are considered.

4.2.1. Imprecision of Spatial Preferences

Let $\underline{X} = \{x_i\}$, $i \in \underline{I}$ (a finite or infinite index set), be the set of all possible alternatives (located goods, for example). Due to market complexities, insufficient information, and subjective valuation, an individual's preference is generally imprecise. Preference intensity is indicated by the degree to which one alternative is preferred or indifferent to another. A consumer's preference relation, R, can then be represented by a fuzzy binary relation

$$\mu_R: \underline{X} \times \underline{X} \longrightarrow [0, 1], \tag{4-1}$$

$$(x_i, x_j) \longrightarrow \mu_R(x_i, x_j), \quad \forall (x_i, x_j) \in \underline{X} \times \underline{X}, \tag{4-2}$$

where $\mu_R(x_i, x_j) \in [0, 1]$ is the degree to which alternative $x_i \in \underline{X}$ is preferred to alternative $x_j \in \underline{X}$.

The set of possible alternatives $\underline{X}$ and the imprecise preference relation R constitute an individual's imprecise spatial preference structure, denoted by $(\underline{X}, R)$. Parallel to conventional spatial choice theories, the imprecise preference structure $(\underline{X}, R)$ can be further broken down into the indifference structure $(\underline{X}, \sim)$ and the strict preference structure $(\underline{X}, \succ)$. Therefore, $(\underline{X}, R)$ can be expressed as $(\underline{X}, \succsim)$. We first examine the properties of such a structure with reference to Ponsard (1981, 1985b).

If $\mu_R(x_i, x_j) > \mu_R(x_j, x_i)$, then $\mu_R(x_i, x_j)$ is called the strong degree of preference for x_i relative to x_j and $\mu_R(x_i, x_j)$ is called the weak degree of preference for x_j relative to x_i.

The imprecise strict preference structure $(\underline{X}, \succ)$ is anti-symmetric. That is

$$\mu_R(x_i, x_j) \neq \mu_R(x_i, x_j), \quad \forall\, (x_i, x_j) \in \underline{X} \times \underline{X}. \tag{4-3}$$

To guarantee that the set of imprecise choices is non-empty, the property of transitivity should be imposed:

$$\mu_R(x_i, x_j) \geq \max_{x_h} \min[\mu_R(x_i, x_h), \mu_R(x_h, x_j)],$$

$$\forall\, (x_i, x_h), (x_h, x_j), (x_i, x_j) \in \underline{X} \times \underline{X}. \tag{4-4}$$

The physical meaning of (4-4) is that the degree of preference for x_i relative to x_j is at least as high as the lowest of the degrees of preference for x_i relative to x_h and for x_h relative to x_j, for all intermediate x_h's.

The imprecise indifference structure $(\underline{X}, \sim)$ is, however, symmetric and reflexive. The property of symmetry:

$$\mu_R(x_i, x_j) = \mu_R(x_j, x_i), \quad \forall\, (x_i, x_j) \in \underline{X} \times \underline{X}, \tag{4-5}$$

implies the property of reflexivity:

$$\mu_R(x_i, x_i) \geq 0, \quad \forall\, (x_i, x_i) \in \underline{X} \times \underline{X}, \tag{4-6}$$

when $x_i = x_j$.

The degree of indifference can be zero. It implies that an individual can either discriminate between any two alternatives which are of no interest to him or it is impossible for him to compare. Under this situation, the set of choices is empty and the values of μ_R is zero. As a consequence, the property of totality of the strict preference relation and the indifference relation is ensured. The conventional axiom of comparability needs not be stated.

Based on the properties of the imprecise strict preference structure $(\underline{X}, \succ)$ and the indifference structure $(\underline{X}, \sim)$, the imprecise preference-indifference structure $(\underline{X}, \succsim)$ is thus a fuzzy total preorder which is reflexive, (4-6), and max-min transitive, (4-4).

Remark. In conventional analysis, the preference structure is also assumed to be reflexive and transitive. Under imprecision, reflexivity does not have to be equal to 1. Such an assumption is important to ensure that the fuzzy total preorder is reducible and the fuzzy utility function is constructable. In relation to the notion of transitive preference in conventional analysis, max-min transitivity in (4-4) is weaker in assumption. By replacing the max- and min-operators in (4-4) by the Boolean sum and Boolean product, the conventional definition of transitivity is obtained. Thus, the fuzzy set formulation, though weaker in assumption, is of a higher level of generalization.

Ponsard (1986a) shows that if reflexivity is defined by:

$$\mu_R(x_i, x_i) = \max_{(x_i, x_j)} \mu_R(x_i, x_j), \ \forall\, x_i \in \underline{X}, \qquad (4\text{-}7)$$

where

$$\mu_R(x_i, x_j) = \mu_R(x_j, x_i), \ \forall\, (x_i, x_j), (x_j, x_i) \in \underline{X} \times \underline{X},$$

then the value of $\mu_R(x_i, x_i)$ is unique. That is, the preference for alternative x_i equals the maximum of the degree of preferences for x_i relative to all other alternatives.

Under this situation, all the fuzzy subsets, with $\mu_R(x_i, x_j) = \mu_R(x_j, x_i)$, form among themselves similitude subrelations, $R(x_i)$, which are reflexive, symmetric, and max-min transitive. For any $x_i \in \underline{X}$, it belongs to one and only one similitude subrelation.

Let $R(x_i)$ corresponds to a equivalence class $C(x_i)$. Let the fuzzy quotient set, $\underline{X}/R(x_i)$, be defined by

$$\underline{X}/R(x_i) = \{C(x_i), \ \forall\, x_i \in \underline{X}\}. \qquad (4\text{-}8)$$

Then, $\underline{X}/R(x_i)$ is isomorphic to an ordinary quotient set $\underline{X}/\sim$ (Negoita and Ralescu, 1975), or equivalently, the following bijection exists:

$$f: \underline{X}/R(x_i) \longrightarrow \underline{X}/\sim, \qquad (4\text{-}9)$$

where $\sim$ is an equivalence relation.

Since the relation formed among the equivalence classes, strongest path from one class to another, is necessarily anti-symmetric, then the equivalence classes form among themselves a fuzzy order relation.

If no constraints are imposed on the decisionmaking process, spatial choice can be directly determined from a given imprecise spatial preference structure. In subsection 4.2.2, decision rules for an imprecise spatial consumption behavior are discussed. Rules for obtaining the most preferrable alternative in the possible consumption space are formulated.

4.2.2. Choice with an Imprecise Spatial Preference Structure

Under no constraints, spatial choice behavior can be solely determined by a given imprecise spatial preference-indifference structure $(\underline{X}, \succsim)$. Based on some optimality criteria, spatial choice behavior can be defined on a finite or infinite set of all possible alternatives $\underline{X}$ (Ponsard, 1985b).

Definition 4.1. Let $(\underline{X}, \succsim)$ be an imprecise preference-indifference structure. Let $J \subset \underline{X}$ be a non-empty fuzzy subset of $\underline{X}$. An alternative $x_i \in J$ is said to be optimal if and only if there does not exist an alternative $x_j \in J$ such that $\mu_J(x_j, x_i) > \mu_J(x_i, x_j)$. That is, to be optimal x_i has to be a maximal element of the structure $(J, \succsim)$.

Having determined a set of optimal alternatives, denoted as N(J), we need to determine the set of best elements in it. Such an alternative is the maximum of the structure $(J, \succsim)$.

Definition 4.2. An alternative $x_i \in J$ is a best alternative in J if and only if $\mu_J(x_i, x_j) \geq \mu_J(x_j, x_i)$, $\forall\, x_j \in J$.

Let

$$S(J) = \{x_i \varepsilon J \mid \mu_J(x_i, x_j) \geq \mu_J(x_j, x_i), \ \forall x_j \varepsilon J\} \quad (4\text{-}10)$$

be the set of imprecise choices consisting of the best alternatives in J. Associated with an imprecise preference-indifference structure $(\underline{X}, \succsim)$, an imprecise spatial choice behavior is defined if and only if the following mapping is defined everywhere:

$$S: \mathscr{F}(\underline{X}) \longrightarrow X, \qquad J \longrightarrow S(J). \quad (4\text{-}11)$$

To guarantee that an imprecise spatial choice behavior is decidable, the structure $(\underline{X}, \succsim)$ has to satisfy specific conditions. Since the set of possible alternatives $\underline{X}$ can be finite or infinite, we should then investigate the situations separately.

(a) Case 1 ($\underline{X}$ finite)

This is quite obvious. Since $\underline{X}$ is finite, then $(\underline{X}, \succsim)$ is a finite fuzzy total preorder. To obtain the best alternatives in J, it is sufficient to make pairwise comparisons of all alternatives in $\underline{X}$. Consequently, S(J) is non-empty and the mapping S in (4-11) is defined. Thus, the imprecise spatial choice behavior is decidable.

(b) Case 2 ($\underline{X}$ is infinite)

Since $\underline{X}$ is infinite, then the structure $(\underline{X}, \succsim)$ is an infinite fuzzy total preorder. Under this situation, the imprecise spatial choice behavior is decidable if and only if every increasing sequence of alternatives, $(x_i)_{(n)}$, in $\underline{X}$ is finite. That is, $(\underline{X},\succsim)$ satisfies the condition of an ascending chain.

It can be seen that if the condition of an ascending chain is not satisfied, then there exists an increasing sequence of alternatives, $(x_i)_{(n)}$, in $\underline{X}$ such that the fuzzy subset J in $\underline{X}$ defined by

$$J = \{x_1, \ldots, x_i, \ldots \mid \mu_J(x_i, x_j) \varepsilon [0, 1], \ \forall x_i, x_j \varepsilon \underline{X}\} \quad (4\text{-}12)$$

has no best alternatives. Thus, S(J) is an empty set and the imprecise spatial choice behavior is not decidable. Therefore, we have a contradiction.

Conversely, if there exists a fuzzy subset $J \varepsilon \mathscr{F}(\underline{X}) - \phi$ such that $S(J) = \phi$, and if we choose $x_j \varepsilon J$ (x_j is not optimal because $S(J) = N(J) = \phi$), then we can find $x_k \varepsilon J$ such that $\mu_J(x_j, x_k) < \mu_J(x_k, x_j)$.

By the same token, since $x_j \varepsilon J$ is not maximal, we can also find a $x_h \varepsilon J$ such that $\mu_J(x_i, x_j) < \mu_J(x_j, x_i) < \mu_J(x_h, x_j)$. Repeating the same argument, it is apparent that we can construct an infinite increasing sequence which contradicts the condition of an ascending chain. Thus, $S(J) \neq \phi$ for each $J \varepsilon \mathscr{F}(\underline{X}) - \phi$.

The above is a general framework of spatial choice behavior with an imprecise spatial preference structure. It applies to cardinal and noncardinal analyses.

Employing slightly different principles, Orlovsky (1978, 1980) proposes to analyze spatial choice behavior via the concept of nondominance. It is, however, a cardinal analysis, and is thus less general.

4.2.3. Imprecise Utility and Spatial Choice

Often, spatial choice is based on the perceived utility of an alternative. The greater is the utility of an alterrnative, the more likely it would be chosen. Since the utility of an alternative is only a reflection of an individual's preference for it. Therefore, it is necessary to determine whether or not an utility function exists in a specific spatial preference structure.

Intuitively, alternatives belonging to the same equivalence class in a reducible imprecise spatial preference structure $(\underline{X}, R)$ should yield the same degree of utility. If an alternative x_i in one equivalence class, $C(x_i)$, is preferred to an alternative x_g in another equivalence class, $C(x_g)$, then the utility of x_i should be greater than x_g.

Let $X \subset \underline{X}$ be a fuzzy subset associated with the structure

$(\underline{X}, \succsim)$ defined by

$$X = \{x_i \mid \mu_X(x_i) \in [0, 1], \ \forall x_i \in \underline{X}\}, \tag{4-13}$$

where

$$\mu_X(x_i) = \mu_R(x_i, x_j) = \mu_R(x_j, x_i) = \mu_X(x_j) = \mu,$$

and $\mu_X(x_i)$ is called the relative degree of preference for x_i relative to other alternatives. The same interpretation applies to $\mu_X(x_j)$.

Then, the fuzzy utility u is a function defined on the set of possible alternatives $\underline{X}$:

$$u: \underline{X} \longrightarrow [0, 1], \tag{4-14}$$

$$x_i \longrightarrow u(x_i) = (f \circ g)(x_i) = f[C(x_i)] = \mu_R(x_i, x_j) = \mu. \tag{4-15}$$

The function

$$f: \underline{X}/\sim \longrightarrow [0, 1], \tag{4-16}$$

$$C(x_i) \longrightarrow f[C(x_i)] = \mu_R(x_i, x_j) = \mu, \tag{4-17}$$

is a strictly increasing function which maps the equivalence classes $C(x_i)$ in the quotient set $\underline{X}/\sim$ to [0, 1]. The function f is strictly increasing if

$$\forall (x_i, x_j) \in C(x_i) \text{ and } \forall (x_g, x_k) \in C(x_g),$$

$$\mu_R(x_i, x_j) > \mu_R(x_g, x_k) \Longrightarrow f[C(x_i)] > f[C(x_g)]. \tag{4-18}$$

The function

$$g: \underline{X} \longrightarrow \underline{X}/\sim, \tag{4-19}$$

$$x_i \longrightarrow g(x_i) = C(x_i), \tag{4-20}$$

is a canonic projection from the possible alternative space $\underline{X}$ to the quotient set $\underline{X}/\sim$.

The existence of such a utility function is guaranteed if the imprecise spatial preference structure $(\underline{X}, R)$ is a completely preordered countable set. The following theorem (Ponsard, 1981) states under what conditions $(\underline{X}, R)$ is a completely preordered countable set.

Theorem 4.1. Let X in an imprecise spatial preference space $(\underline{X}, \succsim, \mathscr{T})$ be convex. If

(a) the fuzzy preference relation is complete and satisfies the convexity condition, and
(b) the fuzzy strict preference relation is max-min transitive,

then $(\underline{X}, \succsim, \mathscr{T})$ is a completely preordered topological preference space.

The above theorem is in fact a generalization of Uzawa's result (1960) in the classical theory of consumption. The symbol $\mathscr{T}$ in Theorem 4.1 denotes a topology on $\underline{X}$ which can be characterized as follows:

Let $\mu_X(x_i)$ be defined in (4-13), and $[\mu_X(x_n)]_n$, $n \in \mathbb{N}$, be an infinite sequence of the relative degrees of preference for the alternatives. The sequence approaches to $\mu_X(x_0)$, $x_0 \in \underline{X}$, as n approaches to infinity. If for any integer n, $[\mu_X(x_n)]_n$ is at most as great (or at least as great) as $\mu_X(x_i)$, then $\mu_X(x_0)$ is at most as great (or at least as great) as $\mu_X(x_i)$. That is, for every x_i and $x_n \in \underline{X}$, the subsets

$$\overline{X}* = \{x_n | \mu_X(x_n) \leq \mu_X(x_i),\ x_n \in \underline{X}\}, \quad (4\text{-}21)$$

and,

$$\underline{X}_* = \{x_n | \mu_X(x_n) \geq \mu_X(x_i),\ x_n \in \underline{X}\}, \quad (4\text{-}22)$$

are closed in $\underline{X}$. Furthermore, the fuzzy preference is continuous if and only if $\overline{X}* \in \mathscr{T}$ and $\underline{X}_* \in \mathscr{T}$, $\forall x_i \in \underline{X}$.

Convexity in Theorem 4.1 means that

$$\mu_X(x_i) \geq \mu_X(x_j) \Longrightarrow \mu_X[\lambda x_i + (1 - \lambda)x_j] \geq \mu_X(x_j),$$
$$\forall x_i \in \underline{X},\ \forall x_j \in \underline{X},\ x_i \neq x_j,\ \lambda \in [0, 1]. \quad (4\text{-}23)$$

That is, the preference for the linear combination of x_i and x_j is greater than or equal to that for x_j if the preference for x_i is strictly greater than or indifferent to x_j.

Therefore, given a countable imprecise spatial preference structures $(\underline{X}, R)$ on which a function of fuzzy utilities exists, this utility function is a continuous homomorphism from $(\underline{X}, \underset{\sim}{\succ}, \underline{\mathscr{T}})$ to $([0, 1], \underline{\geq}, \underline{\mathscr{N}})$ with $\underline{\mathscr{N}}$ being a natural topology of $[0, 1] \subset \mathbb{R}$.

If a utility function on an imprecise spatial preference structure is obtained, spatial choice without constraints can be preformed with reference to the fuzzy utilities of the alternatives. The optimal choice is the set

$$\{x_i \mid u(x_i) \geq u(x_j),\ \forall x_j \in \underline{X}\}. \tag{4-24}$$

Remark. To make the concept of a fuzzy utility function more embracing, Mathieu-Nicot (1986) proposes a concept of fuzzy expected utility. If there is randomness in the decision-making environment a "fuzzy stochastic utility" is obtained. A "fuzzy possibility utility", however, is obtained under imprecision.

Of course, the major purpose of obtaining a function of fuzzy utilities is to perform spatial choice under constraints. The maximization of a function of fuzzy utilities subject to a fuzzy constraint is discussed in section 4.3. For convenience, unless otherwise specified, a function of fuzzy utilities is called a fuzzy utility function.

4.3. CONSTRAINED SPATIAL CHOICE WITH IMPRECISE PREFERENCE STRUCTURES

Often, restrictions are imposed on the possible consumption space in which a spatial choice is determined. Under limited budget, an individual may need to maximize his utility function with respect to a budget constraint. A producer may need to maximize his utility of profit subject to a technological constraint. Maximization of an utility function subject to a budget constraint has thus been a classical problem in spatial choice theories. Treating the utility function as an objective function and the budget or technological constraint as a constraint, the optimal choice problem is in fact a classical constrained optimization problem.

In this section, emphasis is placed on the construction of a general framework for the optimization of a fuzzy objective function subject to a fuzzy constraint. Imprecision of an objective function (utility function, in the context of this chapter) is due to the existence of an imprecise spatial preference structure as discussed in section 4.2. Imprecision of the constraint (budget or technological constraint, in the context of this chapter) is due to a decisionmaker's imprecise cognition or judgement of his technological environment or flexibilities deliberately incorporated into the spatial choice behavior. For example, in place of setting a budget to be β, a real number, we may only be able to specify it to be "*approximately* β", a fuzzy subset.

In subsection 4.3.1, concepts of a fuzzy objective, fuzzy constraint, and fuzzy decision space are discussed. They constitute the basic elements of optimal spatial choice problems under imprecision.

In subsection 4.3.2, selection of the optimal alternative is formulated as an optimization problem with a fuzzy utility function and a fuzzy budget or technological constraint. A general solution procedure is also outlined.

4.3.1. A Notion of Imprecise Decision Spaces

In section 4.2, it is demonstrated that an individual's imprecise utility is a function which maps the alternative space $\underline{X}$ into the set [0, 1]. Thus, if maximizing the utility of an alternative is the motive in a spatial choice problem, the utility function can be treated as a fuzzy objective O defined by

$$\mu_O: \underline{X} \longrightarrow [0, 1], \tag{4-25}$$

$$x_i \longrightarrow \mu_O(x_i), \quad \forall\, x_i \in \underline{X}. \tag{4-26}$$

In a simple constrained spatial choice situation, the fuzzy objective is to be maximized with respect to a fuzzy budget or technological constraint C defined by

$$\mu_C: \underline{X} \longrightarrow [0, 1], \quad (4\text{-}27)$$

$$x_i \longrightarrow \mu_C(x_i), \ \forall\, x_i \in \underline{X}. \quad (4\text{-}28)$$

Since the optimal alternative has to satisfy the fuzzy objective and fuzzy constraint simultaneously, the fuzzy decision space D can then be derived as the intersection of the fuzzy objective O and the fuzzy constraint C (Bellman and Zadeh, 1970), denoted as $D = O \cap C$, defined by

$$\mu_D: \underline{X} \longrightarrow [0, 1], \quad (4\text{-}29)$$

$$x_i \longrightarrow \mu_D(x_i) = \min[\mu_O(x_i), \mu_C(x_i)], \ \forall\, x_i \in \underline{X}. \quad (4\text{-}30)$$

Remark. In the formulation of the decision space D in (4-29) and (4-30), the fuzzy objective function and the fuzzy constraint are treated symmetrically. The differentiation is thus immaterial mathematically. If there are more than one constraint, the decision space becomes $D = O \cap C_1 \cap \ldots \cap C_m$, $m < \infty$, and the defining membership function becomes

$$\mu_D(x_i) = \min\left\{\mu_O(x_i), \min_i[\mu_{C_i}(x_i)]\right\}. \quad (4\text{-}31)$$

4.3.2. Optimal Spatial Choice in an Imprecise Decision Space

In subsection 4.2.3, the optimal alternatives are those with the highest degree of utility in the alternative space. Parallelly, in a constrained situation, the optimal alternatives should be those whose utilities are maximized with respect to a budget or technological constraint.

Due to the symmetry of the imprecise utility function and constraint, the optimal alternatives x_i^* can be defined.

Definition 4.3. The alternative x_i^* is optimal in the possible decision space D if

$$\mu_D(x_i^*) = \sup_{x_i \in \underline{X}} \mu_D(x_i) = \sup_{x_i \in \underline{X}} \min[\mu_O(x_i), \mu_C(x_i)]. \quad (4\text{-}32)$$

For situations involving multiple constraints, the optimal alternative x_i^* should satisfy

$$\mu_D(x_i^*) = \sup_{x_i \in \underline{X}} \mu_D(x_i) = \sup_{x_i \in \underline{X}} \left\{ \min \mu_O(x_i), \min_i[\mu_{C_i}(x_i)] \right\}. \quad (4\text{-}33)$$

That is, an optimal alternative satisfies simultaneously the fuzzy utility function and fuzzy constraints to the highest degree. Thus, μ_D can be interpreted as a constrained fuzzy utility, and $\mu_D(x_i)$ is the utility or degree of satisfaction generated by the alternative x_i. Maximizing $\mu_D(x_i)$ then becomes maximizing the constrained utility or satisfaction of x_i in D.

Generally, the solution of a fuzzy mathematical programming problem is not unique. Strong convexity is usually required. The problem in the following example, however, has an unique solution.

Example 4.1. Let

$$\mu_O(x) = e^{-r_1 x}, \; r_1 > 0, \quad (4\text{-}34)$$

be a single-variable fuzzy objective function.

Let

$$\mu_C(x) = e^{-r_2|x-a|}, \; r_2 > 0, \quad (4\text{-}35)$$

be a single-variable fuzzy constraint, with $r_2 > r_1$.

The functions μ_O and μ_C intersect at

$$x = \left[\frac{r_2}{r_1+r_2}\right]a, \text{ for } 0 < x < a, \quad (4\text{-}36)$$

$$x = \left[\frac{r_2}{r_2-r_1}\right]a, \text{ for } x > a. \quad (4\text{-}37)$$

Thus, the membership function of the imprecise decision space $D = O \cap C$ is

$$\mu_D(x) = \begin{cases} e^{-r_2|a-x|}, & \text{if } x \leq \left[\dfrac{r_2}{r_1+r_2}\right]a, \\ e^{-\left[\frac{r_1 r_2}{r_1+r_2}\right]a}, & \text{if } \left[\dfrac{r_2}{r_1+r_2}\right]a < x \leq \left[\dfrac{r_2}{r_2-r_1}\right]a, \\ e^{-r_2|x-a|}, & \text{if } x > \left[\dfrac{r_2}{r_2-r_1}\right]a. \end{cases} \qquad (4\text{-}38)$$

Applying (4-32), for

$$x^* = \left[\frac{r_2}{r_1+r_2}\right]a, \qquad (4\text{-}39)$$

$\mu_D(x^*) = \sup_x \mu_D(x)$. Thus, the optimal alternative is x^* in (4-39).
(see Fig. 4.1)

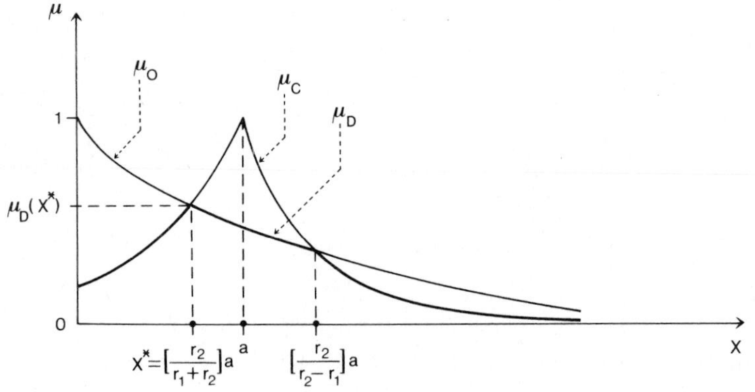

Fig. 4.1 Single-objective, single-constraint fuzzy optimization problem in example 4.1

Algorithmically, the optimization problem in (4-32) can be reduced to the search for the extremum of a scalar function (Tanaka, Okuda, and Asai, 1974; Negoita and Ralescu, 1975,

1977; Negoita, 1979). This result is summarized in the following theorem.

Theorem 4.2. Let $\mu_O: \underline{X} \longrightarrow [0, 1]$, $\mu_C: \underline{X} \longrightarrow [0, 1]$. Let $D = O \cap C$, and $\mu_D: \underline{X} \longrightarrow [0, 1]$. Then

$$\sup_{x_i \in \underline{X}} \mu_D(x_i) = \sup_{\alpha \in [0, 1]} \min[\alpha, \sup_{x_i \in \underline{C}_\alpha} \mu_O(x_i)],$$

where

$$\underline{C}_\alpha = \left\{ x_i \mid \mu_C(x_i) \geq \alpha, x_i \in \underline{X} \right\}, \quad \alpha \in [0, 1],$$

is the α-level set of the fuzzy subset C.

Example 4.2. Consider the optimization problem in Example 4.1. Fig. 4.2 shows how the optimal solution is obtained via theorem 4.2.

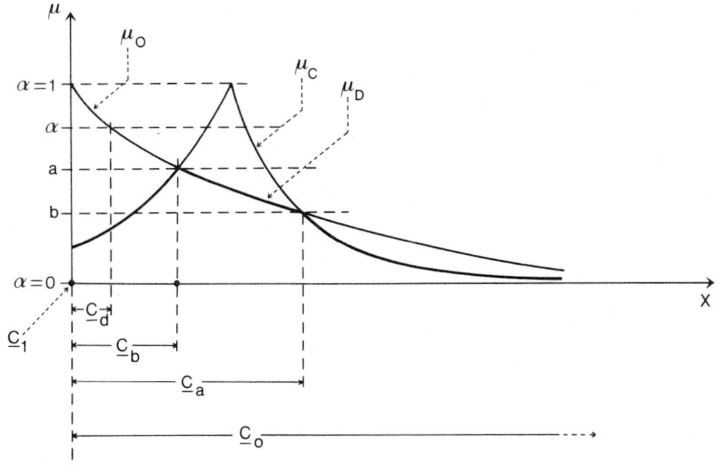

Fig. 4.2 Algorithmical solution of the fuzzy optimization problem in Fig. 4.1

4.4. SPATIAL PARTIAL EQUILIBRIUM UNDER IMPRECISION

In spatial economic analysis, the concept of partial equilibrium plays an important role in analyzing spatial economic behavior of a decisionmaking unit such as a household, a firm, or a public agent. In general, we can classify spatial partial equilibrium analysis into the consumer's spatial partial equilibrium and the producer's spatial partial equilibrium.

In consumer's spatial partial equilibrium, a consumer maximizes his utility with respect to a budget constraint while considering the equilibrium prices as parameters. In producer's spatial partial equilibrium, a producer maximizes his utility of profit subject to a technological constraint while considering the equilibrium prices as parameters.

Due to the imprecise decisionmaking environment previously discussed, consumer's and producer's spatial partial equilibrium may only be obtained with respect to their fuzzy utilities and constraints. The analysis is formally discussed in this section.

In subsection 4.4.1, a consumer's spatial partial equilibrium is determined through the maximization of his fuzzy utility function subject to a fuzzy budget constraint.

In subsection 4.4.2, a producer's spatial partial equilibrium is obtained via the maximization of his fuzzy utility of of profit with respect to a fuzzy technological constraint.

When the utility is precise but the constraints are imprecise, consumer's and producer's spatial partial equilibrium can also be determined. A discussion is provided in subsection 4.4.3.

4.4.1. Consumer's Spatial Partial Equilibrium with a Fuzzy Utility Function and a Fuzzy Constraint

Let $\underline{X}$ be the set of possible consumptions of a consumer i. Let $x_i \in \underline{X}$ be the economic goods in the set of possible consumptions. Let the consumer's imprecise spatial preference-indifference structure be a totally preordered topological space.

Then, by the discussion in subsection 4.2.3, there exists a fuzzy utility function

$$\mu_U: \underline{X} \longrightarrow [0, 1], \tag{4-40}$$

$$x_i \longrightarrow \mu_U(x_i), \quad \forall\, x_i \in \underline{X}. \tag{4-41}$$

By the principle of consumption, a consumer maximizes his imprecise utility, a fuzzy objective function, subject to a budget constraint. Due to the uncertainty over one's budget, the budget constraint is most likely imprecise. Under imprecision, efficiency is a matter of degree. Instead of separating our possible consumption space into sets of efficient and inefficient consumptions, it is more instrumental to delimit the whole possible consumption set by the degree of efficiency. Thus, the budget constraint can be formulated as a fuzzy subset defined by

$$\mu_C: \underline{X} \longrightarrow [0, 1], \tag{4-42}$$

$$x_i \longrightarrow \mu_C(x_i), \quad \forall\, x_i \in \underline{X}. \tag{4-43}$$

Specifically, $\mu_C(x_i)$ depends on the relationship between the available budget w and the price vector, p, of the economic goods x_i. Let the imprecise budget constraint be defined by

$$\mu_C(x_i) = \begin{cases} 1, & \text{if } p \cdot x_i \geq w, \\ \\ \in [0, 1), & \text{if } p \cdot x_i < w. \end{cases} \tag{4-44}$$

Based on theorem 4.2, a consumer's spatial partial equilibrium problem is that, at equilibrium, the optimal consumption is x_i^* such that

$$\begin{aligned} \mu_D(x_i^*) &= \sup_{x_i \in \underline{X}} \mu_D(x_i) \\ &= \sup_{\alpha \in [0, 1]} \min[\alpha, \sup_{x_i \in \underline{C}_\alpha} \mu_U(x_i)] \\ &= \sup_{x_i \in \underline{A}} \mu_U(x_i), \end{aligned} \tag{4-45}$$

where

$D = U \cap C$ is a fuzzy subset defining the imprecise demand space D,

$\underline{C}_\alpha = \{x_i | \mu_C(x_i) \geq \alpha,\ x_i \in U\}$, $\alpha \in [0, 1]$ is the α-level set of the imprecise budget constraint C, and

$\underline{A} = \{x_i | \mu_C(x_i) \geq \mu_U(x_i),\ x_i \in \underline{X}\}$ is the set of economic goods on which the value of the imprecise budget constraint is greater than or equal to that of the imprecise utility.

Obviously, the consumer's spatial partial equilibrium problem is equivalent to the constrained fuzzy spatial optimization problem discussed in section 4.3. By the symmetry of the utility and constraint formulations, the fuzzy utility function and the fuzzy budget constraint are of equal importance in the optimization process. Therefore, conditions such as convexity and continuity can be imposed on the fuzzy utility function or the fuzzy constraint.

Since the function $\sup_{x_i \in \underline{A}} \mu_U(x_i)$ in (4-45) is continuous, U then is strictly convex. Thus, $\sup_{x_i \in \underline{X}} \mu_D(x_i)$ is strictly quasi-concave and D is strictly convex. Moreover, D has to be compact to guarantee a solution. Though the utility and budget constraint are symmetric under the min-operator, convexity is imposed on the utility function here.

4.4.2. Producer's Spatial Partial Equilibrium with a Fuzzy Utility Function and a Fuzzy Constraint

Parallel to the consumer's spatial partial equilibrium problem, a producer's spatial partial equilibrium problem can again be constructed as a fuzzy optimization problem.

Parallel to the definition of $\underline{X}$ in subsection 4.2.1. Let $\underline{Y}$ be the set of possible productions (possible production space of a producer). Let y_r be a production bundle in $\underline{Y}$. For a given price, the producer's profit, a producer attached a utility which is generally imprecise. Therefore, a producer's utility of profit can be formulated as a fuzzy subset P defined by

$$\mu_P: \underline{Y} \longrightarrow [0, 1], \qquad (4\text{-}46)$$

$$y_j \longrightarrow \mu_P(y_j), \ \forall\, y_j \in \underline{Y}. \qquad (4\text{-}47)$$

In general, $\mu_P(y_j)$ depends on the profit, $p \cdot y_j$, of the production bundle y_j.

For a price vector p, the producer's objective P is defined by

$$\mu_P(y_j) = \begin{cases} 1, & \text{if } p \cdot y_j \text{ is maximal,} \\ 0, & \text{if } y_j = 0, \\ \varepsilon\ (0,\ 1), & \text{otherwise.} \end{cases} \qquad (4\text{-}48)$$

In the decisionmaking process, the producer maximizes his utility of profit with respect to a technological constraint which delimits the set of efficient productions. In an imprecise environment, efficiency is only a matter of degree. In place of separating our possible production set into efficient and inefficient production sets, it is more logical to define the whole possible production set by the degree of efficiency. That is, all produced goods are efficient to a certain degree.

Thus, the set of efficient production can be defined as a fuzzy subset E such that

$$\mu_E\colon \underline{Y} \longrightarrow [0,\ 1], \qquad (4\text{-}49)$$

$$y_j \longrightarrow \mu_E(y_j), \ \forall\, y_j \in \underline{Y}, \qquad (4\text{-}50)$$

and,

$$\mu_E(y_j) = \begin{cases} 1, & \text{if } y_j \text{ is of maximal efficiency,} \\ 0, & \text{if } y_j \text{ is of no value,} \\ \varepsilon\ (0,\ 1), & \text{otherwise.} \end{cases} \qquad (4\text{-}51)$$

Based on P in (4-46) and (4-47), E in (4-49) and (4-50), and theorem 4.2, Ponsard (1982a) shows that a producer's spatial partial equilibrium problem is that, at equilibrium, the optimal production is y_j^* such that

$$\mu_D(y_j^*) = \sup_{y_j \in \underline{Y}} \mu_D(y_j)$$

$$= \sup_{\alpha \, \varepsilon \, [0,\, 1]} \min[\alpha, \sup_{y_j \, \varepsilon \, \underline{E}_\alpha} \mu_P(y_j)]$$

$$= \sup_{y_j \, \varepsilon \, \underline{B}} \mu_P(y_j), \tag{4-52}$$

where

$D = P \cap B$ is the imprecise supply space,

$\underline{E} = \left\{ y_j | \mu_E(y_j) \geq \alpha,\ y_j \, \varepsilon \, P \right\}$, $\alpha \, \varepsilon \, [0, 1]$, and

$B = \left\{ y_j | \mu_E(y_j) \geq \mu_P(y_j),\ y_j \, \varepsilon \, \underline{Y} \right\}$ is the set of produced goods on which the value of the fuzzy technological constraint is greater than or equal to that of the fuzzy utility of profit.

The producer's spatial partial equilibrium problem is thus a fuzzy optimization problem with a fuzzy objective function and a fuzzy constraint. Again, the continuity of $\sup_{y_i \, \varepsilon \, \underline{B}} \mu_P(y_j)$ implies that P is strictly convex and $\underline{Y}$ is compact. To free the constraint from convexity, we impose the convexity condition on the objective function instead.

4.4.3. Spatial Partial Equilibrium with a Precise Utility Function and a Fuzzy Constraint

In subsection 4.4.2 and 4.4.3, spatial partial equilibrium problems are based on the existence of a fuzzy utility function and a fuzzy constraint. In some cases, however, consumers and producers may have a precise perception of utility. Nevertheless, their budget and technological constraints may only be imprecisely identified. Under this situation, a consumer needs to maximize his precise utility function subject to a fuzzy budget constraint, or a producer needs to maximize his precise utility of profit with respect to a fuzzy technological constraint. Such a spatial partial equilibrium problem involves the maximization of a precise objective function subject to a fuzzy constraint. Since the formulations are the same for both cases, our discussion only focuses on the producer's spatial partial equilibrium problem.

Let $\underline{Y}$ be the set of possible productions. Let y_j be a production bundle in $\underline{Y}$. For a given spatial pricing system p, a producer's total profit P is a function of y_j defined by

$$f: \underline{Y} \longrightarrow [0, +\infty), \tag{4-53}$$

$$y_j \longrightarrow f(y_j), \ \forall \ y_j \in \underline{Y}. \tag{4-54}$$

The profit function in (4-53) and (4-54) is precise.

Assume that the technological constraint defining the technical efficient productions is imprecise and is defined by (4-49) and (4-50). The producer's spatial partial equilibrium problem becomes the maximization of the precise profit function subject to the fuzzy technological constraint.

Based on a result in Ralescu (1980) (see also Negoita, 1979), Ponsard (1982b) shows that at equilibrium, the optimal production is y_j^* such that

$$f(y_j^*) = \sup_{y_j \in \underline{E}} f(y_j) = \sup_{y_j \in \underline{Y}} \min[f(y_j), \mu_E(y_j)]$$

$$= \sup_{\alpha \in [0, 1]} \min[\alpha, \mu_f\{f \geq \alpha\}]$$

$$= \mu_f\{f \geq \overline{\alpha}\} = \sup_{y_j \in \underline{E}_{\overline{\alpha}}} f(y_j), \tag{4-55}$$

where $\left\{f \geq \alpha\right\} = \left\{y_j | y_j \in \underline{Y}/f(y_j) \geq \alpha\right\}$ is a fuzzy integral; $\sup_{y_j \in \underline{E}_{\overline{\alpha}}} f(y_j)$ is continuous; and $\overline{\alpha} \in [0, 1]$ such that $\sup_{x \in E} f(x) = \sup_{x \in \underline{E}_{\overline{\alpha}}} f(x)$.

Therefore, spatial partial equilibrium problems can be formulated regardless of the precision of the objective function and constraint. The fuzzy set framework appears to be more general and less restrictive. The spatial partial equilibrium analysis can also be extended to examine general equilibrium problems in an imprecise economic decisionmaking environment.

4.5. SPATIAL GENERAL EQUILIBRIUM UNDER IMPRECISION

While partial equilibrium theory concentrates on a

decisionmaking unit and a few selected relationships, spatial general equilibrium problems deal with the spatial economic interdependence of individual decisionmaking units. The analysis in this section focuses on consumers' and producers' spatial general equilibrium problems in a fuzzy economic space. Our discussion is mainly based on Ponsard (1986a).

Before spatial general equilibrium is analyzed in subsection 4.5.3, consumers' and producers' spatial partial equilibriums are first examined in subsections 4.5.1 and 4.5.2 respectively. To facilitate our discussion, basic structure of the economic space is briefly described.

Let there be q economic goods designated by k, k = 1, ..., q, m consumers designated by i, i = 1, ..., m, and n producers designated by j, j = 1, ..., n, in an economic space S. Let q, m, and n be finite and m and n are sufficiently large to constitute a competition in S. Let the indices i and j also indicate the individual locations of the consumers and producers respectively.

The economic space S then has a price system for the located goods. Let the price of good k produced by producer j demanded by consumer i be p_{ijk} defined by

$$p_{ijk} = p_{jk} + p_{ikj}, \tag{4-56}$$

where p_{jk} is the factory price of good k at location j, and p_{ikj} is the transport cost of delivering good k from producer j to consumer i.

The price system in S can then be denoted by $\underline{P}$ with $p = (p_{ijk}) \in \underline{P}$ being a price vector. Now, we are in a position to analyze the spatial general equilibrium in S. Before we do this, we first state the consumers' and producers' partial equilibriums in the following subsections.

4.5.1. Consumers' Spatial Partial Equilibriums

Let $\underline{X}_i$ be the set of possible located consumptions of consumer i, i = 1, ..., m. Let $x_i \in \underline{X}_i$ be a vector (x_{ijk}) in $\mathbb{R}^{mnk}$ indicating the quantities of goods k demanded by consumer i supplied by producer j. Based on the discussion in subsection 4.4.1, for consumer i, his fuzzy utility function μ_{U_i} is defined by

$$\mu_{U_i}: \underline{X}_i \longrightarrow [0, 1], \tag{4-57}$$

$$x_i \longrightarrow \mu_{U_i}(x_i), \ \forall\ x_i \in \underline{X}_i. \tag{4-58}$$

His fuzzy budget constraint is defined by

$$\mu_{C_i}: \underline{X}_i \longrightarrow [0, 1], \tag{4-59}$$

$$x_i \longrightarrow \mu_{C_i}(x_i), \quad \forall\ x_i \in \underline{X}_i. \tag{4-60}$$

Again, $\mu_{C_i}(x_i)$ depends on the available budget w_i of consumer i and is defined by

$$\mu_{C_i}(x_i) = \begin{cases} 1, & \text{if } p \cdot x_i \geq w_i, \\ \varepsilon\ [0, 1), & \text{if } p \cdot x_i < w_i. \end{cases} \tag{4-61}$$

At equilibrium, the optimal consumption is x_i^* such that

$$\mu_{D_i}(x_i^*) = \sup_{x_i \in \underline{X}_i} \mu_{D_i}(x_i),$$

$$= \sup_{x_i \in \underline{A}_i} \mu_{U_i}(x_i), \tag{4-62}$$

where

$D_i = U_i \cap C_i$ is a fuzzy subset in $\underline{X}_i$ defining the demand space of consumer i, and

$A_i = \left\{x_i \mid \mu_C(x_i) \geq \mu_{U_i}(x_i),\ x_i \in \underline{X}_i\right\}$.

Considering all consumers, the set of total possible located consumptions is $\underline{X} = \sum_{i=1}^{m} \underline{X}_i$ whose element $x = \sum_{i=1}^{m} x_i$ is the total consumption in S.

Let X be the corresponding fuzzy subset in $\underline{X}$ characterizing the fuzzy total consumption and is defined by

$$\mu_X\colon \underline{X} \longrightarrow [0,\ 1],$$

$$x \longrightarrow \mu_X(x) = Md(x_i)[\sup_{x_i \in \underline{X}_i} \mu_{D_i}(x_i)], \qquad (4\text{-}63)$$

where $Md(x_i)$ is the median of the grades of membership of the consumers' demands at equilibrium. The median serves as an operator for aggregating the membership functions characterizing the individual demands. The operator is non-additive and is thus more flexible. It gives a central tendency of individual consumers' behavior.

4.5.2. Producers' Spatial Partial Equilibriums

Let $\underline{Y}_j$ be the set of possible productions of producer j, $j = 1, \ldots, n$. Let $y_j \in \underline{Y}_j$ be a vector (y_{ijk}) in $\mathbb{R}^{mnk}$ indicating the quantities of goods k produced by producer j demanded by consumer i. Based on subsection 4.4.2, the imrecise utility of profit of producer j is defined by

$$\mu_{P_j}\colon \underline{Y}_j \longrightarrow [0,\ 1], \qquad (4\text{-}64)$$

$$y_j \longrightarrow \mu_{P_j}(y_j),\ \forall\, y_j \in \underline{Y}_j. \qquad (4\text{-}65)$$

Again, $\mu_{P_j}(y_j)$ depends on the price vector p and is defined by

$$\mu_{P_j}(y_j) = \begin{cases} 1, & \text{if } p\cdot y_j \text{ is maximal,} \\ 0, & \text{if } y_j = 0, \\ \in (0,\ 1), & \text{otherwise.} \end{cases} \qquad (4\text{-}66)$$

Let the imprecise technological constraint of producer j be defined by

$$\mu_{E_j}\colon \underline{Y}_j \longrightarrow [0,\ 1], \qquad (4\text{-}67)$$

$$y_j \longrightarrow \mu_{E_j}(y_j),\ \forall\, y_j \in \underline{Y}_j, \qquad (4\text{-}68)$$

and,

$$\mu_{E_j}(y_j) = \begin{cases} 1, & \text{if } y_j \text{ is of maximal efficiency,} \\ 0, & \text{if } y_j \text{ is of no value,} \\ \varepsilon\ (0,\ 1), & \text{otherwise.} \end{cases} \tag{4-69}$$

At equilibrium, the optimal production is y_j^* such that

$$\mu_{Y_j}(y_j^*) = \sup_{y_j \in \underline{Y}_j} \mu_{D_j}(x_j),$$

$$= \sup_{y_j \in \underline{B}_j} \mu_{P_j}(x_j), \tag{4-70}$$

where

$D_j = P_j \cap E_j$ is a fuzzy subset in $\underline{Y}_j$ defining the supply space of producer j, and

$\underline{B}_j = \{ y_j | \mu_E(y_j) \geq \mu_P(y_j),\ y_j \varepsilon \underline{Y}_j \}$.

Taking all producers into consideration, the set of total possible productions is $\underline{Y} = \sum_{j=1}^{n} \underline{Y}_j$ whose element $y = \sum_{j=1}^{n} y_j$ is the total production in S. Employing the median as an operator, the fuzzy total production Y in $\underline{Y}$ is defined by

$$\mu_Y\colon \underline{Y} \longrightarrow [0,\ 1], \tag{4-71}$$

$$y \longrightarrow \mu_Y(y) = Md(y_j)[\sup_{y_j \varepsilon \underline{Y}_j} \mu_{D_j}(y_j)]. \tag{4-72}$$

Similarly, $Md(y_j)$ is the median of the grades of membership of the producers' supplies at equilibrium. It is an operator for aggregating the membership functions characterizing the individual supplies. The operator is again non-additive and it gives a central tendency of individual suppliers' behavior.

4.5.3. General Spatial Equilibrium

Based on the discussion in subsections 4.5.1 and 4.5.2, the excess demand in the fuzzy economic space S, denoted by h, is

$$h = x - y - w, \tag{4-73}$$

where

$x \in \underline{X}$, $y \in \underline{Y}$,

$w = \sum_{i=1}^{m} w_i$ is the initial resources in S, and

$h \in \underline{H} = \underline{X} - \underline{Y} - \{w\}$.

Since X in $\underline{X}$, and Y in $\underline{Y}$ are fuzzy subsets, then H in $\underline{H}$ is also a fuzzy subset defined by

$$\mu_H(h) = \begin{cases} 1, & \text{if } y = 0 \text{ (excess demand is maximum)}, \\ 0, & \text{if } (x - y) < 0 \text{ (excess demand is non-positive)}, \\ \in (0, 1), & \text{otherwise.} \end{cases} \qquad (4\text{-}74)$$

The function is continuous and monotonically increasing for increasing values of (x - y).

Let the excess demand point-to-set mapping from the set of prices $\underline{P}$ to $\underline{H}$, denoted by h, be defined by

$$h: \underline{P} \longrightarrow \underline{H},$$
$$p \longrightarrow h(p) = x(p) - y(p) - w, \qquad (4\text{-}75)$$

where x(p) is the value of the demand point-to-set mapping for a price p, and y(p) is the value of the supply point-to-set mapping for a price p.

To simplify notation, h is employed to denote the excess demand and the excess demand point-to-set mapping.

Definition 4.4. A fuzzy point-to-set mapping over a set $\underline{X}$ is a mapping, T, from $\underline{X}$ into the fuzzy power set $\mathscr{F}(\underline{X})$ such that

$$T: \underline{X} \longrightarrow \mathscr{F}(\underline{X})$$
$$x \longrightarrow T(x) = A, \ \forall\, x \in \underline{X}, \qquad (4\text{-}76)$$

where

$T(x) = \left\{ y \mid y = f(x) : \mu_X(y) \in \underline{L},\ y \in \underline{X} \right\}$, with L being a lattice, L = [0, 1] in particular.

Based on classical economic theory, it is required that $h \leq 0$ at competitive equilibrium. That is, we have to determine the conditions under which $h(p) \leq 0$ exists. To solve this, we first generalize the Walras Law in the fuzzy economic space S.

It states that excess demand should be non-positive and the goods whose excess demand is negative should have zero price. That is,

$$\sum_{i=1}^{m} p \cdot h_i(p) \leq 0, \ \forall\ p \in \underline{P},\ \forall\ i. \tag{4-77}$$

Thus, if there exists a price vector $p^* \in \underline{P}$ and an excess demand vector $h^* \in h(p^*)$ such that $h^* \leq 0$ and $\mu_H(h^*) = 0$, then the competitive equilibrium exists.

If the generalized Walras Law does not hold, a price adjusting process would take place such that the price, P_{ijk}, would be corrected by a factor ε when $h_i(p) \neq 0$. Thus, at any step of adjusting, we make the following substitution:

$$p'_{ijk} = \begin{cases} p_{ijk}, & \text{if } h_i(p) = 0. \\ p_{ijk} + \varepsilon, & \text{if } h_i(p) > 0. \\ \max(0,\ p_{ijk} - \varepsilon), & \text{if } h_i(p) < 0. \end{cases} \tag{4-78}$$

Let g be a continuous function defined by

$$\begin{aligned} g:\ & \underline{H} \longrightarrow \underline{P}, \\ & h \longrightarrow g(h) = \left\{ p' \middle| p' \cdot h < p \cdot h \right\}. \end{aligned} \tag{4-79}$$

To verify conditions of the fixed point theorems for the subsequent equilibrium analysis, prices are standardized so that $\sum_{ijk} p_{ijk} = 1$.

Let $\underline{\bar{P}} = \left\{ \bar{p} \middle| \sum_{ijk} p_{ijk} = 1,\ p \in \underline{P} \right\}$ be the set of standardized prices. The set $\underline{\bar{P}}$ is then a simplex. Therefore, associated with the function g in (4-79) is a continuous function $\bar{g}$ such that

$$\begin{aligned} \bar{g}:\ & \underline{H} \longrightarrow \underline{P}, \\ & h \longrightarrow \bar{g}(h) = \left\{ \bar{p} \middle| \bar{p}' \cdot h < \bar{p} \cdot h \right\}. \end{aligned} \tag{4-80}$$

Let $\bar{h}$ be a continuous point-to-set mapping associated with the excess demand point-to-set mapping h with $\underline{\bar{P}}$ being the domain and $\underline{H}$ being the range. Then, the generalized Walras Law holds if the following function is continuous:

$$\phi: \underline{\bar{P}} \longrightarrow \underline{\bar{P}},$$

$$\bar{p} \longrightarrow \phi(\bar{p}) = (\bar{g} \circ \bar{h})(\bar{p}). \qquad (4\text{-}81)$$

The function ϕ is apparently continuous because the functions $\bar{g}$ and $\bar{h}$ are continuous.

Since the simplex $\underline{\bar{P}}$ is non-empty, convex, and compact, and the function ϕ is continuous, there exists then a fixed point $\bar{p}^* \varepsilon \underline{\bar{P}}$ which is constant in the transformation $\bar{p} \longrightarrow h(\bar{p}) \longrightarrow \bar{p}'$.

Now we need the following definitions before we state the theorem of spatial general equilibrium.

Definition 4.5. A fixed point of a fuzzy point-to-set mapping T over $\underline{X}$ is an element x* such that $\mu_T(x^*, x^*) \geq \mu_T(x^*, x)$, $\forall\, x \,\varepsilon\, \underline{X}$.

Definition 4.6. An excess demand fuzzy point-to-set mapping, ψ, on $(\underline{\bar{P}} \times \underline{H})$ is defined as a mapping from $(\underline{\bar{P}} \times \underline{H})$ into $\mathscr{F}(\underline{\bar{P}} \times \underline{H})$ such that $\psi(\bar{p}, h) = \{\phi(\bar{p}), \bar{h}(\bar{p})\}$.

Based on the generalization of Kakutani's theorem (Kakutani, 1941) given by Butnariu (1982) and Kaleva (1985), Ponsard (1986a) gives the following theorem for spatial general equilibrium in a fuzzy economic space:

Theorem 4.3. Let $(\underline{\bar{P}} \times \mathbb{R}^{mnk})$ be a real topological vector space which is locally convex and Hausdorff-separated, with $(\underline{\bar{P}} \times \underline{H})$ being a subset of $(\underline{\bar{P}} \times \mathbb{R}^{mnk})$. If the excess demand fuzzy point-to-set mapping ψ on $(\underline{\bar{P}} \times \underline{H})$ is closed and has for its images $\psi(\bar{p}, h)$ which is non-empty, normal, and convex, and verifies the generalized Walras Law, then a competitive equilibrium exists in the fuzzy spatial economy. That is, there exists a price vector $p^* \varepsilon \underline{P}$, and an excess demand vector $h^* \varepsilon h(p^*)$ such that $h^* \leq 0$ and $\mu_H(h^*) = 0$.

The theorem is general enough to cover a variety of spatial behavior in a fuzzy economic space. Spatial equilibrium analyses, partial or general, are more flexible under the imprecision framework. Further investigation may want to focus on the stability of the equilibrium solutions so that fuzzy spatial equilibrium and disequilibrium can be effectively analyzed.

CHAPTER 5

FUZZY MATHEMATICAL PROGRAMMING AND SINGLE OBJECTIVE SPATIAL PLANNING PROBLEMS

5.1. AN EXAMINATION OF SPATIAL PLANNING PROBLEMS

Efficient and effective planning is pertinent to the monitoring and shaping of a spatial system into a desirable path of development. Optimal achievement of an objective or a goal within the limited capacity of a spatial system has long been regarded as an appropriate guideline for tackling spatial planning problems.

Similar to the spatial choice problems discussed in chapter 4, human subjectivity, imperfect or imprecise information, and uncertainty of resource availabilities often prevail in the planning of spatial systems. Objectives and constraints formulated under this situation are most likely fuzzy in nature. With regard to budgetting, a constraint may be specified as: "The total cost should be *approximately* β", where β is a constant. "*Approximately* β" here implies the indeterminancy encountered or flexibility permitted in budget appropriation. Such a formulation may be due to our inability in providing a precise budget or the flexibility we consciously incorporated. In the formulation of an objective, a goal may be stated as "The net return should be *much greater than* γ", where γ is a constant. Here, "*much greater than* γ" implies that an imprecise target value or aspiration level is set for the goal. Apparently, these types of constraints and objectives are imprecise in the sense that the limitation or aspiration levels are fuzzy. Therefore, they constitute classes of alternatives whose boundaries are fuzzily defined.

Thus, efficient and effective planning should take imprecision directly into consideration. The drive for unreasonable level of precision and perfect knowledge in a complex decision-making environment is an impossibility and would make our optimization processes too structured to be instrumental. Instead of treating imprecise information as obsolete and fuzzy cognitive and decisionmaking processes as absurd, planning models should be constructed in such a way that imprecision can be formally incorporated and human valuation, experience, and judgement can be systematically put forth and utilized. By such a practice, greater flexibility and greater insight can be gained in the planning processes.

If objectives and constraints are classified into precise and fuzzy categories, four basic classes of spatial planning problems are obtained in Fig. 5.1 (Leung, 1984c).

	Constraints	
	precise	fuzzy
precise	(1)	(2)
fuzzy	(3)	(4)

Fig. 5.1 A classification of spatial planning problems

In class (1), both the objective and constraints are precise. Literature on this class of conventional spatial planning problems is voluminous and will not be discussed here. However, it should be noticed that this class of problems is in fact a special case of the other three classes in Fig. 5.1.

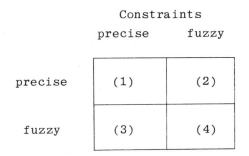

Spatial planning problems in class (2) involve a precise objective and a set of fuzzy constraints. The planning problems are solved by the optimization of a precise objective function subject to a set of fuzzy constraints. The structure of the imprecise constraint set and the optimization over an imprecise feasible region is a subject of analysis in section 5.2.

In class (3), the objective is fuzzy but the constraints are precise. Optimization of a fuzzy objective function over a precisely delimited decision space is investigated in section 5.3.

Class (4) consists of spatial planning problems whose objective and constraints are both fuzzy. Optimization of a fuzzy objective function over a fuzzy constraint set is discussed in section 5.4.

In section 5.5, other types of fuzzy constraints are investigated. Robust programming with fuzzy inclusive constraints is examined.

Unless specified otherwise, our discussion is restricted to planning problems of linear spatial systems. Throughout, fuzzy linear objective functions and fuzzy linear constraints are stated respectively in short as fuzzy objective functions and fuzzy constraints. The word "fuzzy" does not mean that the functional forms of the objective functions and constraints are imprecise. It indicates that the objective functions and constraints consist of fuzzy aspiration levels and fuzzy limitations respectively. In some cases, they comprise fuzzy coefficients. Furthermore, when the term "fuzzy environment" is employed, I mean that the objectives and/or the constraints are fuzzy, but not necessarily the decisionmaking system.

5.2. PRECISE-OBJECTIVE-FUZZY-CONSTRAINTS SPATIAL PLANNING PROBLEMS

A variety of spatial planning problems require the optimization of a precise objective function subject to a set of fuzzy constraints. This is especially common when resource availabilities and other technological conditions are in a state of uncertainty. Our aim is to select an alternative out of a fuzzy class of alternatives to achieve the specified objective within a certain level of tolerance.

In place of the point-valued limits on resource capacities, fuzzy restrictions are in general more realistic and pragmatic. First, fuzzy constraints are a formal representation of the imprecise optimization environment. In some planning processes, decisionmakers may only be able to give a verbal statement, such as "A should be smaller than β as much as possible", which lacks mathematical clarity and precision. Second, fuzzification of technological limitations is a natural way to cope with instability and also a way to reach stability. Uncertainty of the technological environment and human subjectivity often create instabilities in planning. Through fuzzy restrictions, one can evaluate all possible scenarios and select the most appropriate course of action. Third, constraints with fuzzy tolerance can transform a planning problem with incompatible constraints into a new problem having a solution which is close to the initial specification as much as possible.

In subsection 5.2.1, fuzzy constraints are restricted to those having imprecise capacities. Their formulations and the associated structure of the imprecise decision space are examined. In subsection 5.2.2, optimization of a precise objective function over a fuzzy decision space is discussed.

5.2.1. Imprecision of a Constraint Set

The following are three basic types of constraints encountered in conventional spatial planning models:

(a) $\underline{C}_1$: $g(x) \leq b, \; g: \underline{X} \longrightarrow \mathbb{R}, \; x \in \underline{X} \subseteq \mathbb{R}^n, \; b \in \mathbb{R},$ (5-1)
("less than or equal to" constraints);

(b) $\underline{C}_2$: $g(x) \geq b, \; g: \underline{X} \longrightarrow \mathbb{R}, \; x \in \underline{X} \subseteq \mathbb{R}^n, \; b \in \mathbb{R},$ (5-2)
("greater than or equal to" constraints);

(c) $\underline{C}_3$: $g(x) = b, \; g: \underline{X} \longrightarrow \mathbb{R}, \; x \in \underline{X} \subseteq \mathbb{R}^n, \; b \in \mathbb{R},$ (5-3)
("equal to" constraints),

where $\underline{X}$ is the alternative space and g is a $\mathbb{R}^n$ to $\mathbb{R}$ mapping. These constraints are precise because the capacity, b, imposed on g(x) is assumed to be a known constant, and any violation of the constraint would make a solution infeasible.

Due to the imprecision of b, it is more appropriate to specify a tolerance interval with respect to the intended value b. The corresponding constraint becomes fuzzy and lacks a clear-cut boundary. The implication then is that we are allowing for a certain extent of constraint violations. Symbolically, they can be represented as follows:

(a) C_1: $g(x) \lesssim \underline{b}$; $\bar{b}$, g: $\underline{X} \longrightarrow \mathbb{R}$, $x \varepsilon \underline{X} \subseteq \mathbb{R}^n$, $\underline{b}$, $\bar{b} \varepsilon \mathbb{R}$; (5-4)

(b) C_2: $g(x) \gtrsim \bar{b}$; $\underline{b}$, g: $\underline{X} \longrightarrow \mathbb{R}$, $x \varepsilon \underline{X} \subseteq \mathbb{R}^n$, $\underline{b}$, $\bar{b} \varepsilon \mathbb{R}$; (5-5)

(c) C_3: $g(x) \eqsim b$; $\underline{b}$, $\bar{b}$, g: $\underline{X} \longrightarrow \mathbb{R}$, $x \varepsilon \underline{X} \subseteq \mathbb{R}^n$, b, $\underline{b}$, $\bar{b} \varepsilon \mathbb{R}$.

(5-6)

The constraint in (5-4) is a fuzzy "less than or equal to" constraint. The wavy bar, ~, under the symbol, $\leq$, acts as a fuzzifier which transforms the nonfuzzy constraint into a fuzzy constraint which is approximately equal to it. The C_1-type constraint in (5-4) may read "g(x) is essentially less than or equal to $\underline{b}$". The value $\bar{b}$ is the permissible maximal value of g(x). That is, if $g(x) > \underline{b}$, it cannot exceed $\bar{b}$ under any circumstances. Therefore, the interval $[\underline{b}, \bar{b}]$ can be regarded as a tolerance interval with a subjectively chosen length, $d = (\bar{b} - \underline{b})$, indicating the extent to which the deviation from $\underline{b}$ is permissible. Thus, C_1 serves as a fuzzy restriction on the value of g(x). It tries to enforce as much as possible the precise inequality, $\leq \underline{b}$, within a tolerance interval $[\underline{b}, \bar{b}]$.

In the same manner, the C_2-type constraint in (5-5) is the fuzzy "greater than or equal to" constraint which may read "g(x) is essentially greater than or equal to $\bar{b}$". The value $\underline{b}$ is the permissible minimal value of g(x) and the tolerance interval is $[\underline{b}, \bar{b}]$ with length $d = (\bar{b} - \underline{b})$.

Similarly, the C_3-type constraint in (5-6) is the fuzzy "equal to" constraint which may read "g(x) is essentially equal to b". The values $\underline{b}$ and $\bar{b}$ are respectively the permissible minimal and maximal values of g(x). That is, if $g(x) < b$ it cannot be less than $\underline{b}$, and if $g(x) > b$ it cannot be greater than $\bar{b}$. Thus, the length of the tolerance interval $[\underline{b}, \bar{b}]$ is $d = \underline{d} + \bar{d} = (b - \underline{b}) + (\bar{b} - b)$.

To guarantee a rational and logically consistent decision-making process under imprecision, it is important to establish a linear ordering on the set of alternatives $\underline{X}$ in terms of the extent to which a fuzzy constraint is satisfied. A natural way is to define the fuzzy constraint in $\underline{X}$ as a fuzzy subset in $\underline{X}$ (Zimmermann, 1976; Leung, 1983a).

For fuzzy constraint C_1 it can be defined as a fuzzy subset C_1 in $\underline{X}$ (to simplify presentation, C_1 denotes both the constraint and the corresponding fuzzy subset) with a defining membership function:

$$\mu_{C_1}(x) \triangleq \mu_{C_1}(g(x)) = \begin{cases} 1, & \text{if } g(x) \leq \underline{b}, \\ 1 - \dfrac{g(x) - \underline{b}}{d}, & \text{if } \underline{b} < g(x) \leq \bar{b}, \\ 0, & \text{if } g(x) > \bar{b}. \end{cases} \quad (5\text{-}7)$$

(see Fig. 5.2)

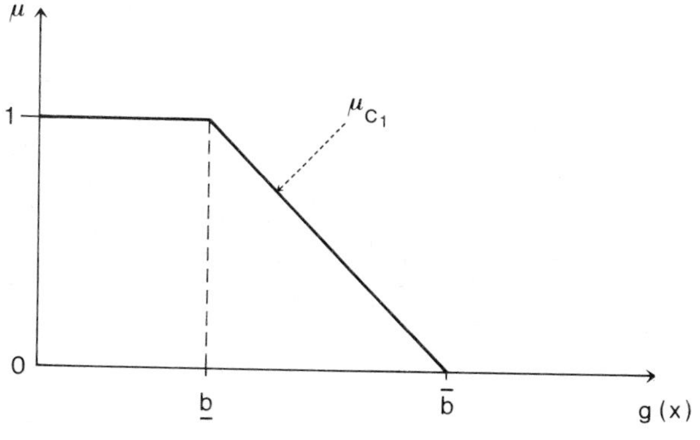

Fig. 5.2 Membership function (satisfaction function) of the fuzzy "less than or equal to" constraint (C_1)

Thus, μ_{C_1} can be treated as a satisfaction function indicating a decisionmaker's degree of satisfaction about the value of g(x). When $\mu_{C_1}(g(x))$ equals 1 and 0 we are completely satisfied and dissatified respectively. Our degree of satisfaction increases with values increasing from 0 to 1. Through this, a linear ordering in $\underline{X}$ is established.

In a similar manner, fuzzy constraint C_2 can be defined as a fuzzy subset C_2 in $\underline{X}$ with a defining membership function:

$$\mu_{C_2}(x) \triangleq \mu_{C_2}(g(x)) = \begin{cases} 1, & \text{if } g(x) > \bar{b}, \\ 1 - \dfrac{\bar{b} - g(x)}{d}, & \text{if } \underline{b} < g(x) \leq \bar{b}, \\ 0, & \text{if } g(x) \leq \underline{b}. \end{cases} \quad (5\text{-}8)$$

(see Fig. 5.3)

Likewise, the fuzzy equality constraint C_3 can be defined by

$$\mu_{C_3}(x) \triangleq \mu_{C_3}(g(x)) = \begin{cases} 0, & \text{if } g(x) < \underline{b}, \\ 1 - \dfrac{b - g(x)}{\underline{d}}, & \text{if } \underline{b} \leq g(x) < b, \\ 1, & \text{if } g(x) = b, \\ 1 - \dfrac{g(x) - b}{\bar{d}}, & \text{if } b < g(x) \leq \bar{b}, \\ 0 & , \text{if } g(x) > \bar{b}. \end{cases} \quad (5\text{-}9)$$

(see Fig. 5.4)

In conventional linear programming models, the constraints delimit a decision space within which the best alternative is sought. Since the fuzzy constraints in the present framework are defined as fuzzy subsets in $\underline{X}$, then the decision space should also be a fuzzy subset in $\underline{X}$. A natural formulation is to construct the decision space as the intersection of the fuzzy constraints.

Referring to fuzzy constraints C_1, C_2, and C_3, the decision space D is a confluence of the fuzzy subsets C_1, C_2, and C_3. Specifically, D is defined by the membership function

$$\mu_D(x) = \mu_{C_1 \cap C_2 \cap C_3}(x) = \min[\mu_{C_1}(x), \mu_{C_2}(x), \mu_{C_3}(x)]. \quad (5\text{-}10)$$

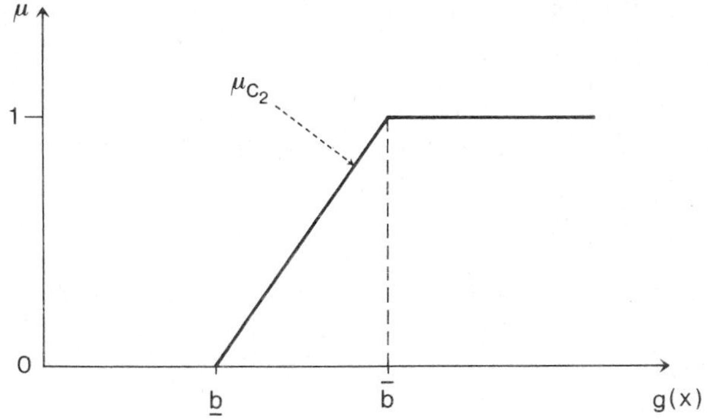

Fig. 5.3 Membership function (satisfaction function) of the fuzzy "greater than or equal to" constraint (C_2)

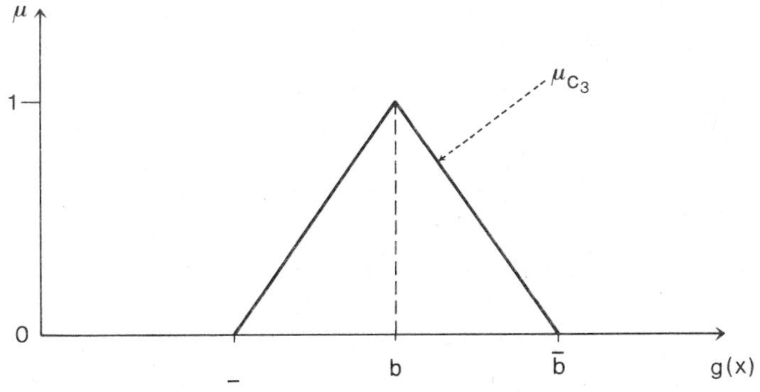

Fig. 5.4 mbe ship function (satisfaction function) f the fuz y "equal to" constraint (C_3)

Here, confluence is treated as intersection which is defined by the min-operator. Of course, as discussed in subsection 2.3.2, it is by no means the only operator. (See also subsection 5.4.1). For consistency, confluence is defined by the min-operator throughout.

Without considering an objective function, the solution x*

for the system $(\underline{X}, \mu_{C_1}, \mu_{C_2}, \mu_{C_3})$ should be the alternative which best satisfies all the fuzzy constraints. That is,

$$\mu_D(x^*) = \sup_{x \in \underline{X}} \mu_D(x) = \sup_{x \in \underline{X}} \min[\mu_{C_1}(x), \mu_{C_2}(x), \mu_{C_3}(x)]. \quad (5\text{-}11)$$

The solution of the optimization problem in (5-11) can be obtained by solving the following conventional linear programming problem:

$$\begin{aligned} &\max \lambda, \\ &\text{s.t. } \mu_{C_1}(x) \geq \lambda \\ &\quad\;\; \mu_{C_2}(x) \geq \lambda \\ &\quad\;\; \mu_{C_3}(x) \geq \lambda \\ &\quad\;\; \lambda \geq 0,\ x \geq \theta, \end{aligned} \quad (5\text{-}12)$$

where λ is the degree of overall constraint-satisfaction which corresponds essentially to (5-10), and θ is the zero vector in $\mathbb{R}^n$.

Based on the definition of μ_{C_1} in (5-7), the constraint $\mu_{C_1}(x) \geq \lambda$ in (5-12), for example, can be rewritten as:

$$\frac{\bar{b} - g(x)}{d} \geq \lambda, \quad (5\text{-}13)$$

or

$$g(x) \leq \underline{b} + d(1 - \lambda). \quad (5\text{-}14)$$

The other constraints in (5-12) can be rewritten in the similar manner.

Example 5.1. Let the following be a system of three fuzzy constraints

$$\begin{aligned} &C_1\colon g_1(x) = 4x_1 + 5x_2 \lesssim 20;\ 24 \\ &C_2\colon g_2(x) = 4x_1 + 2x_2 \lesssim 12;\ 17 \\ &C_3\colon g_3(x) = 3x_1 + 2x_2 \gtrsim 18;\ 12 \\ &x_1 \geq 0,\ x_2 \geq 0. \end{aligned} \quad (5\text{-}15)$$

Note that the original system with precise capacities 20, 12, and 18 has no solution. The fuzzy specification, however, gives a feasible solution by solving

$$\sup_{x=(x_1, x_2)} \min[\mu_{C_1}(x), \mu_{C_2}(x), \mu_{C_3}(x)], \tag{5-16}$$

where

$$\mu_{C_1}(x) = \begin{cases} 1, & \text{if } g_1(x) \le 20, \\ 1 - \dfrac{g_1(x) - 20}{4}, & \text{if } 20 < g_1(x) \le 24, \\ 0, & \text{if } g_1(x) > 24; \end{cases} \tag{5-17}$$

$$\mu_{C_2}(x) = \begin{cases} 1, & \text{if } g_2(x) \le 12, \\ 1 - \dfrac{g_2(x) - 12}{5}, & \text{if } 12 < g_2(x) \le 17, \\ 0, & \text{if } g_2(x) > 17; \end{cases} \tag{5-18}$$

$$\mu_{C_3}(x) = \begin{cases} 1, & \text{if } g_3(x) > 18, \\ 1 - \dfrac{18 - g_3(x)}{6}, & \text{if } 12 < g_3(x) \le 18, \\ 0, & \text{if } g_3(x) \le 12. \end{cases} \tag{5-19}$$

Based on (5-12) and (5-13), to find the solution for (5-16) we solve the following equivalent linear programming problem:

$$\begin{aligned} &\max \lambda, \\ &\text{s.t. } 6 - x_1 - 1.25x_2 \ge \lambda \\ &\quad 3.4 - 0.8x_1 - 0.4x_2 \ge \lambda \\ &\quad -2 + 0.5x_1 + 0.33x_2 \ge \lambda \\ &\quad \lambda \ge 0,\ x_1 \ge 0,\ x_2 \ge 0. \end{aligned} \tag{5-20}$$

The optimal solution is $(\lambda^*, x_1^*, x_2^*) = (0.195, 2.807, 2.398)$.

Negoita and Sularia (1976) demonstrates similarly how to solve a system of incompatible constraints through a fuzzy linear programming algorithm.

5.2.2. Optimization of a Precise Objective Function over a Fuzzy Constraint Set

Assume that the goal of a spatial planning problem is to maximize the value of a precise objective function which depicts, for example, the regional or interregional income (or profit) of a firm. Assume also that the technological constraints are imprecise.

Let $f: \underline{X} \longrightarrow \mathbb{R}$ be a precise objective function with $x \in \underline{X} \subseteq \mathbb{R}^n$ and $f(x) \in \mathbb{R}$. Let $\{\mu_i: X \longrightarrow [0, 1],\ i = 1, \ldots, m\}$ be a set of m fuzzy constraints with $\mu_i(x) \triangleq \mu_i(g_i(x)) \in [0, 1]$ defined by (5-7) or (5-8) or (5-9). (Henceforth, μ_i stands for the fuzzy subset derived from constraint $g_i(x)$).

Based on (5-10), the decision space for the system $(\underline{X}, f, \mu_1, \ldots, \mu_m)$ is defined by

$$\mu_D(x) = \min[f(x), \mu_1(x), \ldots, \mu_m(x)]. \qquad (5\text{-}21)$$

The optimal solution x* should satisfy the relation

$$\mu_D(x^*) = \sup_{x \in \underline{X}} \mu_D(x) = \sup_{x \in \underline{X}} \min[f(x), \mu_1(x), \ldots, \mu_m(x)]. \qquad (5\text{-}22)$$

Since we are maximizing a precise objective function over a fuzzy constraint set, the solution of such an optimization problem is defined by the following fuzzy subset (Orlovsky, 1977; Negoita and Ralescu, 1977; Dubois and Prade, 1980):

$$\psi(x) = \begin{cases} \sup\limits_{x \in \underline{N}_\alpha}, & \text{if } x \in \bigcup\limits_{\alpha > 0} \underline{N}_\alpha, \\ 0, & \text{otherwise.} \end{cases} \qquad (5\text{-}23)$$

where

$$\psi: \underline{X} \longrightarrow [0, 1], \qquad (5\text{-}24)$$

$$\underline{N}_\alpha = \left\{ x \middle| f(x) = \sup_{x \in \underline{C}_\alpha} f(x) \right\}, \tag{5-25}$$

$$\underline{C}_\alpha = \left\{ x \middle| x \in \underline{X},\ \mu_C(x) \geq \alpha \right\},\ \alpha \in (0,\ 1], \tag{5-26}$$

$$\mu_C(x) = \min[\mu_1(x),\ \ldots,\ \mu_m(x)]. \tag{5-27}$$

Thus, the solution concept in (5-23) makes use of the concept of α-level sets of the fuzzy subset μ_C defining the fuzzy constraint set. It, in fact, searches for the highest satisfaction level of the objective function over all α-level sets of the fuzzy constraint set.

To obtain the solution of the optimization problem in (5-23), it is sufficient to solve the following conventional parametric linear programming problem (see also Verdegay, 1982; Chanas, 1983):

$$\begin{aligned} &\max f(x), \\ &\text{s.t. } \mu_i(g_i(x)) \geq \alpha,\ i = 1,\ \ldots,\ m, \\ &\qquad \alpha \geq 0,\ x \geq \theta, \end{aligned} \tag{5-28}$$

or, equivalently

$$\begin{aligned} &\max f(x), \\ &\text{s.t. } g_i(x) \geq \mu_i^{-1}(\alpha),\ i = 1,\ \ldots,\ m, \\ &\qquad \alpha \geq 0,\ x \geq \theta. \end{aligned} \tag{5-29}$$

If $\mu_i(g_i(x))$, for example, is a fuzzy "less than or equal to" constraint defined in (5-7), the optimization problem in (5-29) becomes

$$\begin{aligned} &\max f(x), \\ &\text{s.t. } g_i(x) \leq \underline{b}_i + d(1 - \alpha),\ i = 1,\ \ldots,\ m, \\ &\qquad \alpha \geq 0,\ x \geq \theta. \end{aligned} \tag{5-30}$$

It is apparently a conventional parametric linear programming problem with parameter $(1 - \alpha) \in [0,\ 1]$.

Example 5.2. (Verdegay, 1983)

$$\begin{aligned} &\max\ 2x_1 + x_2, \\ &\text{s.t.}\ 2x_1 \lesssim 80;\ 85 \\ &\quad x_2 \lesssim 30;\ 35 \\ &\quad 4x_1 + 5x_2 \lesssim 200;\ 205 \\ &\quad x_1 \geq 0,\ x_2 \geq 0. \end{aligned} \tag{5-31}$$

By (5-30), the optimal solution is obtained by solving

$$\begin{aligned} &\max\ 2x_1 + x_2, \\ &\text{s.t.}\ 2x_1 \leq 80 + 5(1 - \alpha) \\ &\quad x_2 \leq 30 + 5(1 - \alpha) \\ &\quad 4x_1 + 5x_2 \leq 200 + 5(1 - \alpha) \\ &\quad \alpha \geq 0,\ x_1 \geq 0,\ x_2 \geq 0. \end{aligned} \tag{5-32}$$

Depending on the parameter $(1 - \alpha)$, the optimal solution is expressed as $(x_1^*, x_2^*) = (40 + 2.5(1 - \alpha), 8 - (1 - \alpha))$ with the value of the objective function being: $88 + 4(1 - \alpha)$, $0 \leq \alpha \leq 1$.

5.3. FUZZY-OBJECTIVE-PRECISE-CONSTRAINTS SPATIAL PLANNING PROBLEMS

Often, spatial planning problems involve the achievement of a fuzzy goal within a precisely delimited technological environment. It is especially common when the satisfying principle instead of the maximizing or minimizing principle is employed in a decisionmaking process. Not only a greater flexibility can be achieved through the employment of a fuzzy objective, decisionmakers can also exhaust as much as possible all goal achieving alternatives. Analytically, the planning problem concerns the optimization of a fuzzy objective function over a set of precise constraints.

In subsection 5.3.1, the formulation of a fuzzy objective

function is examined. In subsection 5.3.2, the optimization of a fuzzy objective function over a precise decision space is discussed.

5. 3. 1. Imprecision of an Objective Function

There are three basic types of objective functions in conventional spatial planning models:

(a) O_1: min f(x), f: $\underline{X} \longrightarrow \mathbb{R}$, $x \in \underline{X} \subseteq \mathbb{R}^n$, (minimization); (5-33)

(b) O_2: max f(x), f: $\underline{X} \longrightarrow \mathbb{R}$, $x \in \underline{X} \subseteq \mathbb{R}^n$, (maximization); (5-34)

(c) O_3: f(x) = z, f: $\underline{X} \longrightarrow \mathbb{R}$, $x \in \underline{X} \subseteq \mathbb{R}^n$, (goal setting). (5-35)

The objective functions are precise because f is a well-defined function on $\underline{X}$ and its aspiration level is set for the maximum, the minimum, or a specific value. There are no fuzziness involved.

The optimization of (5-33), (5-34), or (5-35) with respect to a set of precise constraints belongs to the conventional single-objective mathematical programming problems. Their optimization with respect to a set of fuzzy constraints, however, becomes the fuzzy optimization problem in subsection 5.2.2.

As previously argued, objectives may not be precisely stated in some situations. It is especially common in the goal setting environment. For case (c), the aspiration level z may become fuzzy and the objective function may take on the following three basic forms:

(a) O_1: $f(x) \lesssim \underline{z}; \bar{z}$, f: $\underline{X} \longrightarrow \mathbb{R}$, $x \in \underline{X} \subseteq \mathbb{R}^n$, $\underline{z}$, $\bar{z} \in \mathbb{R}$; (5-36)

(b) O_2: $f(x) \gtrsim \bar{z}; \underline{z}$, f: $\underline{X} \longrightarrow \mathbb{R}$, $x \in \underline{X} \subseteq \mathbb{R}^n$, $\underline{z}$, $\bar{z} \in \mathbb{R}$; (5-37)

(c) O_3: $f(x) \approxeq z; \underline{z}, \bar{z}$, f: $\underline{X} \longrightarrow \mathbb{R}$, $x \in \underline{X} \subseteq \mathbb{R}^n$, z, $\underline{z}$, $\bar{z} \in \mathbb{R}$. (5-38)

Fuzzy objective O_1 in (5-36) may read "f(x) should be essentially smaller than $\underline{z}$". Similar to the interpretation of the fuzzy "less than or equal to" constraint in (5-4), O_1 can be represented by a fuzzy subset which may be defined by

$$\mu_{O_1}(x) \triangleq \mu_{O_1}(f(x)) = \begin{cases} 1, & \text{if } f(x) \le \underline{z}, \\ 1 - \dfrac{f(x) - \underline{z}}{d}, & \text{if } \underline{z} < f(x) \le \bar{z}, \\ 0, & \text{if } f(x) > \bar{z}. \end{cases} \tag{5-39}$$

Again, d is the length of the tolerance interval $[\underline{z}, \bar{z}]$ within which f(x) may take its value. μ_{O_1} can be regarded as a satisfaction function indicating a decisionmaker's degree of satisfaction about the value f(x). Parallel to a precise objective function, it defines a linear ordering on the set of alternatives in $\underline{X}$.

By the same argument, fuzzy objective O_2 in (5-37) may read "f(x) should be essentially greater than $\bar{z}$" and can be defined by

$$\mu_{O_2}(x) \triangleq \mu_{O_2}(f(x)) = \begin{cases} 1, & \text{if } f(x) > \bar{z}, \\ 1 - \dfrac{\bar{z} - f(x)}{d}, & \text{if } \underline{z} < f(x) \le \bar{z}, \\ 0, & \text{if } f(x) \le \underline{z}. \end{cases} \tag{5-40}$$

Obviously, fuzzy objective O_3 in (5-38) may read "f(x) should be essentially equal to z" and can be defined by

$$\mu_{O_3}(x) \triangleq \mu_{O_3}(f(x)) = \begin{cases} 0, & \text{if } f(x) < \underline{z}, \\ 1 - \dfrac{z - f(x)}{\underline{d}}, & \text{if } \underline{z} \le f(x) < z, \\ 1, & \text{if } f(x) = z, \\ 1 - \dfrac{f(x) - z}{\bar{d}}, & \text{if } z < f(x) \le \bar{z}, \\ 0, & \text{if } f(x) > \bar{z}. \end{cases} \tag{5-41}$$

Therefore, fuzzy objectives and fuzzy constraints are fuzzy subsets defined on the set of alternatives in $\underline{X}$. If they are of similar types, they can be defined by membership functions of similar formats. To select the most satisfying alternative, one looks for the alternative which maximizes the degree of satisfaction of the fuzzy objective.

5. 3. 2. Optimization of a Fuzzy Objective Function over a Precise Constraint Set

Assume that the objective of a spatial planning problem is fuzzy and is defined by a fuzzy subset $\mu_O: \underline{X} \longrightarrow [0, 1]$. Let $\{g_i: \underline{X} \longrightarrow \mathbb{R} | g_i(x) \leq b_i, \ i = 1, \ldots, m\}$ be a set of m precise constraints.

Then, the decision space for the system $(\underline{X}, \mu_O, g_1, \ldots, g_m)$ is defined by

$$\mu_D(x) = \min[\mu_O(x), g_1(x), \ldots, g_m(x)]. \tag{5-42}$$

The optimal solution is the alternative x* such that

$$\mu_D(x^*) = \sup_{x \in \underline{X}} \min[\mu_O(x), g_1(x), \ldots, g_m(x)]. \tag{5-43}$$

Analytically, x* is obtained by solving the following linear programming problem:

$$\begin{aligned} &\max \lambda, \\ &\text{s.t. } \mu_O(x) \geq \lambda \\ &\qquad g_i(x) \leq b_i, \ i = 1, \ldots, m, \\ &\qquad \lambda \geq 0, \ x \geq \theta, \end{aligned} \tag{5-44}$$

where λ is the degree of objective-satisfaction which corresponds essentially to (5-42).

If μ_O is of the type in (5-37), the optimization problem in (5-44) becomes

$$\max \lambda,$$
$$\text{s.t. } \frac{\bar{z} - f(x)}{d} \geq \lambda$$
$$g_i(x) \leq b_i, \; i = 1, \ldots, m, \qquad (5\text{-}45)$$
$$\lambda \geq 0, \; x \geq \theta.$$

or, equivalently

$$\max \lambda,$$
$$\text{s.t. } f(x) \leq \underline{z} + d(1 - \lambda)$$
$$g_i(x) \leq b_i, \; i = 1, \ldots, m, \qquad (5\text{-}46)$$
$$\lambda \geq 0, \; x \geq \theta.$$

Example 5. 3. Let $f(x) = 3x_1 + 2x_2 \underset{\sim}{\geq} 12; 9$, be a fuzzy objective function. Let the following be a precise constraint set:

$$4x_1 + 5x_2 \leq 0$$
$$4x_1 + 2x_2 \leq 12 \qquad (5\text{-}47)$$
$$x_1 \geq 0, \; x_2 \geq 0.$$

By (5-45), the optimal solution $(\lambda^*, x_1^*, x_2^*) = (0.445, 1.667, 2.667)$ is obtained by solving

$$\max \lambda,$$
$$\text{s.t. } -3 + x_1 + 0.667x_2 \geq \lambda$$
$$4x_1 + 5x_2 \leq 20$$
$$4x_1 + 2x_2 \leq 12 \qquad (5\text{-}48)$$
$$\lambda \geq 0, \; x_1 \geq 0, \; x_2 \geq 0.$$

5. 4. FUZZY-OBJECTIVE-FUZZY-CONSTRAINTS SPATIAL PLANNING PROBLEMS

Spatial planning problems in sections 5.2 and 5.3 are

partially fuzzy optimization problems in which either the objective or the constraints are fuzzy. It is apparent that both the objective and the constraints can be imprecise in a fuzzy planning environment. Since such problems are more general in structure, further elaboration is then necessary.

In subsection 5.4.1, spatial planning problems with a fuzzy objective and a set of fuzzy constraints are formulated via the fuzzy optimization framework proposed by Bellman and Zadeh (1970). In subsection 5.4.2, a connection between interregional equilibrium and fuzzy linear programming is established.

5.4.1. Optimization of a Fuzzy Objective Function over a Fuzzy Constraint Set

Assume that both the objective and constraints of a spatial planning problem are fuzzy and are defined as fuzzy subsets on the set of alternatives in $\underline{X}$. In formulating the decision space, they are structurally identical and can be treated symmetrically.

Let $(\underline{X}, \mu_0, \mu_1, \ldots, \mu_m)$ be the optimization system with μ_0 defining the fuzzy objective and μ_i, $i = 1, \ldots, m$, defining the fuzzy constraint i. The fuzzy decision space D is then defined by

$$\mu_D(x) = \min[\mu_0(x), \mu_1(x), \ldots, \mu_m(x)], \tag{5-49}$$

which is the overall satisfaction function of the fuzzy objective and constraints. The optimal solution is the alternative x* such that

$$\mu_D(x^*) = \sup_{x \varepsilon \underline{X}} \min[\mu_0(x), \mu_1(x), \ldots, \mu_m(x)]. \tag{5-50}$$

The solution concept is based on the rationale that confluence of the fuzzy objective function and fuzzy constraints is a fuzzy subset which defines a linear ordering on the set of alternatives in $\underline{X}$. The best alternative should then be the one which is the highest in order, i.e. the one which gives the highest degree of overall satisfaction.

Based on Zimmermann (1976) and Leung (1983a), to solve the problem in (5-50), it is equivalent to solve

$$\begin{aligned} &\max \lambda, \\ &\text{s.t. } \mu_0(x) \geq \lambda \\ &\quad \mu_i(x) \geq \lambda, \ i = 1, \ldots, m, \\ &\quad \lambda \geq 0, \ x \geq \theta. \end{aligned} \tag{5-51}$$

where λ is the degree of overall satisfaction which corresponds essentially to (5-49).

Specifically, if μ_0 is of the O_1-type in (5-36) and μ_i, i = 1, ..., m, is of the C_1-type in (5-4), the problem in (5-51) becomes

$$\begin{aligned} &\max \lambda, \\ &\text{s.t. } \frac{\bar{z} - f(x)}{d_0} \geq \lambda \\ &\quad \frac{\bar{b}_i - g(x)}{d_i} \geq \lambda \\ &\quad \lambda \geq 0, \ x \geq \theta. \end{aligned} \tag{5-52}$$

The solution x* gives the degree λ* to which both the fuzzy objective and fuzzy constraints are satisfied.

Though the min-operator is employed to aggregate the fuzzy objective and fuzzy constraints in (5-49), it is by no means the only operator which can be used to construct the fuzzy decision space D. If the min-operator is applied, the fuzzy optimization problem is then noncompensating. That is, it always selects the alternative x which gives the lowest degree of satisfaction among the fuzzy objective and fuzzy constraints. Thus, a relatively lower degree of objective satisfaction, for example, cannot be compensated by satisfying the fuzzy constraints to a higher degree. The converse is also true.

To allow for compensation, another "and"-type operator,

such as the algebraic product, •, can be employed to construct the fuzzy decision space D in (5-49). The defining membership function then becomes:

$$\mu_D(x) = \mu_0(x) \cdot \prod_i \mu_i(x), \qquad (5\text{-}53)$$

and, trade-offs among the fuzzy objective and fuzzy constraints are possible.

However, the min- and the algebraic product are both operators which treat the fuzzy objective and fuzzy constraints equally. If they are of unequal importance, operators such as the fuzzy convex combination can be applied. As a result, the fuzzy decision space is defined by:

$$\mu_D(x) = w_0\mu_0(x) + \sum_i w_i\mu_i(x), \quad w_0 + \sum_i w_i = 1, \qquad (5\text{-}54)$$

where w_0 and w_i's are weights attached to the fuzzy objective and fuzzy constraints respectively.

In some optimization problems, conservative strategy imposed by the min-operator may not be desirable. A more optimistic strategy, however, can be made possible by applying some other operators. Among fuzzy set operators, fuzzy decision space defined by the max-operator, corresponding to the connective "or":

$$\mu_D(x) = \max[\mu_0(x), \mu_1(x), \ldots, \mu_m(x)], \qquad (5\text{-}55)$$

would necessitate the alternative x leading to the highest degree of satisfaction among the fuzzy objective and fuzzy constraints to be selected in the fuzzy optimization process. Similar to the min-operator, it is also noncompensatory.

To incorporate compensation, operators such as the algebraic sum, $\hat{+}$, another "or"-type operator, can be employed. Parallel to the algebraic product, it allows for trade-offs among the fuzzy objective and fuzzy constraints. Specifically, the fuzzy decision space D for a single fuzzy objective O and a single fuzzy constraint C would be

$$\mu_D(x) = \mu_O(x) + \mu_C(x) - \mu_O(x) \cdot \mu_C(x). \qquad (5\text{-}56)$$

For multiple fuzzy constraints, as in our present case, the structure of D is too complex to be practical and is not expressed here.

Depending on the purpose of the fuzzy optimization problem, varying operators can be employed to construct the fuzzy decision space. They can be very conservative such as the min- or very optimistic such as the max-, or they can be less restrictive and more compensating such as the algebraic product and the convex combination. Such a variety of choices makes the fuzzy optimization problem structurally less unique (in comparison with the conventional linear programming problem) but more flexible in modelling interregional equilibrium problems.

To facilitate our discussion, I only analyze fuzzy optimization problems based on the min-operator. However, the approach can easily be extended to those involving other operators, albeit some may become computationally more complicated.

Remark. If spatial planning problems have a mixture of fuzzy and precise constraints, the above formulation is still applicable. The solution of the system $(\underline{X}, \mu_0, \mu_1, \ldots, \mu_m, g_{m+1}, \ldots, g_{m+k})$ then is

$$\sup_{x \in \underline{X}} \min[\mu_0(x), \mu_1(x), \ldots, \mu_m(x), g_{m+1}(x), \ldots, g_{m+k}(x)]. \tag{5-57}$$

The corresponding fuzzy linear programming problem becomes

$$\begin{aligned} &\max \lambda, \\ &\text{s.t. } \mu_0(x) \geq \lambda \\ &\qquad \mu_i(x) \geq \lambda, \ i = 1, \ldots, m, \\ &\qquad g_j(x) \leq b_j, \ j = m+1, \ldots, m+k, \\ &\qquad \lambda \geq 0, \ x \geq \theta, \end{aligned} \tag{5-58}$$

where $g_j\colon \underline{X} \longrightarrow \mathbb{R}$, $j = m+1, \ldots, m+k$, for example, is a precise constraint of the "less than or equal to" type.

Example 5.4. (Application to a Regional Resource Allocation Problem)

In this example, the fuzzy linear programming technique is applied to a simlified regional resource allocation problem which involves the communal farming communities, the system of kibbutzim, in Israel.

It is common for groups of kibbutzim to join together as a confederation to share common technical services and to co-ordinate their production. Within the confederation, the over-all planning is carried out by a coordinating technical office. The function of the office is to plan agricultural production of the confederation for the coming year.

The allocation problem concerns a confederation of three kibbutzim. The agricultural output of each kibbutz is limited by the amount of available irrigable land (Table 5.1) and the quantity of water (Table 5.2) (which can only be imprecisely specified) allocated for irrigation by the Water Commissioner, a national government official.

The crops under consideration are sugar beets, cotton, and sorghum. These crops differ in their expected net return and their consumption of water. Imprecise capacities of land that can be devoted to each of the crops are also imposed by the Ministry of Agriculture (see Table 5.3).

The three kibbutzim of the Confederation have agreed that every kibbutz will plant a prespecified proportion of its available irrigable land. Nevertheless, any combination of the crops may be grown at any of the kibbutzim.

For the coming year, the coordinating technical office has to determine the amount of land to be allocated to each crop in the respective kibbutzim so that the above prescribed restrictions are satisfied. In place of taking the maximization of total net return of the confederation as an objective, it is decided that a target value, 350,000 dollars, should be employed and the total net return is required to exceed the target value. In case the target value is too optimistic, the total net return is allowed to fall below it. The bottom line is 250,000 dollars. Thus, the fuzzy goal can be stated as

Table 5. 1 Irrigable land for the confederation of kibbutzim

Kibbutz	Irrigable Land (acres)	
	fuzzy specification*	associated fuzzy interval
1	should be less than 400 or not much greater than 400	[400, 440]
2	should be less than 600 or not much greater than 600	[600, 630]
3	should be less than 300 or not much greater than 300	[300, 320]

*The membership functions are of the C_1-type defined in general by (5-7).

Table 5. 2 Water capacities for the confederation of kibbutzim

Kibbutz	Water Capacities (acre feet)	
	fuzzy specification*	associated fuzzy interval
1	should be less than 600 or not much greater than 600	[600, 660]
2	should be less than 800 or not much greater than 800	[800, 840]
3	should be less than 375 or not much greater than 375	[375, 450]

*The membership functions are of the C_1-type defined in general by (5-7).

Table 5.3 Crop data for the confederation of kibbutzim

Crop	Net Return (dollars/acre)	(acre feet/acre)	Maximum Capacities (acres)	
			fuzzy specification*	associated fuzzy interval
Sugar beets	400	3	should be less than 600 or not much greater than 600	[600, 650]
Cotton	300	2	should be less than 500 or not much greater than 500	[500, 540]
Sorghum	100	1	should be less than 325 or not much greater than 325	[325, 350]

*The membership functions are of the C_1-type defined in general by (5-7).

"Total net return should be greater than 350,000 dollars or not much smaller than 350,000 dollars",

and is defined by the O_2-type membership function in (5-40) with the associated fuzzy interval [250,000, 350,000].

Based on the above information, the resource allocation problem of the Confederation can be formulated as a fuzzy linear programming problem as follows:

Objective:

$$400(x_{11}+x_{12}+x_{13}) + 300(x_{21}+x_{22}+x_{23}) + 100(x_{31}+x_{32}+x_{33}) \gtrsim 350{,}000;\ 250{,}000, \tag{5-59}$$

Constraints:

Water consumption (fuzzy):

$$\begin{aligned} 3x_{11} + 2x_{21} + x_{31} &\lesssim 600;\ 660 \\ 3x_{12} + 2x_{22} + x_{32} &\lesssim 800;\ 840 \\ 3x_{13} + 2x_{23} + x_{33} &\lesssim 375;\ 450 \end{aligned} \tag{5-60}$$

Crop-land (fuzzy):

$$\begin{aligned} x_{11} + x_{12} + x_{13} &\lesssim 600;\ 650 \\ x_{21} + x_{22} + x_{23} &\lesssim 500;\ 540 \\ x_{31} + x_{32} + x_{33} &\lesssim 325;\ 350 \end{aligned} \tag{5-61}$$

Irrigable land (fuzzy):

$$\begin{aligned} x_{11} + x_{21} + x_{31} &\lesssim 400;\ 440 \\ x_{12} + x_{22} + x_{32} &\lesssim 600;\ 630 \\ x_{13} + x_{23} + x_{33} &\lesssim 300;\ 320 \end{aligned} \tag{5-62}$$

Land appropriation (precise):

$$\begin{aligned} \frac{1}{400}(x_{11} + x_{21} + x_{31}) &= \frac{1}{600}(x_{12} + x_{22} + x_{32}) \\ \frac{1}{600}(x_{12} + x_{22} + x_{32}) &= \frac{1}{300}(x_{13} + x_{23} + x_{33}) \\ \frac{1}{300}(x_{13} + x_{23} + x_{33}) &= \frac{1}{400}(x_{11} + x_{21} + x_{31}) \end{aligned} \tag{5-63}$$

Non-negativity:

$$x_{ij} \geq 0, \text{ for } i = 1, 2, 3 \text{ and } j = 1, 2, 3, \qquad (5\text{-}64)$$

where x_{ij} = number of acres of land in kibbutz j to be allocated to plant crop i, i = 1, 2, and 3 (1 = sugar beets, 2 = cotton, 3 = sorghum).

Let the fuzzy constraints be defined by the membership function having the format in (5-7), and the fuzzy objective be defined by the membership function having the format in (5-40). Following the arguments in (5-57) and (5-58), the above fuzzy linear programming problem can then be formulated as a conventional linear programming problem as follows:

$$\begin{aligned}
&\max\ \lambda, \\
&\text{s.t. } -2.5 + 0.004(x_{11} + x_{12} + x_{13}) + 0.003(x_{21} + x_{22} + x_{23}) + \\
&\quad 0.001(x_{31} + x_{32} + x_{33}) \geq \lambda \\
&\quad 11 - 0.05x_{11} - 0.03x_{21} - 0.02x_{31} \geq \lambda \\
&\quad 21 - 0.08x_{12} - 0.05x_{22} - 0.03x_{32} \geq \lambda \\
&\quad 6 - 0.04x_{13} - 0.03x_{23} - 0.01x_{33} \geq \lambda \\
&\quad 13 - 0.02x_{11} - 0.02x_{12} - 0.02x_{13} \geq \lambda \\
&\quad 13.5 - 0.03x_{21} - 0.03x_{22} - 0.03x_{23} \geq \lambda \\
&\quad 14 - 0.04x_{31} - 0.04x_{32} - 0.04x_{33} \geq \lambda \qquad (5\text{-}65) \\
&\quad 11 - 0.025x_{11} - 0.025x_{21} - 0.025x_{31} \geq \lambda \\
&\quad 21 - 0.03x_{12} - 0.03x_{22} - 0.03x_{32} \geq \lambda \\
&\quad 16 - 0.05x_{13} - 0.05x_{23} - 0.05x_{33} \geq \lambda \\
&\quad 3(x_{11} + x_{21} + x_{31}) - 2(x_{12} + x_{22} + x_{32}) = 0 \\
&\quad x_{12} + x_{22} + x_{32} - 2(x_{13} + x_{23} + x_{33}) = 0 \\
&\quad 4(x_{13} + x_{23} + x_{33}) - 3(x_{11} + x_{21} + x_{31}) = 0 \\
&\quad \lambda \geq 0,\ x_{ij} \geq 0, \text{ for } i = 1, 2, 3 \text{ and } j = 1, 2, 3.
\end{aligned}$$

The optimal solution is

$$(x^*_{11}, x^*_{12}, x^*_{13}, x^*_{21}, x^*_{22}, x^*_{23}, x^*_{31}, x^*_{32}, x^*_{33})$$
$$= (152.778, 48.611, 128.472, 104.938, 337.963, 0, 0, 0, 64.815)$$
with $\lambda^* = 0.213$. (5-66)

5.4.2. A Formalization of Interregional Equilibrium in a Fuzzy Environment

Complexity of a spatial system and inexactness of our decisionmaking processes often make the precise specification of objectives and constraints difficult. Inability in accommodating imprecise information tends to decrease the flexibility, on the theoretical and practical level, of conventional linear programming analysis of interregional equilibrium problems. Fuzzy linear programming, as demonstrated in the previous sections, appears to be an effective method for such an analysis.

The purpose of this subsection is to investigate the connection between interregional equilibrium under imprecision and fuzzy linear programming. Emphasis is placed on the basic theoretical properties of the interregional equilibrium solutions and their spatial economic interpretations. It also serves as a formalization of the presentation in subsection 5.4.1.

To keep our discussion general, interregional fuzzy linear programming problems is formulated in general terms. The approach and results, however, are applicable to any spatial economic problems of similar nature. Our discussion is mainly based on Leung (1985d).

Let the interregional equilibrium problem be:

$$\sum_k \sum_j c^k_j \, x^k_j \lesssim \underline{z};\ \bar{z}$$

$$\sum_j a^k_{ij} \, x^k_j \lesssim \underline{b}^k_i;\ \bar{b}^k_i,\quad i = 1, \ldots, m;\ k = 1, \ldots, \ell, \qquad (5\text{-}67)$$

$$x^k_j \geq 0,\quad j = 1, \ldots, n;\ k = 1, \ldots, \ell,$$

where

n = number of activities,
ℓ = number of regions,
m = number of resources,
c_j^k = unit production cost of activity j in region k,
a_{ij}^k = unit consumption of resource i by activity j in region k,
$\underline{z}$ = target value of the total cost,
$\overline{z}$ = maximum tolerated value of the total cost, with $\overline{z} > \underline{z}$,
$\underline{b}_i^k$ = target volume of supply of resource i in region k,
$\overline{b}_i^k$ = maximum tolerated volume of supply of resource i in region k, with $\overline{b}_i^k \geq \underline{b}_i^k$,
x_j^k = production level of activity j in region k.

(Henceforth, activity j in region k is referred to as activity j, k and resource i in region k is referred to as resource i, k)

The objective in (5-67) is to keep the total cost below $\underline{z}$ as much as possible. Should the total cost exceeds $\underline{z}$, it preferrably should not be too much larger than $\underline{z}$ and definitely cannot exceed the maximum tolerated value $\overline{z}$. Thus, the length $d_0 = (\overline{z} - \underline{z})$ of the tolerance interval $[\underline{z}, \overline{z}]$ is the tolerated level set for the total cost.

For each resource i in region k, the tolerated level of the capacity is $d_i^k = (\overline{b}_i^k - \underline{b}_i^k)$. The imprecise constraint then restricts, as much as possible, the total consumption of resource i, k to be less than or equal to $\underline{b}_i^k$ within the tolerated level d_i^k.

To solve the interregional fuzzy linear programming problem in (5-67) in the most satisfactory way, satisfaction functions (fuzzy subsets) for the fuzzy objective and constraints should first be formulated.

Definition 5.1. A fuzzy objective of an interregional equilibrium problem in $\mathbb{R}^{n\ell}$ is a fuzzy subset defined by a satisfaction function (a membership function) in $\mathbb{R}^{n\ell}$.

Let $x = (x_1^1, \ldots, x_j^k, \ldots, x_n^\ell) \in \mathbb{R}^{n\ell}$ be a spatial alternative. Let the satisfaction function of the fuzzy objective be:

$$\mu_0(x) = \begin{cases} 1, & \text{if } \sum_k \sum_j c_j^k x_j^k \leq \underline{z}, \\ 1 - \dfrac{\sum_k \sum_j c_j^k x_j^k - \underline{z}}{d_0}, & \text{if } \underline{z} < \sum_k \sum_j c_j^k x_j^k \leq \bar{z}, \\ 0, & \text{if } \sum_k \sum_j c_j^k x_j^k > \bar{z}, \end{cases} \tag{5-68}$$

where $[\underline{z}, \bar{z}]$ with $d_0 = (\bar{z} - \underline{z})$ is the tolerance interval of the maximum permissible violation of the designated total cost $\underline{z}$.

Definition 5.2. A fuzzy constraint of an interregional equilibrium problem in $\mathbb{R}^{n\ell}$ is a fuzzy subset defined by a satisfaction function in $\mathbb{R}^{n\ell}$.

For resource i, k, i = 1, ..., m; k = 1, ..., ℓ, let the satisfaction function be

$$\mu_i^k(x) = \begin{cases} 1, & \text{if } \sum_j a_{ij}^k x_j^k \leq \underline{b}_i^k, \\ 1 - \dfrac{\sum_j a_{ij}^k x_j^k - \underline{b}_i^k}{d_0^k}, & \text{if } \underline{b}_i^k < \sum_j a_{ij}^k x_j^k \leq \bar{b}_i^k, \\ 0, & \text{if } \sum_j a_{ij}^k x_j^k > b_i, \end{cases} \tag{5-69}$$

where $[\underline{b}_i^k, \bar{b}_i^k]$ with $d_i^k = (\bar{b}_i^k - \underline{b}_i^k)$ is the tolerance interval of the maximum permissible violation of the estimated supply $\underline{b}_i^k$.

Definition 5.3. Let μ_0 and μ_i^k, i = 1, ..., m; k = 1, ..., ℓ, be satisfaction functions defining respectively the fuzzy objective and fuzzy constraints. The fuzzy decision space D of the interregional equilibrium problem in $\mathbb{R}^{n\ell}$ is a fuzzy subset in $\mathbb{R}^{n\ell}$ defined by the membership function

$$\mu_D(x) = \min\left\{\mu_0(x),\ \min_{i,k}[\mu_i^k(x)]\right\},\ x \in \mathbb{R}^{n\ell}. \qquad (5\text{-}70)$$

The membership function in (5-70) is the confluence of the satisfaction functions defining the fuzzy objective and constraints and can thus be interpreted as the overall satisfaction function. The interregional equilibrium problem in (5-67) then becomes a fuzzy optimization problem:

$$\sup_{x \in \mathbb{R}^{n\ell}} \mu_D(x) = \sup_{x \in \mathbb{R}^{n\ell}} \min\left\{\mu_0(x),\ \min_{i,k}[\mu_i^k(x)]\right\}. \qquad (5\text{-}71)$$

The optimal spatial alternative is the one which maximizes the the degree of overall satisfaction.

Let λ be the degree of overall satisfaction which corresponds essentially to (5-70). As discussed in subsection 5.4.1, the optimization problem in (5-71) is equivalent to

$$\begin{aligned} &\max \lambda, \\ &\text{s.t. } \mu_0(x) \geq \lambda \\ &\qquad \mu_i^k(x) \geq \lambda,\ i = 1, \ldots, m;\ k = 1, \ldots, \ell, \\ &\qquad \lambda \geq 0,\ x \geq \theta, \end{aligned} \qquad (5\text{-}72)$$

or, equivalently

$$\begin{aligned} &\max \lambda, \\ &\text{s.t. } \frac{\bar{z} - \sum_k \sum_j c_j^k\, x_j^k}{d_0} \geq \lambda \\ &\qquad \frac{\bar{b}_i^k - \sum_j a_{ij}^k\, x_j^k}{d_i^k} \geq \lambda,\ i = 1, \ldots, m;\ k = 1, \ldots, \ell, \\ &\qquad \lambda \geq 0,\ x_j^k \geq 0,\ j = 1, \ldots, n;\ k = 1, \ldots, \ell. \end{aligned} \qquad (5\text{-}73)$$

Since the problems in (5-71) and (5-73) are equivalent, their solutions should be identical. The existence of a solution to (5-71) [and (5-73)] and its associated properties are of pertinent interest for interregional equilibrium analysis. Let us first examine the general equilibrium solution of the interregional fuzzy linear programming problem in (5-73) via two theorems.

Theorem 5.1. Let $x = (x_1^1, \ldots, x_j^k, \ldots, x_n^\ell)$ be a spatial alternative. If a spatial decision $(\lambda^*, x^*) = (\lambda^*, (x_1^{1*}, \ldots, x_j^{k*}, \ldots, x_n^{\ell*})) \in \mathbb{R}^{n\ell} \times [0, 1]$ is an optimal solution of the following interregional linear programming problem:

$$\max \lambda,$$

$$\text{s.t.} \quad \frac{\bar{z} - \sum_k \sum_j c_j^k x_j^k}{d_0} \geq \lambda$$

$$\frac{\bar{b}_i^k - \sum_j a_{ij}^k x_j^k}{d_i^k} \geq \lambda, \quad i = 1, \ldots, m; \ k = 1, \ldots, \ell,$$

$$\lambda \geq 0, \ x_j^k \geq 0, \ j = 1, \ldots, n; \ k = 1, \ldots, \ell.$$

Then,

$$\lambda^* = \min \left\{ \frac{\bar{z} - \sum_k \sum_j c_j^k x_j^{k*}}{d_0}, \ \min_{i,k} \left[\frac{\bar{b}_i^k - \sum_j a_{ij}^k x_j^{k*}}{d_i^k} \right] \right\}.$$

Theorem 5.2. A spatial alternative $x^* = (x_1^{1*}, \ldots, x_j^{k*}, \ldots, x_n^{\ell*})$ is an optimal solution if and only if the decision $(\lambda^*, x^*) \in \mathbb{R}^{n\ell} \times [0, 1]$ with

$$\lambda^* = \min \left\{ \frac{\bar{z} - \sum_k \sum_j c_j^k x_j^{k*}}{d_0}, \ \min_{i,k} \left[\frac{\bar{b}_i^k - \sum_j a_{ij}^k x_j^{k*}}{d_i^k} \right] \right\}.$$

is a solution of the following interregional linear programming problem:

$$\max \lambda,$$

$$\text{s.t.} \quad \frac{\bar{z} - \sum_k \sum_j c_j^k x_j^k}{d_0} \geq \lambda$$

$$\frac{\bar{b}_i^k - \sum_j a_{ij}^k x_j^k}{d_i^k} \geq \lambda, \quad i = 1, \ldots, m; \; k = 1, \ldots, \ell,$$

$$\lambda \geq 0, \quad x_j^k \geq 0, \quad j = 1, \ldots, n; \; k = 1, \ldots, \ell.$$

Theorems 5.1 and 5.2 state that the optimal solution of an interregional fuzzy linear programming problem is the spatial alternative which maximizes simultaneously the degrees of satisfaction of the fuzzy objective function and fuzzy constraints over the imprecise fuzzy decision space. To obtain the optimal solution for problem (5-71), it is sufficient to obtain the optimal solution for problem (5-73).

To put the interregional fuzzy linear programming problem in line with the general fuzzy optimization problems. Corollary 5.1 shows that the interregional equilibrium solution can actually take on the expression of the solution of a general fuzzy mathematical programming problem, specifically an unconstrained extremum.

Corollary 5.1. If (λ^*, x^*) is an optimal solution of the interregional linear programming problem in (5-71), then

$$\lambda^* = \sup_{\alpha \in [0,1]} \mu_D(x) = \sup_{\alpha \in [0,1]} \min \left\{ \alpha, \sup_{x \in \underline{C}_\alpha} \left[\frac{\bar{z} - \sum_k \sum_j c_j^k x_j^k}{d_0} \right] \right\},$$

where

$$\mu_D(x) = \min \left\{ \mu_0(x), \min_{i,k} [\mu_i^k(x)] \right\},$$

and,

$$\underline{C}_\alpha = \bigcap_{i,k} \underline{C}_{\alpha_i^k} \text{ with } \underline{C}_{\alpha_i^k} = \left\{x \in \mathbb{R}^{n\ell} \mid \mu_i^k(x) \geq \alpha_i^k\right\},\ \alpha_i^k \in [0, 1].$$

Parallel to conventional linear programming problems, associated with the equilibrium analysis of any interregional fuzzy linear programming problem is a dual optimal problem from which interesting mathematical results and spatial economic interpretations can be drawn. In what to follow, a discussion of the basic duality arguments is provided.

Treating the interregional problem in (5-73) as the primal problem, it can be rewritten as:

$$\max \lambda,$$

$$\text{s.t. } \sum_k \sum_j c_j^k x_j^k + d_0 \lambda \leq \bar{z}$$

$$\sum_j a_{ij}^k x_j^k + d_i^k \leq \bar{b}_i^k,\ i = 1, \ldots, m;\ k = 1, \ldots, \ell, \quad (5\text{-}74)$$

$$\lambda \geq 0,\ x_j^k \geq 0,\ j = 1, \ldots, n;\ k = 1, \ldots, \ell.$$

The corresponding dual problem is:

$$\min \bar{z} y_0 + \sum_k \sum_i \bar{b}_i^k y_i^k$$

$$\text{s.t. } d_0 y_0 + \sum_k \sum_i d_i^k y_i^k \geq 1$$

$$c_j^k y_0 + \sum_k \sum_i a_{ij}^k y_i^k \geq 0,\ j = 1, \ldots, n;\ k = 1, \ldots, \ell,$$

$$y_0 \geq 0,\ y_i^k \geq 0,\ i = 1, \ldots, m;\ k = 1, \ldots, \ell. \quad (5\text{-}75)$$

The dual variable y_0 denotes the relative change of the degree of overall satisfaction due to the marginal variation of the total cost. In brief, it can be defined as the marginal satisfaction of total cost.

Similarly, the dual variable y_i^k, $i = 1, \ldots, m;$ $k = 1,$

..., ℓ, indicates the relative change of the degree of overall satisfaction due to the marginal variation of the supply of resource i, k. In brief, it can be defined as the marginal satisfaction of resource i, k.

A fundamental relation between the primal and the dual problem can be stated by the theorem of duality:

Theorem 5.3. (Fundamental Theorem of Duality)

Given a pair of primal and dual interregional fuzzy linear programming problems in (5-74) and (5-75), one and only one of the following statements is true:

(a) both problems possess optimal solutions and

$$\max \lambda = \min[\bar{z}y_0 + \sum_k \sum_i \bar{b}_i^k \, y_i^k];$$

(b) one problem has no feasible solutions and the other has at least one feasible solution but no bounded optimal solutions;

(c) both problems are infeasible.

Based on Theorem 5.3, the existence of an equilibrium solution for problem (5-74) and (5-75) can be established:

Theorem 5.4. (Existence Theorem)

Given a pair of primal and dual interregional fuzzy linear programming problems in (5-74) and (5-75), the necessary and sufficient conditions that both problems have optimal solutions is that both have feasible solutions.

Since problem (5-71) is equivalent to problem (5-74), the corresponding existence theorem then becomes:

Theorem 5.5. (Existence Theorem)

The system of satisfaction functions in problem (5-71) has an optimal solution in $\mathbb{R}^{n\ell}$ if D is non-vacuous.

Apparently, if D in (5-71) is non-vacuous, then a feasible solution exists in both (5-74) and (5-75). Specifically, let the fuzzy objective, O, and the fuzzy constraints, C_i^k, i = 1, ..., m; k = 1, ..., ℓ, be fuzzy subsets defined respectively by μ_0 and μ_i^k, i = 1, ..., m; k = 1, ..., ℓ. If $D = O \cap [\bigcap_{i,k} C_i^k] \neq \phi$, then an optimal solution exists.

To be optimal, a feasible solution has to satisfy certain conditions. They are stated in theorem 5.6.

Theorem 5.6. (Weak Theorem of Complementary Slackness)

Given a pair of primal and dual interregional fuzzy linear programming problems in canonical form in (5-74) and (5-75), the necessary and sufficient conditions for feasible solutions $(\lambda^*, x_1^{1*}, \ldots, x_j^{k*}, \ldots, x_n^{\ell*})$ and $(y_0^*, y_1^{1*}, \ldots, y_i^{k*}, \ldots, y_m^{\ell*})$ to be optimal are that they satisfy relations (a) to (d):

$$\text{(a)}\quad y_0^*(\bar{z} - \sum_k \sum_j c_j^k x_j^{k*} - d_0 \lambda^*) = 0,$$

$$\text{(b)}\quad y_i^{k*}(\bar{b}_i^k - \sum_j a_{ij}^k x_j^{k*} - d_i^k \lambda^*) = 0,$$

$$i = 1, \ldots, m;\ k = 1, \ldots, \ell;$$

and

$$\text{(c)}\quad \lambda^*(d_0 y_0^* + \sum_k \sum_i d_i^k y_i^{k*} - 1) = 0,$$

$$\text{(d)}\quad x_j^{k*}(c_j^k y_0^* + \sum_k \sum_i a_{ij}^k y_i^{k*}) = 0,\quad j = 1, \ldots, n;\ k = 1, \ldots, \ell.$$

Based on theorem 5.6, an interesting corollary having significant spatial economic interpretations can be derived.

Corollary 5.2. (Corollary of Complementary Slackness)

Given a pair of primal and dual interregional fuzzy linear programming problems in canonical form in (5-74) and (5-75), the necessary and sufficient conditions for feasible solutions

$(\lambda^*, x_1^{1*}, \ldots, x_j^{k*}, \ldots, x_n^{\ell*})$ and $(y_0^*, y_1^{1*}, \ldots, y_i^{k*}, \ldots, y_m^{\ell*})$ to be optimal are that they satisfy relations (a) to (h):

(a) $y_0^* > 0 \Longrightarrow \sum_k \sum_j c_j^k x_j^{k*} + d_0 \lambda^* = \bar{z}$,

(b) $\sum_k \sum_j c_j^k x_j^{k*} + d_0 \lambda^* < \bar{z} \Longrightarrow y_0^* = 0$,

(c) $y_i^{k*} > 0 \Longrightarrow \sum_j a_{ij}^k x_j^{k*} + d_i^k \lambda^* = \bar{b}_i^k$,

$i = 1, \ldots, m; \; k = 1, \ldots, \ell$,

(d) $\sum_k \sum_j a_{ij}^k x_j^{k*} + d_i^k \lambda^* < \bar{b}_i^k \Longrightarrow y_i^{k*} = 0$,

$i = , \ldots, m; \; k = , \ldots, \ell$.

Similarly,

(e) $\lambda^* > 0 \Longrightarrow d_0 y_0^* + \sum_k \sum_i d_i^k y_i^{k*} = 1$,

(f) $d_0 y_0^* + \sum_k \sum_i d_i^k y_i^{k*} > 1 \Longrightarrow \lambda^* = 0$,

(g) $x_j^{k*} > 0 \Longrightarrow c_j^k y_0^* + \sum_k \sum_i a_{ij}^k y_i^{k*} = 0$,

$j = 1, \ldots, n; \; k = 1, \ldots, \ell$,

(h) $c_j^k y_0^* + \sum_k \sum_i a_{ij}^k y_i^{k*} > 0 \Longrightarrow x_j^{k*} = 0$,

$j = 1, \ldots, n; \; k = 1, \ldots, \ell$.

In general, corrollary 5.2 states that whenever a variable in one problem is positive, the corresponding constraint in the other problem must be tight. On the other hand, whenever a constraint in one problem is not tight, the corresponding variable in the other problem must be zero.

Specifically, relation (a) states that whenever the

marginal satisfaction of total cost, y_0^*, is positive, the spatial system operates at the maximum tolerated level of total cost, $\bar{z}$. That is, a higher degree of overall satisfaction, λ, can be achieved by decreasing the total cost from the maximum tolerated level.

Similarly, relation (b) states that the marginal satisfaction of total cost is zero whenever the system operates below the maximum tolerated level of total cost.

Likewise, relation (c) states that whenever the marginal satisfaction of resource i, k, y_i^{k*}, is positive, the maximum tolerated supply of resource i, k, $\bar{b}_i^k$, is exhausted. That is, a higher level of satisfaction can be obtained by decreasing the supply of resource i, k, b_i^k, from the maximum tolerated level.

Similarly, relation (d) says that the marginal satisfaction of resource i, k is zero whenever resource i, k is consumed below the maximum tolerated level.

Relations (e) to (h) can be interpreted in the similar manner.

Relation (e) says that whenever the system operates at a positive degree of satisfaction, the marginal satisfaction of total cost and resources it consumes equals one.

Relation (f) states that whenever the marginal satisfaction of total cost and resources is greater than one, the degree of satisfaction equals zero.

Relation (g) states that whenever activity j, k, x_j^{k*}, operates at a strictly positive level, the marginal satisfaction of total cost and resources it consumes must be zero.

Relation (h) says that whenever the marginal satisfaction of total cost and resources is greater than zero, the corresponding activity must not be produced.

Therefore, the existence of an optimal solution is crucial to our understanding of interregional equilibrium under imprecision. Similar to ordinary equilibrium analysis, it is possible that more than one equilibrium exist in an interregional fuzzy linear programming problem. Depending on the satisfaction functions defining the fuzzy objective and fuzzy constraints and the decision space thus constructed, the equilibrium solution may not be unique.

Once the existence of the equilibrium solution is confirmed, its stability can be analyzed. Changes in unit production costs, unit consumptions of resources, target and tolerated values of the total cost, target and tolerated volumes of supply of resources, and the shifts or changes of the satisfaction functions may lead to instability of the equilibrium solution. It is well known that within the linear programming framework, stability can be analyzed through duality theory and sensitivity analysis. Parallel treatment can also be made within the imprecise system and is not elaborated here.

The present framework can be applied to a variety of spatial equilibrium problems. Location-allocation problems and spatial price equilibrium are typical examples. Within our highly complex spatial economic systems, interregional demand and supply of commodities and interregional transport costs are often imprecise. Spatial price equilibrium, for example, can then be formulated as an interregional fuzzy linear programming problem having imprecise supply, demand, and transport-cost functions. The existence of an interregional equilibrium price, its uniqueness and stability can thus be rigorously analyzed through the arguments presented in this paper.

5.5. ROBUST PROGRAMMING FOR SPATIAL PLANNING PROBLEMS

In our discussion so far, fuzziness of a constraint is due to the imprecision on the capacity of a resource, namely the right hand side of the constraint. In many spatial planning problems, however, imprecision may occur in the estimation of coefficients. Technological production coefficients, for example, are often indeterministic in regional or interregional programming problems. Point-valued estimation of the unit usages of resources by a given activity is often difficult or inappropriate. Probabilistic or stochastic argument seems to pick up what the deterministic models left off. Nevertheless, as argued previously, uncertainty over the technological production coefficients may have nothing to do with randomness. It may be a consequence of our imprecise knowledge of the market and the production process or the tolerance intentionally imposed by the decisionmakers. To take imprecision into consideration and to increase planning robustness, interval within which technological coefficient may take its value with certain membership grade should be incorporated in the programming models.

In subsection 5.5.1, constraints with fuzzy coefficients and fuzzy capacities are discussed. The corresponding decision space is also examined.

In subsection 5.5.2, robust programming with fuzzy constraints are investigated. Emphasis is placed on the optimization of a precise objective function over a fuzzy decision space delimited by constraints with fuzzy coefficients and fuzzy capacities. Robust optimization with objective functions having fuzzy coefficients are briefly outlined.

5.5.1. Technological Constraints with Fuzzy Coefficients and Fuzzy Capacities

In this subsection, two methods of treating fuzzy coefficients and fuzzy capacities in a constraint is discussed.

Let

$$\begin{aligned} \underline{g}(x) &\lesssim B, \\ x &\geq \theta, \end{aligned} \tag{5-76}$$

be the constraint set. Let $g(x) \lesssim B$ be a fuzzy constraint taking the specific form:

$$g(x) = A_1x_1 \oplus A_2x_2 \oplus \ldots \oplus A_nx_n \lesssim B, \tag{5-77}$$

where A_j, $j = 1, \ldots, n$ and B are fuzzy subsets, and $x_j \in \mathbb{R}$, $j = 1, \ldots, n$. The symbol $\oplus$ denotes the addition of fuzzy subsets.

Fuzziness of the constraint is due to the existing imprecisions of the coefficients A_j's and the capacity B. Our task is to reformulate this type of fuzzy constraint into one which is in line with the conventional constraints.

In the discussion to follow, two methods of handling the type of constraint in (5-77) are presented. The first method treats the constraint as a fuzzy inclusive constraint, (5-80), while the second method treats it as it is in (5-77).

Method 1. Let A_j, $j = 1, \ldots, n$, be a fuzzy subset defined by a membership function

$$\mu_{A_j}: \underline{U}_j \longrightarrow [0, 1], \qquad (5\text{-}78)$$

where $\mu_{A_j}(a_j)$ indicates the possibility of consuming a specific amount of resource by activity j.

Let B be a fuzzy subset indicating the possible availability of resource and is defined by

$$\mu_B: \underline{V} \longrightarrow [0, 1]. \qquad (5\text{-}79)$$

Since both A_j's and B are fuzzy subsets, they then impose fuzzy restrictions on the corresponding base variables $\underline{U}_j$ and $\underline{V}$. Thus, to state it precisely, the fuzzy constraint in (5-77) can be regarded as a fuzzy inclusive constraint:

$$A_1x_1 \oplus A_2x_2 \oplus \ldots \oplus A_nx_n \subseteq B. \qquad (5\text{-}80)$$

Employing the concept of level sets and the representation theorem, Negoita, Minoiu, and Stan (1976) show that the the constraint in (5-77) can be written as

$$(\underline{A}_1)_\alpha x_1 \oplus (\underline{A}_2)_\alpha x_2 \oplus \ldots \oplus (\underline{A}_n)_\alpha x_n \subseteq \underline{B}_\alpha, \quad \alpha \,\varepsilon\, [0, 1], \qquad (5\text{-}81)$$

where

$$(\underline{A}_j)_\alpha = \left\{ a_j \,\varepsilon\, \underline{U}_j \,|\, \mu_{A_j}(a_j) \geq \alpha \right\}, \qquad (5\text{-}82)$$

and,

$$\underline{B}_\alpha = \left\{ b \,\varepsilon\, \underline{V} \,|\, \mu_B(b) \geq \alpha \right\}. \qquad (5\text{-}83)$$

Assume that the fuzzy subsets in (5-80) is finite and possess the property

$$\left\{\mu_{A_j}(a_j) \mid a_j \varepsilon \underline{U}_j\right\} = \left\{\alpha_1, \alpha_2, \ldots, \alpha_k\right\},$$

$$0 \leq \alpha_1 \leq \alpha_2 \leq \ldots \leq \alpha_k \leq 1. \qquad (5\text{-}84)$$

Then, for each value of α, say α_k, the constraint in (5-81) becomes

$$(\underline{A}_1)_{\alpha_k}x_1 + (\underline{A}_2)_{\alpha_k}x_2 + \ldots + (\underline{A}_n)_{\alpha_k}x_n \subseteq \underline{B}_{\alpha_k}, \qquad (5\text{-}85)$$

where $(\underline{A}_j)_{\alpha_k}$, $j = 1, \ldots, n$, and $\underline{B}_{\alpha_k}$ constitute convex and non-empty fuzzy intervals.

Thus, the vector $\{x_j\}$, $j = 1, \ldots, n$, is admissible for (5-85) if and only if for each possible vector $\left\{a_j \varepsilon (\underline{A}_j)_{\alpha_k}\right\}$, $j = 1, \ldots, n$, the following is obtained:

$$a_1x_1 + a_2x_2 + \ldots + a_nx_n \ \varepsilon \ \underline{B}_{\alpha_k}. \qquad (5\text{-}86)$$

Based on Soyster (1973), the constraint in (5-85) can be replaced by

$$\begin{aligned} \bar{a}_1^k x_1 + \bar{a}_2^k x_2 + \ldots + \bar{a}_n^k x_n &\leq \bar{b}^k, \\ \underline{a}_1^k x_1 + \underline{a}_2^k x_2 + \ldots + \underline{a}_n^k x_n &\geq \underline{b}^k, \end{aligned} \qquad (5\text{-}87)$$

where

$$\bar{a}_j^k = \sup_{a_j^k \varepsilon (\underline{A}_j)_{\alpha_k}} a_j^k, \qquad (5\text{-}88)$$

$$\underline{a}_j^k = \inf_{a_j^k \varepsilon (\underline{A}_j)_{\alpha_k}} a_j^k, \qquad (5\text{-}89)$$

$$\bar{b}^k = \sup_{b^k \varepsilon \underline{B}_{\alpha_k}} b^k, \qquad (5\text{-}90)$$

$$\underline{b}^k = \inf_{b^k \varepsilon \underline{B}_{\alpha_k}} b^k. \qquad (5\text{-}91)$$

Therefore, for k levels of cut, the fuzzy constraint in (5-81) is replaced by 2k precise constraints:

$$\bar{a}_1^h x_1 + \bar{a}_2^h x_2 + \ldots + \bar{a}_n^h x_n \leq \bar{b}^h, \; h = 1, \ldots, k,$$
$$\underline{a}_1^h x_1 + \underline{a}_2^h x_2 + \ldots + \underline{a}_n^h x_n \geq \underline{b}^h, \; h = 1, \ldots, k. \quad (5\text{-}92)$$

In a spatial planning problem with m fuzzy constraints of the type in (5-80), our decision space is delimited by, together with the decision variables, the following fuzzy inclusive constraints

$$A_{i1}x_1 \oplus A_{i2}x_2 \oplus \ldots \oplus A_{in}x_n \subseteq B_i, \; i = 1, \ldots, m. \quad (5\text{-}93)$$

Represented in terms of α-level sets, they become

$$(\underline{A}_{i1})_\alpha x_1 + (\underline{A}_{i2})_\alpha x_2 + \ldots + (\underline{A}_{in})_\alpha x_n \subseteq \underline{B}_{i\alpha},$$
$$i = 1, \ldots, m, \; \alpha \in [0, 1]. \quad (5\text{-}94)$$

If

$$\left\{ \mu_{A_{ij}}(a_{ij}) \mid a_{ij} \in \underline{U}_{ij} \right\} = \left\{ \alpha_{i1}, \alpha_{i2}, \ldots, \alpha_{ik} \right\},$$
$$0 \leq \alpha_{i1} \leq \alpha_{i2} \leq \ldots \leq \alpha_{ik} \leq 1, \; i = 1, \ldots, m, \quad (5\text{-}95)$$

then, following the arguments in (5-85) to (5-92) the decision space is delimited by the following constraints:

$$\bar{a}_{i1}^h x_1 + \bar{a}_{i2}^h x_2 + \ldots + \bar{a}_{in}^h x_n \leq \bar{b}_i^h, \; i = 1, \ldots, m; \; h = 1, \ldots, k,$$
$$\underline{a}_{i1}^h x_1 + \underline{a}_{i2}^h x_2 + \ldots + \underline{a}_{in}^h x_n \geq \underline{b}_i^h, \; i = 1, \ldots, m; \; h = 1, \ldots, k,$$
$$x_j \geq 0, \; j = 1, \ldots, n. \quad (5\text{-}96)$$

The number of constraints involved is $2 \cdot k \cdot m$.

Apparently, this method increases the robustness of a spatial planning problem but also increases the problem dimension. Existence of fuzzy coefficients substantially increase the number of constraints. For a problem of reasonable size, it is not too demanding a task. For extremely large problem, such a transformation may become noneconomical or impractical. However, the enlargement of problem dimension is a price we have to pay for the enhancement of programming robustness in a fuzzy environment. If (5-77) is not considered as a fuzzy inclusive constraint, a more parsimony method can be employed.

Method 2. Consider the fuzzy constraint in (5-77). Let A_j, $j = 1, \ldots, n$, and B be fuzzy numbers of the L-R type (definition 2.25), with

$$A_j = (a_j, \underline{\underline{a}}_j, \bar{\bar{a}}_j) \text{ and } B = (b, \underline{\underline{b}}, \bar{\bar{b}}), \tag{5-97}$$

where a_j, $\underline{\underline{a}}_j$, and $\bar{\bar{a}}_j$ are respectively the mean value, the left and the right spreads of A_j, and b, $\underline{\underline{b}}$, and $\bar{\bar{b}}$ are respectively the mean value, the left and the right spreads of B.

Since $x_j \geq 0$, $j = 1, \ldots, n$, then by the property of multiplying a L-R fuzzy number by a real number, (2-100), the constraint in (5-77) is equivalent to:

$$(a_1x_1, \underline{\underline{a}}_1x_1, \bar{\bar{a}}_1x_1) + (a_2x_2, \underline{\underline{a}}_2x_2, \bar{\bar{a}}_2x_2) + \ldots + (a_nx_n, \underline{\underline{a}}_nx_n, \bar{\bar{a}}_nx_n) \lesssim (b, \underline{\underline{b}}, \bar{\bar{b}}). \tag{5-98}$$

By the properties of addition, (2-94), and ordering of fuzzy numbers in terms of $\leq$, (2-105), the constraint in (5-98) can be replaced by a system of 3 precise constraints:

$$\begin{aligned} a_1x_1 + a_2x_2 + \ldots + a_nx_n &\leq b \\ \underline{\underline{a}}_1x_1 + \underline{\underline{a}}_2x_2 + \ldots + \underline{\underline{a}}_nx_n &\geq \underline{\underline{b}} . \\ \bar{\bar{a}}_1x_1 + \bar{\bar{a}}_2x_2 + \ldots + \bar{\bar{a}}_nx_n &\leq \bar{\bar{b}} \end{aligned} \tag{5-99}$$

For a system of m fuzzy constraints taking the format of (5-77), we would have the following representation:

$$(a_{i1}x_1, \underline{\underline{a}}_{i1}x_1, \bar{\bar{a}}_{i1}x_1) + (a_{i2}x_2, \underline{\underline{a}}_{i2}x_2, \bar{\bar{a}}_{i2}x_2) + \ldots + (a_{in}x_n, \underline{\underline{a}}_{in}x_n, \bar{\bar{a}}_{in}x_n) \lesssim (b_i, \underline{\underline{b}}_i, \bar{\bar{b}}_i),$$
$$i = 1, \ldots, m, \tag{5-100}$$

In the similar manner, it is equivalent to the following system of 3m precise constraints

$$\begin{aligned} a_{i1}x_1 + a_{i2}x_2 + \ldots + a_{in}x_n &\leq b_i, \; i = 1, \ldots, m, \\ \underline{\underline{a}}_{i1}x_1 + \underline{\underline{a}}_{i2}x_2 + \ldots + \underline{\underline{a}}_{in}x_n &\geq \underline{\underline{b}}_i, \; i = 1, \ldots, m, \\ \bar{\bar{a}}_{i1}x_1 + \bar{\bar{a}}_{i2}x_2 + \ldots + \bar{\bar{a}}_{in}x_n &\leq \bar{\bar{b}}_i, \; i = 1, \ldots, m. \end{aligned} \tag{5-101}$$

Remark. Obviously, the decision space defined in this subsection is more robust. It is more realistic in dealing with indeterministic spatial planning problems in which knowledge on the unit consumptions and capacities of resources is imprecise. Such a formulation is especially instrumental in regional input-output analysis when the technological production coefficients are imprecisely known and we have to approximate them by fuzzy numbers (Leung, 1987b). By taking fuzzy intervals as their values, the defined feasible region permits us to consider all possible situations of our production capabilities.

5.5.2. Robust Optimization with Fuzzy-Coefficient Constraints

As discussed previously, spatial planning problems may involve the optimization of a precise objective function over an imprecise decision space. In subsection 5.2.2, fuzziness of the decision space is due to the imprecision of resource capacities. In this subsection, the fuzzy decision space is constituted by the fuzzy inclusive constraints examined in the preceding subsection.

Let the following be a spatial optimization problem:

$$\begin{aligned} \max \quad & c_1x_1 + c_2x_2 + \ldots + c_nx_n, \\ \text{s.t.} \quad & A_{i1}x_1 \oplus A_{i2}x_2 \oplus \ldots \oplus A_{in}x_n \lesssim B_i, \quad i = 1, \ldots, m, \\ & x_j \geq 0, \quad j = 1, \ldots, n, \end{aligned} \tag{5-102}$$

where the objective function is precise and the constraints are fuzzy with fuzzy coefficients and fuzzy capacities.

Based on the formulation of the constraint set in subsection 5.5.1, the above maximization problem can be solved by two methods.

Method 1. Based on method 1 in subsection 5.5.1 and Negoita, Minoia and Stan (1976), if (5-95) holds, the maximization problem in (5-102) is equivalent to

$$\max \quad c_1x_1 + c_2x_2 + \ldots + c_nx_n,$$

$$\text{s.t.} \quad (\underline{A}_{i1})_\alpha x_1 + (\underline{A}_{i2})_\alpha x_2 + \ldots + (\underline{A}_{in})_\alpha x_n \subseteq \underline{B}_{i\alpha}, \quad i = 1, \ldots, m,$$

$$x_j \geq 0, \quad j = 1, \ldots, n, \quad \alpha \in [0, 1]. \qquad (5\text{-}103)$$

Thus, (5-102) becomes a conventional linear programming problem:

$$\max \quad c_1x_1 + c_2x_2 + \ldots + c_nx_n,$$

$$\text{s.t.} \quad \bar{a}^h_{i1}x_1 + \bar{a}^h_{i2}x_2 + \ldots + \bar{a}^h_{in}x_n \leq \bar{b}^h_i, \quad i = 1, \ldots, m; \quad h = 1, \ldots, k,$$

$$\underline{a}^h_{i1}x_1 + \underline{a}^h_{i2}x_2 + \ldots + \underline{a}^h_{in}x_n \geq \underline{b}^h_i, \quad i = 1, \ldots, m; \quad h = 1, \ldots, k,$$

$$x_j \geq 0, \quad j = 1, \ldots, n. \qquad (5\text{-}104)$$

Example 5.5. Let the following be a robust spatial programming problem:

$$\max \; 3x_1 + 2x_2,$$

$$\text{s.t.} \quad \underset{\sim}{4}x_1 \oplus \underset{\sim}{5}x_2 \subseteq \underset{\sim}{22}$$

$$\underset{\sim}{4}x_1 \oplus \underset{\sim}{2}x_2 \subseteq \underset{\sim}{14.5} \qquad (5\text{-}105)$$

$$x_1 \geq 0, \; x_2 \geq 0.$$

The symbol "$\sim$" means "approximately". Thus, $\underset{\sim}{4}$, $\underset{\sim}{5}$, $\underset{\sim}{2}$, $\underset{\sim}{22}$, and $\underset{\sim}{14.5}$ are fuzzy subsets labelled "*approximately* 4", "*approximately* 5", "*approximately* 2", "*approximately* 22", and "*approximately* 14.5" respectively.

Let the fuzzy subsets be defined by membership functions having the following general format (of course, they can be defined by any asymmetric quasi-concave functions):

$$\mu_A(y) = \begin{cases} 0, & \text{if } y < \underline{a} \text{ and } y > \bar{a}, \\ 1 - \dfrac{|A - y|}{\frac{1}{2}(\bar{a} - \underline{a})}, & \text{if } \underline{a} \leq y \leq \bar{a}, \end{cases} \qquad (5\text{-}106)$$

where $[\underline{a}, \bar{a}]$ is the fuzzy interval imposed by fuzzy subset A.

For example, "*approximately* 4" is defined by

$$\mu_{\underset{\sim}{4}}(y) = \begin{cases} 0, & \text{if } y < 2 \text{ and } y > 6, \\ 1 - \dfrac{|4 - y|}{2}, & \text{if } 2 \le y \le 6. \end{cases} \quad (5\text{-}107)$$

Let the fuzzy subsets be evaluated at the following points $\alpha_1 = 0$, $\alpha_2 = 0.2$, $\alpha_3 = 0.5$, and $\alpha_4 = 0.8$. Then,

$$\underset{\sim}{4} = \{(0, 2), (0.2, 2.4), (0.5, 3), (0.8, 3.6), (0.8, 4.4), (0.5, 5), (0.2, 5.6), (0, 6)\},$$

$$\underset{\sim}{5} = \{(0, 4), (0.2, 4.2), (0.5, 4.5), (0.8, 4.8), (0.8, 5.8), (0.5, 5.5), (0.2, 5.8), (0, 6)\},$$

$$\underset{\sim}{2} = \{(0, 0), (0.2, 0.2), (0.5, 1), (0.8, 1.6), (0.8, 2.4), (0.5, 3), (0.2, 3.6), (0, 4)\},$$

$$\underset{\sim}{22} = \{(0, 20), (0.2, 20.4), (0.5, 21), (0.8, 21.6), (0.8, 22.4), (0.5, 23), (0.2, 23.6), (0, 24)\},$$

$$\underset{\sim}{14.5} = \{(0, 12), (0.2, 12.5), (0.5, 13.25), (0.8, 14), (0.8, 15), (0.5, 15.75), (0.2, 16.5), (0, 17)\}.$$

Based on (5-103) and (5-104), the equivalent problem for (5-105) becomes

$$\begin{aligned} \max\ & 3x_1 + 2x_2, \\ \text{s.t. } & 6x_1 + 6x_2 \le 24 \\ & 2x_1 + 4x_2 \ge 20 \\ & 5.6x_1 + 5.8x_2 \le 23.6 \\ & 2.4x_1 + 4.2x_2 \ge 20.4 \\ & 5x_1 + 5.5x_2 \le 23 \\ & 3x_1 + 4.5x_2 \ge 21 \\ & 4.4x_1 + 5.8x_2 \le 22.4 \\ & 3.6x_1 + 4.8x_2 \ge 21.6 \end{aligned}$$

$$6x_1 + 4x_2 \leq 17$$

$$2x_1 \geq 12$$

$$5.6x_1 + 3.6x_2 \leq 16.5$$

$$2.4x_1 + 0.2x_2 \geq 12.5$$

$$5x_1 + 3x_2 \leq 15.75$$

$$3x_1 + x_2 \geq 13.25$$

$$4.4x_1 + 2.4x_2 \leq 15$$

$$3.6x_1 + 1.6x_2 \geq 14$$

$$x_1 \geq 0,\ x_2 \geq 0. \qquad (5\text{-}108)$$

Thus, the robust programming problem with fuzzy inclusive constraints in (5-105) is transformed into a conventional linear programming problem consisting of 16 precise constraints. The solution of (5-108) is $(x_1^*, x_2^*) = (0.523, 3.465)$ with the value of the objective function being 8.5.

Method 2. Without further elaboration, if the A_{ij}'s and B_i's are fuzzy numbers of the L-R type, then by method 2 in subsection 5.5.1, the maximization problem in (5-102) is equivalent to

$$\begin{aligned}
\max\ & c_1x_1 + c_2x_2 + \dots + c_nx_n, \\
\text{s.t.}\ & a_{i1}x_1 + a_{i2}x_2 + \dots + a_{in}x_n \leq b_i,\ i = 1, \dots, m, \\
& \underline{\underline{a}}_{i1}x_1 + \underline{\underline{a}}_{i2}x_2 + \dots + \underline{\underline{a}}_{in}x_n \geq \underline{\underline{b}}_i,\ i = 1, \dots, m, \qquad (5\text{-}109) \\
& \bar{\bar{a}}_{i1}x_1 + \bar{\bar{a}}_{i2}x_2 + \dots + \bar{\bar{a}}_{in}x_n \leq \bar{\bar{b}}_i,\ i = 1, \dots, m, \\
& x_j \geq 0,\ j = 1, \dots, n.
\end{aligned}$$

Example 5.6. Let the following be a fuzzy linear programming problem:

$$\begin{aligned} &\max\ 3x_1 + 5x_2, \\ &\text{s.t. } \underset{\sim}{1}x_1 \lesssim \underset{\sim}{4} \\ &\quad \underset{\sim}{3}x_1 \oplus \underset{\sim}{2}x_2 \lesssim \underset{\sim}{18} \\ &\quad x_1 \geq 0,\ x_2 \geq 0. \end{aligned} \tag{5-110}$$

Let $\underset{\sim}{1}$, $\underset{\sim}{2}$, $\underset{\sim}{3}$, $\underset{\sim}{4}$, and $\underset{\sim}{18}$ be L-R fuzzy numbers such that $\underset{\sim}{1}$ = (1, 1, 0), $\underset{\sim}{2}$ = (2, 0, 1), $\underset{\sim}{3}$ = (3, 1, 1), $\underset{\sim}{4}$ = (4, 1, 2), $\underset{\sim}{18}$ = (18, 2, 7).

Based on (5-109), the solution $(x_1^*, x_2^*) = (2, 5)$ is obtained by solving

$$\begin{aligned} &\max\ 3x_1 + 5x_2, \\ &\text{s.t. } x_1 \leq 4 \\ &\quad x_1 \geq 1 \\ &\quad 3x_1 + 2x_2 \leq 18 \\ &\quad x_1 \geq 2 \\ &\quad x_1 + x_2 \leq 7 \\ &\quad x_1 \geq 0,\ x_2 \geq 0. \end{aligned} \tag{5-111}$$

Notice that the constraint $x_1 \geq 1$ is redundant in (5-111).

In some spatial planning problems, coefficients in the objective function and constraint may both be fuzzy. If, for example, the objective function represents total profit, unit profit of a product may not be precisely known. It may take on fuzzy numbers as its value. Under this situation, our optimization problems can be formulated as:

$$\begin{aligned} &\underset{\sim}{\max}\ C_1x_1 \oplus C_2x_2 \oplus \ldots \oplus C_nx_n, \\ &\text{s.t. } A_{i1}x_1 \oplus A_{i2}x_2 \oplus \ldots \oplus A_{in}x_n \lesssim B_i,\ i = 1, \ldots, m, \\ &\quad x_j \geq 0,\ j = 1, \ldots, n, \end{aligned} \tag{5-112}$$

Let C_j, j = 1, ..., n, be a L-R fuzzy number such that C_j

$= (c_j, \underline{\underline{c}}_j, \bar{\bar{c}}_j)$. Let A_{ij} and B_i, $i = 1, \ldots, m$, $j = 1, \ldots, n$, be defined as before. Then, by the approximation of $\underset{\sim}{\max}$, (5-113) is approximately equivalent to the following 3-objective linear programming problem:

$$
\begin{aligned}
&\max\ c_1x_1 + c_2x_2 + \ldots + c_nx_n, \\
&\min\ \underline{\underline{c}}_1x_1 + \underline{\underline{c}}_2x_2 + \ldots + \underline{\underline{c}}_nx_n, \\
&\max\ \bar{\bar{c}}_1x_1 + \bar{\bar{c}}_2x_2 + \ldots + \bar{\bar{c}}_nx_n, \\
&\text{s.t. } a_{i1}x_1 + a_{i2}x_2 + \ldots + a_{in}x_n \le b_i,\ i = 1, \ldots, m, \qquad (5\text{-}113) \\
&\quad \underline{\underline{a}}_{i1}x_1 + \underline{\underline{a}}_{i2}x_2 + \ldots + \underline{\underline{a}}_{in}x_n \ge \underline{\underline{b}}_i,\ i = 1, \ldots, m, \\
&\quad \bar{\bar{a}}_{i1}x_1 + \bar{\bar{a}}_{i2}x_2 + \ldots + \bar{\bar{a}}_{in}x_n \le \bar{\bar{b}}_i,\ i = 1, \ldots, m, \\
&\quad x_j \ge 0,\ j = 1, \ldots, n.
\end{aligned}
$$

The solution to (5-113) can be obtained by applying the method to be discussed in section 6.2.

CHAPTER 6

FUZZY MATHEMATICAL PROGRAMMING AND MULTIOBJECTIVE SPATIAL PLANNING PROBLEMS

6.1. A REMARK ON MULTIOBJECTIVE SPATIAL PLANNING PROBLEMS

Within a complex spatial system, planning problems often involve more than one objective. In determining the optimal levels of production for diverse economic activities, objectives such as the maximization of profits, the minization of total costs, the maximization of long term growth, the maximization of employment opportunities, and the minization of environmental pollution may need to be achieved simultaneously. In general, objectives are conflicting and simultaneous optimals of individual objectives are generally impossible to achieve. To resolve conflicts, decisionmakers need to search for a compromise.

As argued in chapter 5, information within a multiobjective optimization problem may be imprecise and objectives and constraints may not be precisely specified. In general, conflict resolution is carried out in a more or less fuzzy environment. Contrary to common belief, imprecision is in fact not a deterrent but an instrumental mechanism in resolving conflicts among incommensurable objectives and constraints. It makes room for compromise.

The purpose of this chapter is to examine multiobjective spatial optimization problems under varying contexts of imprecision. In section 6.2, spatial planning problems with precise objectives and precise constraints are analyzed within a multiobjective fuzzy linear programming framework. Multiobjective interregional equilibrium is also scrutinized.

In section 6.3, three major types of multiobjective fuzzy optimization problems are discussed. Problems with fuzzy objectives and precise constraints are first analyzed. Problems with fuzzy objectives and fuzzy constraints are then discussed. Lastly, planning with precise objectives and fuzzy constraints is examined.

In section 6.4, solution of fuzzy multiobjective planning problems is discussed from the dynamic point of view. Iterative and interactive procedures as well as the ideal displacement method for conflict resolution are analyzed.

6.2. MULTIOBJECTIVE SPATIAL PLANNING PROBLEMS UNDER CERTAINTY

Conventional multiobjective spatial planning problems consist of a precisely delimited decision space over which a set of precisely specified objectives are optimized. The vector maximum methods (see for example Charnes and Cooper, 1961; Yu and Zeleny, 1975; and Ecker and Kouada, 1975), the goal programming methods (see for example Charnes and Cooper, 1961; Lee, 1972; and Ignizio, 1976), and the interactive methods (see for example Geoffrion, Dyer, and Feinberg, 1972; Dyer, 1972; and Zionts and Wallenius, 1976) have been developed to solve general programming problems with multiple objectives. All these methods require precise information and precisely stated objectives and constraints. They have been applied to solve a variety of spatial planning problems (see for example Delft and Nijkamp, 1979; and Nijkamp, 1977). I demonstrate in this section that by fuzzifying the precise objectives, planning problems with precise objectives and constraints can be efficiently solved via a multiobjective fuzzy linear programming framework. The implication then is that conventional multiobjective programming problems can be solved within the fuzzy optimization framework.

In subsection 6.2.1, a general fuzzy set formulation of programming problems with precise objectives and constraints is outlined. In subsection 6.2.2, a formal treatment of interregional equilibrium problem with precise objectives and constraints is presented.

6.2.1. Precise-Objectives-Precise-Constraints Spatial Planning Problems

Let the following be an optimization problem with q precisely stated objectives:

$$\max F(x) = [f_1(x), \ldots, f_h(x), \ldots, f_q(x)],$$
$$\text{s.t. } g_i(x) \leq b_i, \ i = 1, \ldots, m, \tag{6-1}$$
$$x \geq \theta,$$

where $x \in \underline{X} \subseteq \mathbb{R}^n$, and θ is the zero vector in $\mathbb{R}^n$.

Assume that the individual objectives cannot be simultaneously maximized. That is, some or all of the objectives are conflicting. The resolution of conflict depends then on whether or not a satisfactory compromise can be attained. If the technological constraints are fixed, a compromise may be attained if a certain degree of deviation from the optimal of each objective can be tolerated. In other words, room for compromising needs to be created. A simple way is to specify tolerance intervals for each objective. The idea is to reach a compromise with minimal tolerance. Since tolerance intervals are involved, the logic employed in chapter 5 can then be applied to the multiobjective situation.

Let $x_h^* \in \mathbb{R}^n$, $h = 1, \ldots, q$, be the optimal solution for the following single-objective optimization problem:

$$\max f_h(x),$$
$$\text{s.t. } q_i(x) \leq b_i, \ i = 1, \ldots, m, \tag{6-2}$$
$$x \geq \theta.$$

Let $\bar{f}_h = f_h(x_h^*)$ be the optimal value of objective h when it is individually optimized in (6-2).

Based on (6-1), $\bar{f}_h$ may most likely not be attained when other objective functions are simultaneously maximized over the same decision space. To reach a compromise, it is then necessary to determine the worst possible value for objective h when the maximization of all other objectives are considered.

Let $\{x_h^*\}$, h = 1, ..., q, be the set of individual optimal solutions obtained by solving (6-2) q times. Then

$$\underline{f}_h = \min[f_h(x_1^*), \ldots, f_h(x_h^*), \ldots, f_h(x_q^*)], \; h = 1, \ldots, q, \tag{6-3}$$

can be defined as the worst possible value for objective h. That is, in the search for compromise, it is the worst permissible value to which $\bar{f}_h$ can be lowered.

Based on the tolerance interval $[\underline{f}_h, \bar{f}_h]$, a satisfaction function can be specified for objective h as follows:

$$\mu_h(x) \triangleq \mu_h(f_h(x)) = \begin{cases} 1, & \text{if } f_h(x) > \bar{f}_h, \\ 1 - \dfrac{\bar{f}_h - f_h(x)}{d_h}, & \text{if } \underline{f}_h < f_h(x) \le \bar{f}_h, \\ 0, & \text{if } f_h(x) \le \underline{f}_h, \end{cases} \tag{6-4}$$

where $d_h = (\bar{f}_h - \underline{f}_h)$ is the length of the tolerance interval.

Employing the min-operator, the satisfaction function for all objectives can be defined by

$$\min_h[\mu_h(x)]. \tag{6-5}$$

The multiobjective optimization problem in (6-1) can then be solved by determining an alternative x* which maximizes the overall satisfaction function in (6-5) as follows (Zimmermann, 1978; Leung, 1985e):

$$\begin{aligned} &\max \min_h[\mu_h(x)], \\ &\text{s.t. } g_i(x) \le b_i, \; i = 1, \ldots, m, \\ &\quad x \ge \theta. \end{aligned} \tag{6-6}$$

Equivalently, the compromise solution x* is obtained by solving

$$\begin{aligned} &\max \lambda, \\ &\text{s.t } \mu_h(x) \ge \lambda, \; h = 1, \ldots, q, \\ &\quad g_i(x) \le b_i, \; i = 1, \ldots, m, \\ &\quad \lambda \ge 0, \; x \ge \theta, \end{aligned} \tag{6-7}$$

or,

$$\begin{aligned} &\max \lambda, \\ &\text{s.t. } f_h(x) - \lambda d_h \geq \underline{f}_h, \; h = 1, \ldots, q, \\ &\quad g_i(x) \leq b_i, \; i = 1, \ldots, m, \\ &\quad \lambda \geq 0, \; x \geq \theta, \end{aligned} \tag{6-8}$$

where λ is the degree of overall objective-satisfaction which corresponds essentially to (6-5).

Therefore, the multiobjective optimization problem with precise objectives and constraints in (6-1) is transformed into a single-objective fuzzy optimization problem with precise constraints in (6-7) and can be solved by conventional mathematical programming algorithm.

Example 6.1. Let the following be a simple two-objective optimization problem with precise constraints:

$$\begin{aligned} \max F(x) &= [f_1(x), f_2(x)] \\ &= [2x_1 + x_2, -3x_1 + 2x_2], \\ \text{s.t. } x_1 \quad &\leq 5 \\ x_2 &\leq 6 \\ x_1 + x_2 &\leq 9 \\ x_1 \geq 0, \; x_2 &\geq 0. \end{aligned} \tag{6-9}$$

Based on (6-2), the individual maximization of f_1 and f_2 gives $\bar{f}_1 = 14$ and $\bar{f}_2 = 12$. Based on (6-3), the worst possible values for objectives f_1 and f_2 are respectively $\underline{f}_1 = 6$ and $\underline{f}_2 = -7$. Therefore, the tolerance intervals for the optimal values of f_1 and f_2 are respectively $[\underline{f}_1, \bar{f}_1] = [6, 14]$ and $[\underline{f}_2, \bar{f}_2] = [-7, 12]$ with $d_1 = 8$ and $d_2 = 19$.

Applying the arguments in (6-5) to (6-7), the compromise solution $(\lambda^*, x_1^*, x_2^*) = (0.628, 2.51, 6)$ is obtained by solving

$$\begin{aligned}
&\max\ \lambda,\\
&\text{s.t. } 0.25\,x_1 + 0.125\,x_2 - 0.75 \geq \lambda\\
&\quad -0.16\,x_1 + 0.11\,x_2 + 0.37 \geq \lambda\\
&\quad x_1 \leq 5\\
&\quad x_2 \leq 6\\
&\quad x_1 + x_2 \leq 9\\
&\quad \lambda \geq 0,\ x_1 \geq 0,\ x_2 \geq 0.
\end{aligned} \tag{6-10}$$

Though the two objectives cannot be fully satisfied, the compromise solution gives an overall degree of satisfaction $\lambda^* = 0.628$. (See Fig. 6.1).

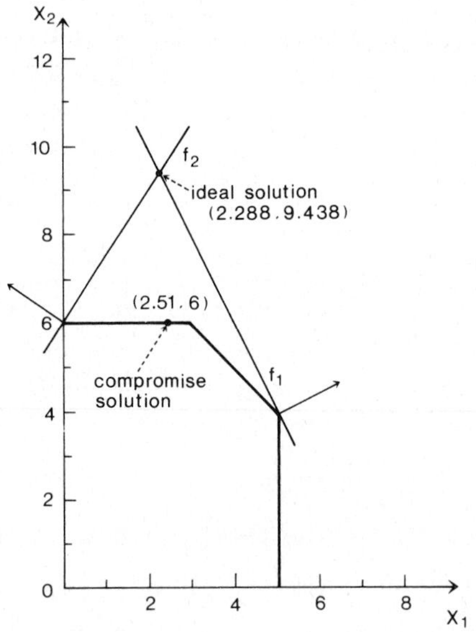

Fig. 6.1 The two-objective optimization problem in (6-9) in the decision space

Example 6. 2. (A simple application to a regional development problem)

In a simplified regional planning problem for the development of a new industrial area, Meuse-flat, in the Netherlands, Nijkamp (1977, pp. 200-204) applies several conventional multi-objective programming methods to obtain an optimal allocation of land to seven industrial activities. I demonstrate here that the problem can also be solved within the fuzzy programming framework.

In the study, two conflicting objectives are employed as the optimization criteria. They are respectively, f_1: maximization of additional employment, and f_2: minimization of additional pollution. A total area of 900 hectares is earmarked for new industrial uses, and the maximum and minimum land requirement for each industrial uses are imposed as constraints (see Table 6.1).

Table 6.1 Data of the Netherlands regional planning problem

Industrial Activities	h_j	k_j	$\underline{b}_j$	$\overline{b}_j$
1. Integrated steel plant (x_1)	24.0	82.57	250	500
2. Tank storage plant (x_2)	1.0	0.00	50	100
3. Ores transshipment and processing plang (x_3)	2.0	48.23	100	200
4. Container terminal (x_4)	1.2	0.00	50	100
5. Oil refinery (x_5)	6.0	510.11	100	200
6. Petro-chemical plant (x_6)	20.0	42.89	50	100
7. Tanker-cleaning and ship-repair yard (x_7)	8.0	0.00	20	40

The two-objective optimization problem is formulated as

$$\max f_1(x) = \sum_{j=1}^{7} h_j x_j,$$

$$\min f_2(x) = \sum_{j=1}^{7} k_j x_j,$$

$$\text{s.t.} \sum_{j=1}^{7} x_j \le 900 \qquad (6\text{-}11)$$

$$\underline{b}_j \le x_j \le \overline{b}_j, \ j = 1, \ldots, 7,$$

$$x_j \ge 0, \ j = 1, \ldots, 7,$$

where

x_j = hectares to be allocated to industry j, and $x = \{x_j\}$, $j = 1, \ldots, 7$,

h_j = employment generated per hectare of industry j (measured in man-year),

k_j = pollutant emitted per hectare of industry j (coal monoxide and sulfur dioxide measured in tons per hectare per year),

$\underline{b}_j$ = minimum land requirement of industry j,

$\overline{b}_j$ = maximum land requirement of industry j.

Maximizing f_1 and minimizing f_2 separately, the tolerance intervals for f_1 and f_2 are respectively obtained as $[\underline{f}_1, \overline{f}_1]$ = [8,070; 14,670] and $[\underline{f}_2, \overline{f}_2]$ = [78,621, 100,550]. If the satisfaction function of f_1 is defined by (6-4) and that of f_2 is defined by the one similar in format to (5-39), then applying (6-7) we have the following fuzzy optimization problem:

$$\max \lambda,$$

$$\text{s.t. } -1.223 + 0.004x_1 + 0.0002x_2 + 0.0003x_3 + 0.0002x_4 + 0.001x_5 + 0.003x_6 + 0.001x_7 \ge \lambda$$

$$4.585 - 0.004x_1 - 0.002x_3 - 0.023x_5 - 0.002x_6 \ge \lambda$$

$$x_1 + x_2 + x_3 + x_4 + x_5 + x_6 + x_7 \le 900$$

$$250 \le x_1 \le 500$$

$$50 \leq x_2 \leq 100$$

$$100 \leq x_3 \leq 200$$

$$50 \leq x_4 \leq 100$$

$$100 \leq x_5 \leq 200$$

$$50 \leq x_6 \leq 100$$

$$20 \leq x_7 \leq 40$$

$$\lambda \geq 0,\ x_j \geq 0,\ j = 1, \ldots, 7. \tag{6-12}$$

The compromise solution is $(\lambda^*, x_1^*, x_2^*, x_3^*, x_4^*, x_5^*, x_6^*, x_7^*) = (0.586, 324.75, 100, 100, 100, 100, 100, 40)$ with $(f_1^*, f_2^*) = (11{,}134, 86{,}914.875)$. Thus, the individual optimals $(\overline{f}_1, \overline{f}_2) = (14{,}670, 78{,}621)$ cannot be technologically achieved but the compromise solution satisfies the two objective functions at 0.586 degree of satisfaction.

Example 6.3. (A formulation of facility location problems)

Facility location problems in real life are often multiobjective in structure. In addition to monetary objectives, there are intangible criteria on which a location decision is based. In locating a production plant, for example, objectives such as minimization of cost, minimization of risk, maximization of accessibility, minimization of negative externalities (such as emission of pollutants), and maximization of growth may need to be considered. Generally, these objectives are mutually conflicting. A higher achievement of one may result in a lower achievement of the others. To reconcile the differences, a compromise solution is sought so that the optimal location can accomplish various objectives to a certain degree. In this example, a fuzzy optimization approach is employed to solve a general class of multiobjective facility location problems.

Suppose we need to decide on the locations of a certain number of production plants among m possible sites. Assume that all plants produce homogeneous products. Let there be n demand points for the products with a volume of demand d_j, j = 1, ..., n. Let the volume of supply at plant i be s_i, i = 1, ..., m.

Assume that there are q objectives taking the following general format:

$$z_r(x, y) = \sum_{i=1} (\sum_{j=1} c^r_{ij} x_{ij} + f^r_i y_i), \quad r = 1, \ldots, q, \tag{6-13}$$

where

c_{ij} = unit amount (can be cost, profit, or pollutants) of producing a unit of product at plant i and shipping it to demand point j;

f_i = a fixed amount (can be fixed cost, fixed in-plant profit, or fixed amount of inplant emitted pollutants);

x_{ij} = volume of production (integers);

$$y_i = \begin{cases} 1, & \text{if } x_{ij} > 0, \ i = 1, \ldots, m, \\ 0, & \text{if } x_{ij} = 0, \ i = 1, \ldots, m, \end{cases}$$

y_i can be interpreted as plant i being constructed, albeit a redundant information solutionwise;

$x = (x_{11}, \ldots, x_{ij}, \ldots, x_{mn})$ and $y = (y_1, \ldots, y_i, \ldots, y_m)$.

Without loss of generality, let all of the objective functions be minimized. Then, the multiobjective facility location problem can be formulated as:

$$\min Z_1(x, y) = \sum_i (\sum_j c^1_{ij} x_{ij} + f^1_i y_i),$$
$$\vdots$$
$$\min Z_q(x, y) = \sum_i (\sum_j c^q_{ij} x_{ij} + f^1_i y_i),$$

$$\text{s.t.} \sum_j x_{ij} \leq s_i, \quad i = 1, \ldots, m, \tag{6-14}$$

$$\sum_i x_{ij} \geq d_j, \quad j = 1, \ldots, n,$$

$$x_{ij} \geq 0 \text{ and integers}, \quad i = 1, \ldots, m; \ j = 1, \ldots, n,$$

$$y_i = (0, 1), \quad i = 1, \ldots, m.$$

The problem in (6-14) is now a multiobjective mixed-integer programming problem. To determine the solutions in a more efficient way, we can transform (6-14) into a zero-one integer programming problem.

Since for any integer x_{ij}, if $0 \leq x_{ij} \leq 2^p$, then x_{ij} can be uniquely represented by

$$x_{ij} = \sum_h 2^h x_{ijh}, \tag{6-15}$$

where $x_{ijh} \varepsilon \{0, 1\}$, and $h = 0, 1, 2, \ldots, p$.

Substituting x_{ij} in (6-14) by (6-15), the facility location problem in (6-14) becomes

$$\min Z_1(x_h, y) = \sum_i [\sum_j c_{ij}^1 (\sum_h 2^h x_{ijh}) + f_i^1 y_i],$$

$$\vdots$$

$$\min Z_q(x_h, y) = \sum_i [\sum_j c_{ij}^q (\sum_h 2^h x_{ijh}) + f_i^k y_i],$$

$$\text{s.t.} \sum_j \sum_h 2^h x_{ijh} \leq s_i, \ i = 1, \ldots, m, \tag{6-16}$$

$$\sum_i \sum_h 2^h x_{ijh} \geq d_j, \ j = 1, \ldots, n,$$

$$x_{ijh} = (0, 1), \ i = 1, \ldots, m; \ j = 1, \ldots, n; \ h = 1, \ldots, p,$$

$$y_i = (0, 1), \ i = 1, \ldots, m.$$

To solve (6-16), we first solve k single-objective zero-one integer programming problem via, for example, the generalized network algorithm. Let $\underline{Z}_r$, $r = 1, \ldots, q$, be the optimal value when $Z_r(x_h, y)$ is minimized over the constraint set in (6-16). Let $\overline{Z}_r$ be the maximal permissible value obtained by:

$$\overline{Z}_r = \max[Z_r(x_h^{1*}, y^{1*}), \ldots, Z_r(x_h^{q*}, y^{q*})], \tag{6-17}$$

where (x_h^{r*}, y^{r*}), $r = 1, \ldots, q$, is the optimal solution for the r-th single-objective problem.

Thus, $[\underline{Z}_r, \overline{Z}_r]$ is the tolerance interval for objective Z_r.

Let

$$\mu_r(Z_r(x_h, y)) = \begin{cases} 1, & \text{if } Z_r(x_h, y) < \underline{Z}_r, \\ 1 - \dfrac{Z_r(x_h, y) - \underline{Z}_r}{\overline{Z}_r - \underline{Z}_r}, & \text{if } \underline{Z}_r \leq Z_r(x_h, y) < \overline{Z}_r, \\ 0, & \text{if } Z_r(x_h, y) \geq \overline{Z}_r, \end{cases} \tag{6-18}$$

be the satisfaction function for objective Z_r, $r = 1, \ldots, q$.

Then, the compromise solution of problem (6-14) can be obtained by solving the following single-objective problem:

$$\begin{aligned} & \max \lambda, \\ & \text{s.t. } \frac{\overline{Z}_r - Z_r(x_h, y)}{\overline{Z}_r - \underline{Z}_r} \geq \lambda, \; r = 1, \ldots, q, \\ & \sum_j \sum_h 2^h x_{ijh} \leq s_i, \; i = 1, \ldots, m, \\ & \sum_i \sum_h 2^h x_{ijh} \geq d_j, \; j = 1, \ldots, n, \\ & x_{ijh} = (0, 1), \; i = 1, \ldots, m; \; j = 1, \ldots, n; \\ & \qquad h = 1, \ldots, p, \\ & y_i = (0, 1), \; i = 1, \ldots, m. \end{aligned} \tag{6-19}$$

Since $\lambda \in [0, 1]$ is a real number, then (6-19) is a mixed integer programming problem. To be able to apply the generalized network algorithm, we need to transform (6-19) into a zero-one integer programming problem. Since can be approximated by:

$$\lambda = \sum_{\ell=1} 2^{-\ell} \lambda_\ell, \tag{6-20}$$

where $\lambda_\ell \in \{0, 1\}$, and $\ell = 1, 2, \ldots, q$, then (6-19) can be transformed into:

$$\max \sum_{\ell} 2^{-\ell}\lambda_{\ell},$$

$$\text{s.t. } \left\{\underline{Z}_r - \sum_i \left[\sum_j c^r_{ij}\left(\sum_h 2^h x_{ijh}\right) + f^r_i y_i\right]\right\} /$$

$$(\overline{Z}_r - \underline{Z}_r) \geq \sum_{\ell} 2^{-\ell}\lambda_{\ell},\ r = 1, \ldots, q,$$

$$\sum_j \sum_h 2^h x_{ijh} \leq s_i,\ i = 1, \ldots, m, \tag{6-21}$$

$$\sum_i \sum_h 2^h x_{ijh} \geq d_j,\ j = 1, \ldots, n,$$

$$x_{ijh} = (0, 1),\ i = 1, \ldots, m;\ j = 1, \ldots, n;\ h = 1, \ldots, p,$$

$$y_i = (0, 1),\ i = 1, \ldots, m,$$

$$\lambda_{\ell} = (0, 1),\ \ell = 1, \ldots, q.$$

Though problem (6-21) is larger in dimension than problem (6-19), it can be solved efficiently by the generalized network algorithm.

Therefore, multiobjective facility location problems can in general be expressed as a zero-one integer programming problem. Efficient algorithm is also available to solve such a large scale integer program.

Apparently, multiobjective problems with precise objectives and precise constraints can be solved within a fuzzy optimization framework. A formalization of the multiobjective interregional equilibrium with precise linear objectives and constraints is given in the following subsection.

6.2.2. A Formalization of Multiobjective Interregional Equilibrium under Certainty

In subsection 5.4.2, single-objective interregional equilibrium in a fuzzy environment is formalized within a fuzzy linear programming framework. Major theoretical results of fuzzy linear programming in the spatial context are also examined. In this subsection, multiobjective interregional equilibrium is to be formalized within a multiobjective fuzzy linear programming framework. The decisionmaking environment, however, is precise. Our discussion is mainly based on Leung (1985e).

Without loss of generality, let the multiobjective interregional equilibrium problem be formulated in general terms as follows:

$$\max F(x) = [f_1(x), \ldots, f_h(x), \ldots, f_q(x)],$$

$$\text{s.t. } \sum_j a_{ij}^k x_j^k \leq b_i^k, \; i = 1, \ldots, m; \; k = 1, \ldots, \ell, \qquad (6\text{-}22)$$

$$x_j^k \geq 0, \; j = 1, \ldots, n; \; k = 1, \ldots, \ell,$$

where

$x = (x_1^1, \ldots, x_j^k, \ldots, x_n^\ell) \in \mathbb{R}^{n\ell}$ is a spatial alternative,

$f_h(x) = \sum_k \sum_j c_{hj}^k x_j^k$, $h = 1, \ldots, q$, is an objective function,

n = number of activities,
ℓ = number of regions,
m = number of resources,
c_{hj}^k = unit return of producing activity j in region k under objective h,
a_{ij}^k = unit consumption of resource i by activity j in region k,
b_i^k = volume of supply of resource i in region k,
x_j^k = production level of activity j in region k.

(Henceforth, activity j in region k is referred to as activity j, k and resource i in region k is referred to as resource i, k),

Based on subsection 6.2.1, interregional equilibrium is to be achieved by maximizing all individual objectives as much as possible. In general, not all objectives can simultaneously achieve their individual optimal values. Interregional equilibrium is usually achieved by searching for an efficient solution (Pareto-optimal) for (6-22).

Definition 6.1. Let $\underline{X}$ be the decision space bounded by the constraints in (6-22), a spatial alternative $x^* \in \underline{X}$ is efficient if there does not exist an alternative $x^{**} \in \underline{X}$ such that $f_h(x^{**}) \geq f_h(x^*)$ and $f_h(x^{**}) > f_h(x^*)$, $h = 1, \ldots, q$, for at least one h.

To obtain an efficient solution for (6-22), a satisfaction function for all objectives needs to be determined first.

Let $x_h^* = (x_{h1}^{1*}, \ldots, x_{hj}^{k*}, \ldots, x_{hn}^{\ell*})$, $h = 1, \ldots, q$, be the optimal solution obtained by solving the single-objective interregional equilibrium problem:

$$\max f_h(x) = \sum_k \sum_j c_{hj}^k \, x_j^k,$$

$$\text{s.t.} \sum_j a_{ij}^k \, x_j^k \leq b_i^k, \quad i = 1, \ldots, m; \; k = 1, \ldots, \ell, \qquad (6\text{-}23)$$

$$x_j^k \geq 0, \quad j = 1, \ldots, n; \; k = 1, \ldots, \ell.$$

Let $\bar{f}_h$ be the optimal value of objective h in (6-23), and $\underline{f}_h$, by (6-3), be the worst possible value of objective h. Let the satisfaction function for objective h be defined by

$$\mu_h(x) = \begin{cases} 1, & \text{if } \sum_k \sum_j c_{hj}^k \, x_j^k > \bar{f}_h, \\ 1 - \dfrac{\bar{f}_h - \sum_k \sum_j c_{hj}^k \, x_j^k}{d_h}, & \text{if } \underline{f}_h < \sum_k \sum_j c_{hj}^k \, x_j^k \leq \bar{f}_h, \\ 0, & \text{if } \sum_k \sum_j c_{hj}^k \, x_j^k \leq \underline{f}_h. \end{cases} \qquad (6\text{-}24)$$

Considering all objectives, the aggregated satisfaction function becomes

$$\min_h \left[\frac{\sum_k \sum_j c_{hj}^k x_j^k - \underline{f}_h}{d_h} \right]. \tag{6-25}$$

The equilibrium solution $(\lambda^*, x^*) = (\lambda^*, (x_1^{1*}, \ldots, x_j^{k*}, \ldots, x_n^{\ell*}))$ of the multiobjective interregional equilibrium problem in (6-22) is obtained by solving

$$\max \min_h \left[\frac{\sum_k \sum_j c_{hj}^k x_j^k - \underline{f}_h}{d_h} \right],$$

$$\text{s.t. } \sum_j a_{ij}^k x_j^k \leq b_i^k, \; i = 1, \ldots, m; \; k = 1, \ldots, \ell, \tag{6-26}$$

$$x_j^k \geq 0, \; j = 1, \ldots, n; \; k = 1, \ldots, \ell,$$

or equivalently,

$$\max \lambda,$$

$$\text{s.t. } \frac{\sum_k \sum_j c_{hj}^k x_j^k - \underline{f}_h}{d_h} \geq \lambda, \; h = 1, \ldots, q,$$

$$\sum_j a_{ij}^k x_j^k \leq b_i^k, \; i = 1, \ldots, m; \; k = 1, \ldots, \ell, \tag{6-27}$$

$$\lambda \geq 0, \; x_j^k \geq 0, \; j = 1, \ldots, n; \; k = 1, \ldots, \ell,$$

where λ is the degree of overall objective-satisfaction which corresponds essentially to (6-25).

Theorem 6.1. If a spatial alternative $(\lambda^*, x^*) = (\lambda^*, (x_1^{1*}, \ldots, x_j^{k*}, \ldots, x_n^{\ell*}))$ is an optimal solution of (6-27), then x^* is an efficient solution of (6-22), and $f_h(x^*) \in [\underline{f}_h, \overline{f}_h]$, $h = 1, \ldots, q$.

Theorem 6.1 says that the multiobjective interregional equilibrium is in fact an efficient solution which best satisfies, with λ^* degree of satisfaction, all objectives under the given technological constraints. Therefore, it is a compromise solution.

The basic idea of the above fuzzy linear programming approach is that individual objectives are to be maximized to their fullest extent within tolerable intervals of compromising. Such a tactic is in fact similar to finding a compromise solution with respect to the ideal solution of (6-22).

Let $(\bar{f}_1, \ldots, \bar{f}_h, \ldots, \bar{f}_q)$ be the ideal solution of (6-22) with $(\mu_1(\bar{f}_1), \ldots, \mu_2(\bar{f}_2), \ldots, \mu_q(\bar{f}_q)) = (1, \ldots, 1, \ldots, 1)$. That is, the ideal solution is the one which maximizes simultaneously all objective functions. Then, the maximal distance between the ideal alternative and any alternative x is

$$d_\infty = \max_h \left\{ 1 - \frac{\sum_k \sum_j c_{hj}^k \, x_j^k - \underline{f}_h}{d_h} \right\}. \tag{6-28}$$

Thus, a compromise solution can be obtained through the minimization of the maximal distance d_∞. That is, we need to solve

$$\begin{aligned} &\min d_\infty, \\ &\text{s.t. } \frac{\bar{f}_h - \sum_k \sum_j c_{hj}^k \, x_j^k}{d_h}, \le d_\infty, \quad h = 1, \ldots, q, \\ &\sum_k \sum_j a_{ij}^k \, x_j^k \le b_i^k, \quad i = 1, \ldots, m; \; k = 1, \ldots, \ell, \\ &x_j^k \ge 0, \quad j = 1, \ldots, n; \; k = 1, \ldots, \ell. \end{aligned} \tag{6-29}$$

The physical meaning of the interregional equilibrium problem in (6-29) is that to reach a compromise we search for an efficient solution which minimizes the maximal distance between a spatial alternative and the ideal alternative.

Theorem 6. 2. If (λ^*, x^*) is a solution of (6-27), then x^* is also a solution of (6-29).

Example 6.4. Consider the two-objective optimization problem in (6-9). Based on (6-29), the efficient solution $(d_\infty^*, x_1^*, x_2^*) = (0.372, 2.51, 6)$ is obtained by solving

$$
\begin{aligned}
&\min d_\infty, \\
&\text{s.t. } 1.75 - 0.25x_1 - 0.125x_2 \le d_\infty \\
&\qquad 0.63 + 0.16x_1 - 0.11x_2 \le d_\infty \\
&\qquad x_1 \le 5 \\
&\qquad x_2 \le 6 \\
&\qquad x_1 + x_2 \le 9 \\
&\qquad x_1 \ge 0, \ x_2 \ge 0.
\end{aligned}
\tag{6-30}
$$

Observe that $(d_\infty^*, x_1^*, x_2^*) = (1 - \lambda^*, x_1^*, x_2^*) = (0.372, 2.51, 6)$.
(See Fig. 6.1 and 6.2).

Parallel to the single-objective interregional equilibrium problem analyzed in subsection 5.4.2, associated with every multiobjective interregional equilibrium problem there is a dual optimal problem rendering important spatial economic interpretations. Taking the problem in (6-27) as the primal problem, the corresponding dual problem becomes

$$
\begin{aligned}
&\min \sum_h f_h y_h + \sum_k \sum_i b_i^k y_i^k, \\
&\text{s.t. } 1 + \sum_h d_h y_h \ge 0 \\
&\qquad \sum_h c_{hj}^k y_h + \sum_k \sum_i a_{ij}^k y_i^k \ge 0, \ i = 1, \ldots, n; \\
&\qquad\qquad k = 1, \ldots, \ell, \\
&\qquad y_h \ge 0, \ h = 1, \ldots, q, \\
&\qquad y_i^k \ge 0, \ i = 1, \ldots, m; \ k = 1, \ldots, \ell.
\end{aligned}
\tag{6-31}
$$

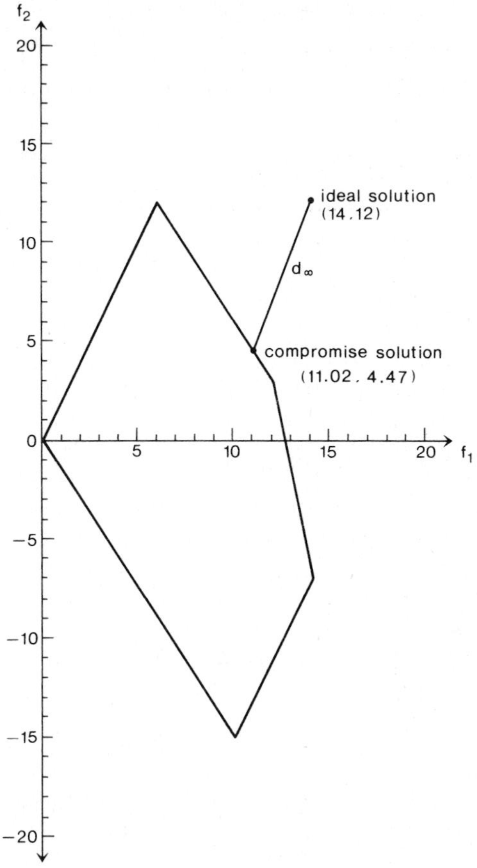

Fig. 6.2 The two-objective optimization problem in (6-9) via (6-30) in the objective space

The dual variable y_h denotes the relative change of the degree of overall satisfaction due to the marginal variation of the value of objective h. It sheds some light on the way a compromise should be made by varying the value of an objective function within its tolerance interval. In brief, it can be defined as the marginal satisfaction of profit under objective h. The dual variable y_i^k, however, is the conventionally defined shadow price of resource i, k.

Since both problems in (6-27) and (6-31) are conventional

single-objective linear programming problems, based on the fundamental theorem of duality, if (6-27) and (6-31) have an optimal solution, the following condition is satisfied:

$$\max \lambda = \min[\sum_{h} \underline{f}_h y_h + \sum_{k}\sum_{i} b_i^k y_i^k]. \tag{6-32}$$

Parallel to the discussion in subsection 5.4.2, useful theorems on the complementarity of (6-27) and (6-31) can be easily derived from the duality arguement and are not elaborated here.

Therefore, interregional equilibrium with precise objectives and constraints can be analyzed within a multiobjective fuzzy linear programming framework. As demonstrated in theorem 6.2, it has a close relationship with a conventional compromise programming framework. However, the usefulness of the present framework is not restricted to solve equilibrium problem under precision. In the section to follow, it is extended to solve interregional equilibrium problem with fuzzy objectives and precise or fuzzy constraints.

6.3. MULTIOBJECTIVE SPATIAL PLANNING PROBLEMS IN A FUZZY ENVIRONMENT

Planning under certainty is almost an exception rather than a rule. In a complex decisionmaking environment, objectives and constraints are often imprecise. Similar to the single-objective problem, three major types of fuzzy environments can be classified as those depicted in Fig. 5.1. That is, multiobjective optimization may involve fuzzy objectives and precise constraints, fuzzy objectives and fuzzy constraints, as well as precise objectives and fuzzy constraints. These are situations which often occur in real-world planning environments.

In subsection 6.3.1, the framework discussed in section 6.2 is extended to solve planning problems involving fuzzy objectives and precise constraints. Fuzzy goal programming methods are also examined.

In subsection 6.3.2, planning problems with fuzzy objectives and fuzzy constraints are discussed. A method extended from section 6.2 is examined.

In subsection 6.3.3, the planning environment consists of precise objectives and fuzzy constraints. The optimal solution is obtained through a fuzzy compromise programming procedure.

6.3.1. Fuzzy-Objectives-Precise-Constraints Spatial Planning Problems

Let the following be a planning problem with q fuzzy objectives and $m \times \ell$ precise constraints:

$$\max \underset{\sim}{F}(x) = [\underset{\sim}{f}_1(x), \ldots, \underset{\sim}{f}_h(x), \ldots, \underset{\sim}{f}_q(x)],$$

$$\text{s.t. } \sum_j a_{ij}^k x_j^k \leq b_i^k, \; i = 1, \ldots, m; \; k = 1, \ldots, \ell, \tag{6-33}$$

$$x_j^k \geq 0, \; j = 1, \ldots, n; \; k = 1, \ldots, \ell,$$

where

$$f_h(x) = \sum_k \sum_j c_{hj}^k x_j^k \gtrsim \bar{f}_h; \underline{f}_h, \; h = 1, \ldots, q, \tag{6-34}$$

is a fuzzy objective with $\bar{f}_h$ and $\underline{f}_h$ being respectively the target value and the lowest permissible value specified for $f_h(x)$. The other symbols in (6-33) are defined as those in (6-22).

In this type of problem, a fuzzy goal is set and the motive is to achieve the target value on a precisely delimited technological frontier as much as possible. Apparently, the best and the worst alternative for objective h would be the one which gives the optimal value $\bar{f}_h$ and the compromising value $\underline{f}_h$ respectively. Based on this rationale, a linear ordering can thus be imposed on the possible alternatives via a satisfaction function.

Let the satisfaction function be defined by (6-24). Then, by the argument in subsection 6.2.2, problem (6-33) becomes

$$\max \lambda,$$

$$\text{s.t.} \quad \frac{\sum_k \sum_j c_{hj}^k x_j^k - \underline{f}_h}{d_h} \geq \lambda, \quad h = 1, \ldots, q,$$

$$\sum_j a_{ij}^k x_j^k \leq b_i^k, \quad i = 1, \ldots, m; \ k = 1, \ldots, \ell, \qquad (6\text{-}35)$$

$$\lambda \geq 0, \ x_j^k \geq 0, \ j = 1, \ldots, n; \ k = 1, \ldots, \ell.$$

The solution of (6-35) is stated in the following theorem (Leung, (1985e):

Theorem 6.3. If a spatial alternative $(\lambda^*, x^*) = (\lambda^*, (x_1^{1*}, \ldots, x_j^{k*}, \ldots, x_n^{\ell*}))$ is an optimal solution of (6-35), then λ^* is an efficient solution and $f_h(x^*) \in [\underline{f}_h, \overline{f}_h]$, $h = 1, \ldots, q$.

Therefore, spatial planning problems with fuzzy objectives and precise constraints are in fact similar in format to that with precise objectives and precise constraints. Moreover, if $\overline{f}_h$ and $\underline{f}_h$ in (6-35) are defined as their counterparts in (6-24), the solution of (6-35) is identical to that of (6-27). However, the two approaches are built on different rationales. In the fuzzy-objective case, target values are set for individual objectives and the idea is to achieve as much as possible the target values within the tolerated region. In the precise-objective case, no target values are set and the aim is to maximize individual objectives to the fullest extent.

Thus, spatial planning problems with fuzzy objectives may be viewed as problems with predetermined fuzzy goals. They can also be addressed as fuzzy goal programming problems (Narasimhan, 1980; Rubin and Narasimhan, 1984). In what to follow, fuzzy goal programming methods are briefly examined.

Consider the multiobjective planning problem in (6-33). Let z_h be a fuzzy goal, aspiration level, for objective h, h = 1, ..., q. For example, z_h can be of the fuzzy "less than or equal to" type, fuzzy "greater than or equal to" type, or fuzzy "equal to" type depicted in subsection 5.3.1.

Let the fuzzy goal z_h be defined by a satisfaction function similar to (6-24). Then the fuzzy goal programming formulation of (6-33) becomes

$$\max \lambda,$$

$$\text{s.t.} \frac{\sum_k \sum_j c_{hj}^k x_j^k - \underline{z}_h}{d_h} \geq \lambda, \; h = 1, \ldots, q,$$

$$\sum_k \sum_j c_{hj}^k x_j^k + d_h^- - d_h^+ = z_h, \; h = 1, \ldots, q,$$

$$\sum_j a_{ij}^k x_j^k \leq b_i^k, \; i = 1, \ldots, m; \; k = 1, \ldots, \ell, \qquad (6\text{-}36)$$

$$\lambda \geq 0, \; x_j^k \geq 0, \; j = 1, \ldots, n; \; k = 1, \ldots, \ell,$$

$$d_h^-, \; d_h^+ > 0, \; h = 1, \ldots, q.$$

Observe that the overall degree of satisfaction of the fuzzy objectives and fuzzy goals is maximized in (6-35) and (6-36) respectively. The only difference is in (6-36), deviational variables d_h^- and d_h^+ representing the under- and over-achievement of the h-th goal z_h are incorporated.

In both (6-35) and (6-36), however, priorities of objectives or goals are not considered. Therefore, the optimal solution generally leads to similar degree of satisfaction for all objectives. If objectives are of varying degrees of importance, the satisfaction function in (6-24) can always be modified by a priority function μ_{P_h} such that

$$\mu_{P_h}(\mu_h(x)): \underline{X} \longrightarrow [0, 1]. \qquad (6\text{-}37)$$

The function μ_{P_h} is in fact a composite membership function which gives the degree of satisfaction about the value $\mu_h(x) = \beta_h$. Thus, instead of requiring $\mu_h(x) \geq \lambda$ in (6-35) and (6-36), we require $\mu_{P_h}(\mu_h(x)) \geq \lambda$. The purpose of μ_{P_h} is then to appreciate or depreciate the degree of satisfaction with respect to the importance of an objective.

Of course, composite membership function is not the only way to assign weights to objectives according to their priorities. Explicit weights, real numbers, can be attached to objective functions so that objectives of higher priorities receive larger weights in the optimization process (see for example Hannan, 1981). The optimal alternative is then selected through the maximization of the weighted overall degree of satisfaction

$$\sum_h w_h(x)\ \mu_h(x), \tag{6-38}$$

where w_h are membership functions such that

$$\sum_h w_h(x) = 1, \tag{6-39}$$

or, weights such that

$$\sum_h w_h = 1. \tag{6-40}$$

Sometimes, composite functions or weights may be difficult to determine. Hierarchical procedures can be constructed so that compromise solutions are obtained through a sequential optimization procedure which optimizes the objectives by their order of priorities (see subsection 6.4.1).

6. 3. 2. Fuzzy-Objectives-Fuzzy-Constraints Spatial Planning Problems

So far, the multiobjective planning problems only involve fuzzy objectives. As discussed in chapter 5, imprecision may also occur in the formulation of technological constraints. I demonstrate in this subsection that planning problems involving fuzzy objectives and fuzzy constraints can also be analyzed by extending the method in section 6.2.

Let the following be a planning problem with q fuzzy objectives and m fuzzy constraints:

$$\max \underset{\sim}{F}(x) = [\underset{\sim}{f}_1(x), \ldots, \underset{\sim}{f}_h(x), \ldots, \underset{\sim}{f}_q(x)]$$

$$\text{s.t. } \underset{\sim}{g}_i^k(x) \leq b_i, \ i = 1, \ldots, m; \ k = 1, \ldots, \ell, \qquad (6\text{-}41)$$

$$x \geq \theta,$$

where $x \in \mathbb{R}^{n\ell}$, $f_h(x)$ assumes the form in (6-34), and

$$g_i^k(x) = \sum_j a_{ij}^k x_j^k \underset{\sim}{\leq} \underline{b}_i^k; \bar{b}_i^k, \ i = 1, \ldots, m; \ k = 1, \ldots, \ell.$$

Since objectives and constraints in a fuzzy optimization problem are treated symmetrically, then a satisfaction function for $g_i(x)$ needs to be established. Let it be defined by

$$\mu_i(x) \triangleq \mu_i(g_i(x)) = \begin{cases} 1, & \text{if } \sum_j a_{ij}^k x_j^k \leq \underline{b}_i^k, \\ 1 - \dfrac{\sum_j a_{ij}^k x_j^k - \underline{b}_i^k}{d_i^k}, & \text{if } b_i^k < \sum_j a_{ij}^k x_j^k \leq b_i^k, \\ 0, & \text{if } \sum_j a_{ij}^k x_j^k > \bar{b}_i^k. \end{cases}$$

(6-42)

Let the satisfaction functions of the fuzzy objectives be defined by (6-24). Considering all objectives and constraints, the overall satisfaction can then be defined by

$$\min\left\{\min_h\left[\frac{\sum_k \sum_j c_{hj}^k x_j^k - \underline{f}_h}{d_h}\right], \min_{i,k}\left[\frac{\bar{b}_i^k - \sum_j a_{ij}^k x_j^k}{d_i^k}\right]\right\}. \qquad (6\text{-}43)$$

Such a satisfaction function imposes a linear ordering on the spatial alternatives. Apparently, the best solution is the one which maximizes (6-43) (Leung, 1985e).

Therefore, the multiobjective planning problems in (6-41) becomes

$$\sup_{x_j^k} \min \left\{ \min_h \left[\frac{\sum_k \sum_j c_{hj}^k x_j^k - \underline{f}_h}{d_h} \right], \min_{i,k} \left[\frac{\bar{b}_i^k - \sum_j a_{ij}^k x_j^k}{d_i^k} \right] \right\}. \quad (6\text{-}44)$$

Obviously, it is a fuzzy optimization problem whose solution is obtained by solving:

$$\max \lambda,$$

$$\text{s.t. } \frac{\sum_k \sum_j c_{hj}^k x_j^k - \underline{f}_h}{d_h} \geq \lambda, \; h = 1, \ldots, q,$$

$$\frac{\bar{b}_i^k - \sum_j a_{ij}^k - x_j^k}{d_h^k} \geq \lambda, \; i = 1, \ldots, m; \; k = 1, \ldots, \ell, \quad (6\text{-}45)$$

$$\lambda \geq 0, \; x_j^k \geq 0, \; j = 1, \ldots, n; \; k = 1, \ldots, \ell.$$

Since (6-45) is now similar in format to the case of the single fuzzy objective and fuzzy constraints problem, all theoretical arguments in subsection 5.4.2 can be parallelly applied and are not elaborated here. The corresponding dual problem is

$$\min \sum_h \underline{f}_h y_h + \sum_k \sum_i \bar{b}_i^k y_i^k,$$

$$\text{s.t. } 1 + \sum_h d_h y_h + \sum_k \sum_i d_i^k y_i^k \geq 0$$

$$\sum_h c_{hj}^k y_h + \sum_k \sum_i a_{ij}^k y_i^k \geq 0, \; j = 1, \ldots, n; \; k = 1, \ldots, \ell,$$

$$y_h \geq 0, \; h = 1, \ldots, q,$$

$$y_i^k \geq 0, \; i = 1, \ldots, m; \; k = 1, \ldots, \ell. \quad (6\text{-}46)$$

The dual variable y_h, as those defined in (6-31), is the marginal satisfaction of profit under objective h. The dual variable y_i^k denotes the degree of satisfaction in achieving the fuzzy constraint due to the marginal variation of the supply of resource i, k.

Example 6.4. (A simple application to a regional development problem)

Consider the regional planning problem in example 6.2. Assume that in place of maximizing additional employment and minimizing additional pollution, the policy-makers prefer to set a target value for the two objectives which the optimal development policy of the Meuse-flat should achieve as much as possible.

Let the two fuzzy objectives be specified as:

$$\underset{\sim}{f}_1(x) = \sum_{j=1}^{7} h_j\, x_j \gtrsim 14{,}670;\ 8{,}070, \qquad (6\text{-}47)$$

$$\underset{\sim}{f}_2(x) = \sum_{j=1}^{7} k_j\, x_j \lesssim 78{,}621;\ 100{,}550. \qquad (6\text{-}48)$$

Thus, 14,670 and 78,621 are the target values for f_1 and f_2 respectively.

To allow for more flexibility, the total area for new industrial uses is set for 800 hectares. It, however, can be extended to 1,000 hectares if necessary, albeit less desirable. Thus, we have the following fuzzy constraint:

$$\underset{\sim}{g}(x) = \sum_{j=1}^{7} x_j \lesssim 800;\ 1{,}000. \qquad (6\text{-}49)$$

Let the satisfaction function of f_1 takes on the format in (6-24), and that of f_2 and g(x) in (6-42). If the maximum and minimum land requirements for each industrial uses are same as those in example 6.2. Then, the regional planning problem becomes an optimization problem with two fuzzy objectives and a set of fuzzy and precise constraints.

The optimal solution $(\lambda^*, x_1^*, x_2^*, x_3^*, x_4^*, x_5^*, x_6^*, x_7^*)$ =

(0.454, 339.222, 50, 100, 50, 100, 50, 20) is obtained by solving

$$\begin{aligned}
&\max \lambda, \\
&\text{s.t. } -1.223 + 0.004x_1 + 0.0002x_2 + 0.0003x_3 + 0.0002x_4 + \\
&\quad 0.001x_5 + 0.003x_6 + 0.001x_7 \geq \lambda \\
&\quad 4.585 - 0.004x_1 - 0.002x_3 - 0.023x_5 - 0.002x_6 \geq \lambda \\
&\quad 4 - 0.005x_1 - 0.005x_2 - 0.005x_3 - 0.005x_4 - 0.005x_5 - \\
&\quad 0.005x_6 - 0.005x_7 \geq \lambda \\
&\quad 250 \leq x_1 \leq 500 \\
&\quad 50 \leq x_2 \leq 100 \\
&\quad 100 \leq x_3 \leq 200 \\
&\quad 50 \leq x_4 \leq 100 \\
&\quad 100 \leq x_5 \leq 200 \\
&\quad 50 \leq x_6 \leq 100 \\
&\quad 20 \leq x_7 \leq 40 \\
&\quad \lambda \geq 0,\ x_j \geq 0,\ j = 1, \ldots, 7.
\end{aligned} \tag{6-50}$$

Thus, both fuzzy objectives and the fuzzy constraint are satisfied at 0.454 degree of satisfaction via the compromise solution x^*. By varying the target values and tolerance intervals, different scenarios can be provided for decision-making.

Though the present application is a single-region problem, it is apparent that the framework can be similarly applied to multiregion problems. (See for example Leung, 1985e).

6. 3. 3. Precise-Objectives-Fuzzy-Constraints Spatial Planning Problems

In some spatial planning problems, objectives are precise but the technological environment may be more or less fuzzy.

Decisionmakers intend to maximize or minimize precise objective functions over a fuzzy decision space. Flexibility is achieved by relaxing or tightening constraints within permissible region of deviation. Under these circumstances, precise objectives and fuzzy constraints are involved in the planning process and require different optimization procedure.

Let the following be a planning problem with q precise objectives and m fuzzy constraints:

$$\max F(x) = [f_1(x), \ldots, f_h(x), \ldots, f_q(x)],$$
$$\text{s.t. } \underset{\sim}{g}_i^k(x) \leq b_i, \; i = 1, \ldots, m, \; k = 1, \ldots, \ell, \qquad (6\text{-}51)$$
$$x \geq \theta,$$

where $x \in \mathbb{R}^{n\ell}$ and $f_h(x)$, $h = 1, \ldots, q$, are the same as that defined in (6-22), and $g_i^k(x)$, $i = 1, \ldots, m$; $k = 1, \ldots, \ell$, is the same as that defined in (6-41).

Then, problem (6-51) tries to maximize q objective functions over a set of fuzzy constraints. Though the problem can be solved through different approaches, the fuzzy compromise programming framework (Leung, 1984b) discussed in the following appears to be more revealing.

If the constraints are precise in problem (6-51), by employing the conventional compromise programming method, a point-valued ideal solution can be determined as a reference point and the corresponding compromise solution set can be derived by minimizing a family of distance functions. However, the resource capacities in (6-51) are imprecise, i.e. they may be stretched from $\underline{b}_i^k$ to $\overline{b}_i^k$ for all i, k. Therefore, the ideal solution may be stretched accordingly and is in general fuzzy. In place of a point-value, it becomes a fuzzy region with bounds derived from extending the limits of the constraints.

To obtain the lower bound of the fuzzy ideal solution, we first solve separately the following q linear programming problems:

$$\max f_h(x),$$
$$\text{s.t. } \sum_j a_{ij}^k x_j^k \leq \underline{b}_i^k, \; i = 1, \ldots, m; \; k = 1, \ldots, \ell,$$

$$x_j^k \geq 0, \; j = 1, \ldots, n; \; k = 1, \ldots, \ell. \tag{6-52}$$

Let $\underline{x}_h^* = (\underline{x}_{h1}^{1*}, \ldots, \underline{x}_{hj}^{k*}, \ldots, \underline{x}_{hn}^{\ell*})$, $h = 1, \ldots, q$, be the optimal solution of problem (6-52). Let $\underline{f}_h = f_h^*(\underline{x}_h)$ be the optimal value of objective h. Then, the lower bound of the fuzzy ideal solution $\underline{x}^* = (x_1^{1*}, \ldots, x_j^{k*}, \ldots, x_n^{\ell*})$ with $\underline{f}^* = (\underline{f}_1^*, \ldots, \underline{f}_h^*, \ldots, f_q^*)$ is obtained by solving the following system of simultaneous equations:

$$f_h(x) = \underline{f}_h^*, \; h = 1, \ldots, q. \tag{6-53}$$

Since $\underline{x}^*$ is generally infeasible, compromise solutions are then sought so that the individual optimals $\underline{f}_h^*$'s can be preserved or reached as much as possible. To determine all possible compromise solutions, let

$$\underline{d}_p = \left\{ \sum_h \left| \frac{\underline{f}_h^* - f_h(x)}{\underline{f}_h^*} \right|^p \right\}^{\frac{1}{p}}, \; p = 1, 2, \ldots, \infty, \tag{6-54}$$

be a family of distance measures with parameter p. The function $\underline{d}_p$ represents the distance between any alternative x and the ideal solution $\underline{x}^*$. A compromise solution is then the solution which minimizes $\underline{d}_p$. Considering all values of p, a set of compromise solutions can be determined.

Specifically, for p = 1, the compromise solution $\underline{x}^p = \underline{x}^1$ is obtained by solving

$$\min \underline{d}_1 = \max \sum_h \frac{f_h(x)}{\underline{f}_h^*},$$

$$\text{s.t.} \sum_j a_{ij}^k x_j^k \leq \underline{b}_i^k, \; i = 1, \ldots, m; \; k = 1, \ldots, \ell, \tag{6-55}$$

$$x_j^k \geq 0, \; j = 1, \ldots, n; \; k = 1, \ldots, \ell.$$

The solution $\underline{x}^1$ minimizes the sum of the individual distances to the lower bound of the fuzzy ideal solution.

For $p = \infty$, the compromise solution $\underline{x}^\infty$ is obtained by solving

$$\min \underline{d}_\infty,$$

$$\text{s.t.} \quad \frac{f_h^* - f_h(x)}{f_h^*} \le \underline{d}_\infty, \quad h = 1, \ldots, q,$$

$$\sum_j a_{ij}^k x_j^k \le \underline{b}_i^k, \quad i = 1, \ldots, m; \; k = 1, \ldots, \ell, \qquad (6\text{-}56)$$

$$x_j^k \ge 0, \quad j = 1, \ldots, n; \; k = 1, \ldots, \ell.$$

Such a solution minimizes the maximal distance a feasible spatial alternative may have to the lower bound of the fuzzy ideal solution.

These two compromise solutions serve as bounds of the compromise solution set for all p, $1 \le p \le \infty$, with respect to the ideal solution $\underline{x}^*$. In particular, for $p = 2$, compromise solution $\underline{x}^2$ is the one which minimizes the physical distance to the lower bound of the fuzzy ideal solution. It can be obtained by interpolating from $\underline{x}^1$ and $\underline{x}^\infty$ or by solving the corresponding quadratic programming problem.

Therefore, corresponding to the lower bound of the fuzzy ideal solution, a set of compromise solutions can be obtained. Such a set constitutes the lower bound of the fuzzy compromise solution set. To derive its upper bound, the upper bound of the fuzzy ideal solution needs to be determined first.

To obtain the upper bound of the fuzzy ideal solution, we need to solve the following q linear programming problems:

$$\max f_h(x),$$

$$\text{s.t.} \quad \sum_j a_{ij}^k x_j^k \le \bar{b}_i^k, \quad i = 1, \ldots, m; \; k = 1, \ldots, \ell, \qquad (6\text{-}57)$$

$$x_j^k \ge 0, \quad j = 1, \ldots, n; \; k = 1, \ldots, \ell.$$

Let $\bar{x}_h^* = (\bar{x}_{h1}^{1*}, \ldots, \bar{x}_{hj}^{k*}, \ldots, \bar{x}_{hn}^{\ell*})$, $h = 1, \ldots, q$, be the optimal solution of problem (6-57). Let $\bar{f}_h^* = f_h(\bar{x}_h^*)$ be the optimal value of objective h. Then, the upper bound of the fuzzy ideal solution $\bar{x}^* = (\bar{x}_1^{1*}, \ldots, \bar{x}_j^{k*}, \ldots, \bar{x}_n^{\ell*})$ with $\bar{f}^* = (\bar{f}_1^*, \ldots, \bar{f}_h^*, \ldots, \bar{f}_q^*)$ is obtained by solving the following system of simuiltaneous equations:

$$f_h(x) = \bar{f}_h^*, \; h = 1, \ldots, q. \tag{6-58}$$

Based on $\bar{x}^*$, a set of compromise solutions can be determined. Let

$$\bar{d}_p = \left\{ \sum_h \left| \frac{\bar{f}_h^* - f_h(x)}{\bar{f}_h^*} \right|^p \right\}^{\frac{1}{p}}, \; p = 1, 2, \ldots, \infty, \tag{6-59}$$

be a family of distance measures with parameter p. The compromise solution set can again be obtained by minimizing $\bar{d}_p$ for all p.

Specifically, for $p = 1$, the compromise solution $\bar{x}^1$ is obtained by solving

$$\min \bar{d}_1 = \max \sum_h \frac{f_h(x)}{\bar{f}_h^*},$$

$$\text{s.t. } \sum_j a_{ij}^k x_j^k \leq \bar{b}_i^k, \; i = 1, \ldots, m; \; k = 1, \ldots, \ell, \tag{6-60}$$

$$x_j^k \geq 0, \; j = 1, \ldots, n; \; k = 1, \ldots, \ell.$$

For $p = \infty$, the compromise solution $\bar{x}^\infty$ is obtained by solving

$$\min \bar{d}_\infty,$$

$$\text{s.t. } \frac{\bar{f}_h^* - f_h(x)}{\bar{f}_h^*} \leq \bar{d}_\infty, \; h = 1, \ldots, q,$$

$$\sum_j a_{ij}^k x_j^k \leq \bar{b}_i^k, \; i = 1, \ldots, m; \; k = 1, \ldots, \ell, \tag{6-61}$$

$$x_j^k \geq 0, \; j = 1, \ldots, n; \; k = 1, \ldots, \ell.$$

Thus, by stretching the capacities of the fuzzy constraints from $\underline{b}_i^k$ to $\overline{b}_i^k$, $i = 1, \ldots, m$; $k = 1, \ldots, \ell$, the optimal values of the individual objectives change from $\underline{f}_h^*$ to $\overline{f}_h^*$, $h = 1, \ldots, q$. As a result, the ideal solution moves from f^* to $\overline{f}^*$. Considering all variations within the intervals $[\underline{b}_i^k, \overline{b}_i^k]$, the fuzzy ideal solution can be expressed as

$$\left\{(f_1, \ldots, f_h, \ldots, f_q) \middle| \underline{f}_h^* \leq f_h \leq \overline{f}_h^*, \ h = 1, \ldots, q\right\}. \tag{6-62}$$

The corresponding compromise solution is then imprecise.

To determine the fuzzy compromise solution, the imprecision of the constraints and the objectives should be explicitly incorporated. Since each objective function can assume its value within the interval $[\underline{f}_h^*, \overline{f}_h^*]$. The objective can in fact be treated as a fuzzy objective having the following expression:

$$f_h(x) \gtrsim \overline{f}_h^*;\ \underline{f}_h^*, \quad h = 1, \ldots, q. \tag{6-63}$$

To derive the most appropriate ideal solution, we first need to solve separately q single-objective fuzzy programming problems as follows:

$$\begin{aligned} & f_h(x) \gtrsim \overline{f}_h^*;\ \underline{f}_h^* \\ & \sum_j a_{ij}^k\, x_j^k \lesssim \underline{b}_i^k;\ \overline{b}_i^k, \quad i = 1, \ldots, m;\ k = 1, \ldots, \ell, \\ & x_j^k \geq 0, \quad j = 1, \ldots, n;\ k = 1, \ldots, \ell. \end{aligned} \tag{6-64}$$

Since problem (6-64) is an optimization problem with a fuzzy objective and a set of fuzzy constraints, by subsection 5.4.1, the optimal solution $x_h^* = (x_{h1}^{1*}, \ldots, x_{hj}^{k*}, \ldots, x_{hn}^{\ell*})$ of problem (6-64) is obtained by solving

$$\max \lambda,$$

$$\text{s.t.} \quad \frac{f_h(x) - \underline{f}_h^*}{\overline{f}_h^* - \underline{f}_h^*} \geq \lambda$$

$$\frac{\bar{b}_i^k - \sum_j a_{ij}^k x_j^k}{d_i^k} \geq \lambda, \; i = 1, \ldots, m; \; k = 1, \ldots, \ell,$$

$$\lambda \geq 0, \; x_j^k \geq 0, \; j = 1, \ldots, n; \; k = 1, \ldots, \ell. \tag{6-65}$$

Taking all q problems in (6-64) into consideration, the most appropriate ideal solution is $x^* = (x_1^{1*}, \ldots, x_j^{k*}, \ldots, x_n^{\ell *})$ with $f^* = (f_1^*, \ldots, f_h^*, \ldots, f_q^*)$ which is obtained by solving the system of simultaneous equations:

$$f_h(x) = f_h^*, \; h = 1, \ldots, q. \tag{6-66}$$

With respect to λ^*, the decision space on which a compromise solution is obtained is bounded by

$$\sum_j a_{ij}^k x_j^k \leq \underline{b}_i^k + (1 - \lambda^*) d_i^k, \; i = 1, \ldots, m; \; k = 1, \ldots, \ell,$$

$$x_j^k \geq 0, \; j = 1, \ldots, n; \; k = 1, \ldots, \ell, \tag{6-67}$$

where $d_i^k = (\bar{b}_i^k - \underline{b}_i^k)$ is the length of the tolerance interval $[\underline{b}_i^k, \bar{b}_i^k]$.

Therefore, when x^* is treated as the ideal solution, the most appropriate compromise solution can be determined by minimizing the distance

$$d_p = \left\{ \sum_h \left| \frac{f_h^* - f_h(x)}{f_h^*} \right|^p \right\}^{\frac{1}{p}}, \; p = 1, 2, \ldots, \infty, \tag{6-68}$$

For $p = 1$, the most appropriate compromise solution is obtained by solving

$$\min d_1 = \max \sum_h \frac{f_h(x)}{f_h^*},$$

$$\text{s.t.} \quad \sum_j a_{ij}^k x_j^k \leq \underline{b}_i^k + (1 - \lambda^*) d_i^k, \; i = 1, \ldots, m; \; k = 1, \ldots, \ell, \tag{6-69}$$

$$x_j^k \geq 0, \; j = 1, \ldots, n; \; k = 1, \ldots, \ell.$$

For $p = \infty$, the most appropriate compromise solution is obtained by solving

$$\min d_\infty,$$

$$\text{s.t. } \frac{f_h^* - f_h(x)}{f_h^*} \le d_\infty, \; h = 1, \ldots, q,$$

$$\sum_j a_{ij}^k x_j^k \le b_i^k + (1 - \lambda^*) d_i^k, \; i = 1, \ldots, m; \quad k = 1, \ldots, \ell, \qquad (6\text{-}70)$$

$$x_j^k \ge 0, \; j = 1, \ldots, n; \; k = 1, \ldots, \ell.$$

Thus, the most appropriate compromise solution set with $1 \le p \le \infty$ is bounded by the solutions in (6-69) and (6-70).

Since λ denotes the degree of satisfaction of the fuzzy objectives and constraints, then $(1 - \lambda)$ can be interpreted as the degree of dissatisfaction. Therefore, by varying the value of λ, we are changing the reference point in the fuzzy ideal solution and the compromise solution set changes accordingly. If $\theta = 1 - \lambda$ is treated as a parameter in (6-70), the entire compromise solution set with parameter p can be generated via the parametric programming method.

Therefore, within permitted limits of tolerance, the fuzzy compromise programming method allows us to generate alternatives to resolve conflicts which may not be resolved or dissolved otherwise. The imprecision of constraints free decision-makers from restricting themselves to a single point-valued ideal solution.

Example 6. 6. Let the following be a two-objective maximization problem with four fuzzy constraints:

$$\max F(x) = [f_1(x_1, x_2), f_2(x_1, x_2)]$$

$$= [2x_1 + x_2, -3x_1 + 2x_2],$$

$$\text{s.t. } x_1 \lesssim 5;\ 6$$

$$x_2 \lesssim 6;\ 8$$

$$x_1 + x_2 \lesssim 9;\ 10$$

$$2x_1 + 3x_2 \gtrsim 8;\ 6$$

$$x_1 \geq 0,\ x_2 \geq 0. \qquad (6\text{-}71)$$

By (6-52) and (6-53), the separate maximization of f_1 and f_2 with respect to the constraint set

$$\begin{aligned} x_1 \quad\quad & \leq 5 \\ x_2 & \leq 6 \\ x_1 + x_2 & \leq 9 \\ 2x_1 + 3x_2 & \geq 8 \\ x_1 \geq 0,\ x_2 & \geq 0, \end{aligned} \qquad (6\text{-}72)$$

gives the ideal solution $\underline{x}^* = (2.288,\ 9.438)$ with $\underline{f}^* = (14,\ 12)$.

Employing the distance measure in (6-54), for $p = 1$, the compromise solution $\underline{x}^1 = (0,\ 6)$ with $\underline{f}^1 = (6,\ 12)$ is obtained by solving

$$\begin{aligned} \max\ & -0.107x_1 + 0.238x_2, \\ \text{s.t.}\ & x_1 \quad\quad \leq 5 \\ & x_2 \leq 6 \\ & x_1 + x_2 \leq 9 \\ & 2x_1 + 3x_2 \geq 8 \\ & x_1 \geq 0,\ x_2 \geq 0. \end{aligned} \qquad (6\text{-}73)$$

For $p = \infty$, the compromise solution $\underline{x}^\infty = (1.466,\ 6)$ with $\underline{f}^\infty = (0.932,\ 7.602)$ is obtained by solving

$$\begin{aligned} \min\ & \underline{d}_\infty, \\ \text{s.t.}\ & x_1 \quad\quad \leq 5 \\ & x_2 \leq 6 \\ & x_1 + x_2 \leq 9 \end{aligned}$$

$$2x_1 + 3x_2 \geq 8$$

$$0.143x_1 + 0.071x_2 + \underline{d}_\infty \geq 1$$

$$-0.25x_1 + 0.167x_2 + \underline{d}_\infty \geq 1$$

$$x_1 \geq 0,\ x_2 \geq 0. \qquad (6\text{-}74)$$

Thus for $1 \leq p \leq \infty$, the minimization of $\underline{d}_p$ generates a compromise solution set with respect to the ideal solution $\underline{x}^*$ (see Fig. 6.3 and 6.4).

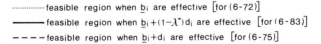

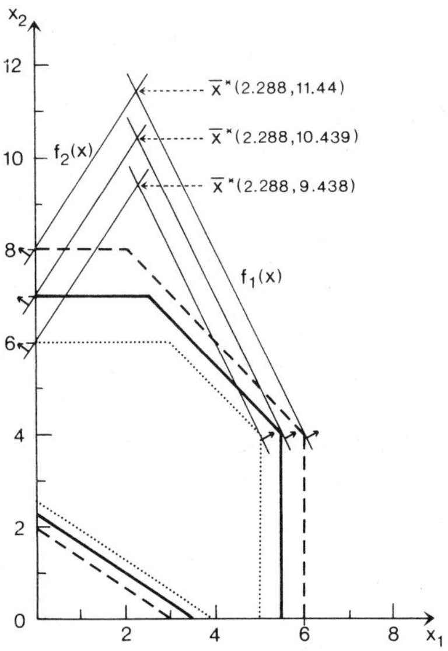

.... feasible region when $\underline{b}_i$ are effective [for (6-72)]

——— feasible region when $\underline{b}_i + (1-\lambda^*)d_i$ are effective [for (6-83)]

- - - - feasible region when $\underline{b}_i + d_i$ are effective [for (6-75)]

Fig. 6.3 The ideal solutions in the fuzzy decision space for the two-objective problem in (6-71)

By stretching the constraints in (6-71) to the other end of the tolerance intervals, the ideal solution $\bar{x}^* = (2.288, 11.44)$ with $\bar{f}^* = (16, 16)$ is obtained by maximizing separately f_1 and f_2 with respect to the following constraint set:

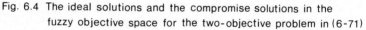

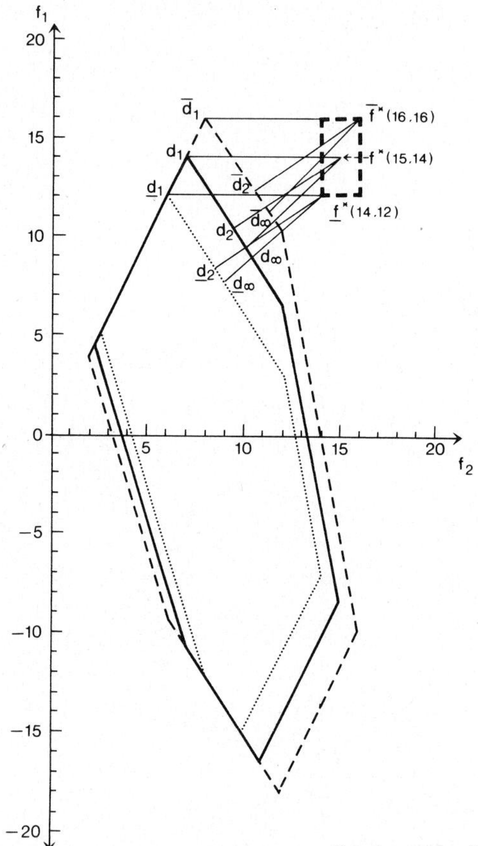

Fig. 6.4 The ideal solutions and the compromise solutions in the fuzzy objective space for the two-objective problem in (6-71)

$$
\begin{aligned}
x_1 \qquad\quad &\leq 6 \\
x_2 &\leq 8 \\
x_1 + \ x_2 &\leq 10 \qquad\qquad (6\text{-}75)\\
2x_1 + 3x_2 &\geq 6 \\
x_1 \geq 0,\ x_2 &\geq 0.
\end{aligned}
$$

Employing the distance measure in (6-59), the compromise solution $\overline{x}^1 = (0, 8)$ with $\overline{f}^1 = (8, 16)$ is obtained by solving

$$\begin{aligned}
\max\ & 0.063x_1 + 0.188x_2, \\
\text{s.t.}\ & x_1 \le 6 \\
& x_2 \le 8 \\
& x_1 + x_2 \le 10 \\
& 2x_1 + 3x_2 \ge 16 \\
& x_1 \ge 0,\ x_2 \ge 0.
\end{aligned} \tag{6-76}$$

Similarly, the compromise solution $\overline{x}^{\infty} = (1.585, 8)$ with $\overline{f}^{\infty} = (11.17, 11.245)$ is obtained by solving

$$\begin{aligned}
\min\ & d_{\infty}, \\
\text{s.t.}\ & 0.125x_1 + 0.063x_2 + \overline{d}_{\infty} \ge 1 \\
& -0.188x_1 + 0.125x_2 + \overline{d}_{\infty} \ge 1 \\
& x_1 \le 6 \\
& x_2 \le 8 \\
& x_1 + x_2 \le 10 \\
& 2x_1 + 3x_2 \ge 6 \\
& x_1 \ge 0,\ x_2 \ge 0.
\end{aligned} \tag{6-77}$$

(See Fig. 6.3 and 6.4).

Thus, by (6-62), the fuzzy ideal can be expressed as

$$\left\{ (f_1, f_2) \,|\, 14 \le f_1 \le 16,\ 12 \le f_2 \le 16 \right\}. \tag{6-78}$$

Considering the above tolerances, objectives f_1 and f_2 in (6-71) can be respectively respecified as

$$f_1(x_1, x_2) = 2x_1 + x_2 \gtrsim 16;\ 14, \tag{6-79}$$

and,

$$f_2(x_1, x_2) = -3x_1 + 2x_2 \gtrsim 16;\ 12. \tag{6-80}$$

By (6-65), the solution $x_1^* = (5.5, 4)$ with $f_1^* = (15, -8.5)$ and $\lambda^* = 0.5$ is obtained by solving

$$\begin{aligned} &\max \lambda, \\ &\text{s.t} \quad -8 + x_1 + 0.5x_2 \geq \lambda \\ &\qquad 5 - x_1 \geq \lambda \\ &\qquad 3 - 0.5x_2 \geq \lambda \\ &\qquad 9 - x_1 - x_2 \geq \lambda \\ &\qquad -4 + x_1 + 1.5x_2 \geq \lambda \\ &\qquad \lambda \geq 0,\ x_1 \geq 0,\ x_2 \geq 0. \end{aligned} \tag{6-81}$$

Similarly, the solution $x_2^* = (0, 7)$ with $f_2^* = (7, 14)$ and $\lambda^* = 0.5$ is obtained by solving

$$\begin{aligned} &\max \lambda, \\ &\text{s.t.} \quad -4 - 0.75x_1 + 0.5x_2 \geq \lambda \\ &\qquad 5 - x_1 \geq \lambda \\ &\qquad 3 - 0.5x_2 \geq \lambda \\ &\qquad 9 - x_1 - x_2 \geq \lambda \\ &\qquad -4 + x_1 + 1.5x_2 \geq \lambda \\ &\qquad \lambda \geq 0,\ x_1 \geq 0,\ x_2 \geq 0. \end{aligned} \tag{6-82}$$

By (6-66), the most appropriate ideal solution of (6-71) is $x^* = (2.288, 10.439)$ with $f^* = (15, 14)$. The most appropriate compromise solution is then the one which minimizes the distance measure in (6-68).

For $p = 1$, the most appropriate compromise solution $x^1 = (0, 7)$ with $f^1 = (7, 14)$ is obtained by solving

$$\begin{aligned} &\max -0.081x_1 + 0.21x_2, \\ &\text{s.t.} \quad x_1 \leq 5.5 \\ &\qquad x_2 \leq 7 \\ &\qquad x_1 + x_2 \leq 9.5 \\ &\qquad x_1 + 1.5x_2 \geq 3.5 \end{aligned}$$

$$x_1 \geq 0,\ x_2 \geq 0. \qquad (6\text{-}83)$$

For p = ∞, the most appropriate compromise solution x^∞ = (1.533, 7) with f^∞ = (10.066, 9.401) is obtained by solving

$$\begin{aligned} &\min\ d_\infty, \\ &\text{s.t.}\ x_1 \leq 5.5 \\ &\qquad x_2 \leq 7 \\ &\qquad x_1 + x_2 \leq 9.5 \\ &\qquad x_1 + 1.5x_2 \geq 3.5 \\ &\qquad 0.133x_1 + 0.067x_2 + d_\infty \geq 1 \\ &\qquad -0.214x_1 + 0.143x_2 + d_\infty \geq 1 \\ &\qquad x_1 \geq 0,\ x_2 \geq 0. \end{aligned} \qquad (6\text{-}84)$$

(See Fig. 6.3 and 6.4).

6.4. DYNAMIC CONFLICT RESOLUTION IN A FUZZY ENVIRONMENT

So far, conflicts among objectives are resolved in a one-stage noninteractive optimization process. In real-life, conflicts are usually resolved or dissolved by an iterative and/or interactive procedure necessitated by additional information acquired, changes in decisionmakers' valuations and/or judgements and/or the planning environment. It is especially crucial if the original compromise solution fails to yield a settlement and a new alternative is required.

In subsection 6.4.1, interactive procedures are employed to solve planning problems with hierarchical fuzzy objectives. In subsection 6.4.2, resolution of conflicts is examined through a theory of displaced fuzzy ideal. Conflicts among objectives are resolved within a recursive and interactive framework.

6.4.1. Interactive Approach to Hierarchical Fuzzy Objectives Problems

In subsection 6.3, spatial planning problems involving multiple objectives are solved by some one-stage fuzzy optimization methods. The process is noninteractive. Once the decisionmakers have provided all necessary information, the optimization procedure is self-fulfilling.

In some situations, however, the one-stage process is too mechanical and restrictive. It is more instrumental, albeit more taxing, if it is on an interactive mode. The method is especially suitable to solve problems involving hierarchical objectives. As discussed in the preceding section, weights can be preassigned to objectives according to their order of priorities. A compromise solution is then obtained by optimizing the composite objective function.

In this subsection, a stepwise optimization procedure is formulated (Leung, 1985c). In place of prespecifying weights, trade-off between objectives are obtained from decisionmakers in each step of the interactive process. In brief, the stepwise procedure optimize the most important objective first. The next most important objective is then optimized within a tolerable trade-off specified for the optimal value of the preceding objective. By the same token, the third most important objective is optimized within the restriction imposed by the tolerable trade-offs of the previous optimal solutions. The process continues until the least important objective is optimized and the final compromise solution is reached. Strictly speaking, it is an interactive but noniterative procedure. It becomes iterative only if the whole or part of the procedure is repeated a number of times before the final decision is made. In what to follow, a stepwise procedure is outlined.

Let

$$
\begin{aligned}
& f_h(x) \gtrsim \overline{f}_h;\ \underline{f}_h,\ h = 1, \ldots, q, \\
& g_i(x) \gtrsim \overline{b}_i;\ \underline{b}_i,\ i = 1, \ldots, m, \qquad (6\text{-}85) \\
& x \geq \theta,
\end{aligned}
$$

be an optimization problem with q fuzzy objectives and m fuzzy constraints, $x \in \mathbb{R}^n$.

Let the satisfaction functions of $f_h(x)$ and $g_i(x)$ be respectively defined by

$$\mu_h(x) \triangleq \mu_h(f_h(x)) = \begin{cases} 1, & \text{if } f_h(x) > \bar{f}_h, \\ 1 - \dfrac{\bar{f}_h - f_h(x)}{d_h}, & \text{if } \underline{f}_h < f_h(x) \leq \bar{f}_h, \\ & h = 1, \ldots, q, \\ 0, & \text{if } f_h(x) \leq \underline{f}_h; \end{cases} \quad (6\text{-}86)$$

$$\mu_i(x) \triangleq \mu_i(g_i(x)) = \begin{cases} 1, & \text{if } g_i(x) > \bar{b}_i, \\ 1 - \dfrac{\bar{b}_i - g_i(x)}{s_i}, & \text{if } \underline{b}_i < g_i(x) \leq \bar{b}_i, \\ & i = 1, \ldots, m, \\ 0, & \text{if } g_i(x) \leq \underline{b}_i, \end{cases} \quad (6\text{-}87)$$

where $d_h = (\bar{f}_h - f_h)$ and $s_i = (\bar{b}_i - \underline{b}_i)$ are the lengths of the respective tolerance intervals $[\underline{f}_h, \bar{f}_h]$ and $[\underline{b}_i, \bar{b}_i]$.

Suppose that the fuzzy objectives can be ranked in decending order of priorities as

$$f_1 \succ f_2 \succ \cdots \succ f_q, \quad (6\text{-}88)$$

with f_1 being the most important objective and f_m the least important objective.

In place of assigning weights to the q objectives, sequential optimization of individual objectives by their decending order of priorities can be performed. That is, the satisfaction function $\mu_1(f_1)$ is first optimized with respect to the fuzzy constraints $\mu_i(g_i)$'s. A tolerable level of deviation from the optimal value f_1^* is then employed as a restraint on the optimization of $\mu_2(f_2)$ in the next step. Along the same line of reasoning, $\mu_3(f_3)$ is optimized subject to $\mu_i(g_i)$'s and the tolerable levels of deviation from the optimal values f_1^* and f_2^* obtained from the previous steps. The procedure continues until $\mu_q(f_q)$ is optimized.

The following are the major steps:

Step 1. Obtain the optimal solution $(\lambda_1^*, x_1^*, f_1^*)$, the subscript in x indicates step number, by solving the following single objective optimization problem:

$$f_1(x) \gtrsim \bar{f}_1; \underline{f}_1$$

$$g_i(x) \gtrsim \bar{b}_i; \underline{b}_i, \; i = 1, \ldots, m, \qquad (6\text{-}89)$$

$$x \geq \theta.$$

That is, we solve

$$\max \lambda_1,$$

$$\text{s.t.} \; \frac{f_1(x) - \underline{f}_1}{d_1} \geq \lambda_1$$

$$\frac{g_i(x) - \underline{b}_i}{s_i} \geq \lambda_1, \; i = 1, \ldots, m, \qquad (6\text{-}90)$$

$$\lambda_1 \geq 0, \; x \geq \theta.$$

Step 2. Obtain from the decisionmaker a trade-off coefficient β_1^2 indicating the extent to which λ_1^* can be compromised in order to best achieve f_2 by maximizing $\mu_2(f_2)$. Thus, β_1^2 may be regarded as the tolerance level for the permissible deviation from λ_1^* when f_2 is maximized.

Determine the optimal solution $(\lambda_2^*, x_2^*, f_2^*) \Longleftrightarrow (\lambda_{1,2}^*, \lambda_2^*, x_2^*, f_{1,2}^*, f_2^*)$, the second subscript indicates the current step number for λ_1^* and f_1^*, by solving

$$\max \lambda_2,$$

$$\text{s.t.} \; \frac{f_2(x) - \underline{f}_2}{d_2} \geq \lambda_2$$

$$\frac{g_i(x) - \underline{b}_i}{s_i} \geq \lambda_2, \; i = 1, \ldots, m, \qquad (6\text{-}91)$$

$$\frac{f_1(x) - \underline{f}_1}{d_1} \geq \beta_1^2 \lambda_1^*$$

$$\lambda_2 \geq 0, \; x \geq \theta,$$

where $0 \leq \beta_1 \leq \beta_1^2 \leq 1$, with β_1 being the maximal tolerable trade-off of f_1^* for f_2^*. For instance, β_1 can be set in such a way that the degree of satisfaction of f_1 would always be greater than or equal to that of f_2.

Step 3. Based on λ_1^* and λ_2^*, obtain from the decisionmaker trade-off coefficients β_1^3 and β_2^3 to serve as restraints when f_3 is maximized.

Determine the optimal solution $(\lambda_3^*, x_3^*, f_3^*) \Longleftrightarrow (\lambda_{1,3}^*, \lambda_{2,3}^*, \lambda_3^*, x_3^*, f_{1,3}^*, f_{2,3}^*, f_3^*)$ by solving

$$\begin{aligned} &\max \lambda_3, \\ &\text{s.t. } \frac{f_3(x) - \underline{f}_3}{d_3} \geq \lambda_3 \\ &\qquad \frac{g_i(x) - \underline{b}_i}{s_i} \geq \lambda_3, \; i = 1, \ldots, m, \\ &\qquad \frac{f_1(x) - \underline{f}_1}{d_1} \geq \beta_1^3 \lambda_1^* \\ &\qquad \frac{f_2(x) - \underline{f}_2}{d_2} \geq \beta_2^3 \lambda_2^* \\ &\qquad \lambda_3 \geq 0, \; x \geq \theta, \end{aligned} \tag{6-92}$$

where $0 \leq \beta_1 \leq \beta_1^3 \leq \beta_1^2 \leq 1$ and $0 \leq \beta_2 \leq \beta_2^3 \leq 1$, with β_1 and β_2 respectively the maximal permissible trade-offs of f_1^* and f_2^* for f_3^*.

Step q. Applying the same method throughout the sequential optimization procedure, the compromise solution $(\lambda_q^*, x_q^*, f_q^*) \Longleftrightarrow (\lambda_{1,q}^*, \ldots, \lambda_q^*, x_q^*, f_{1,q}^*, \ldots, f_q^*)$ in the qth step is obtained by solving

$$\begin{aligned} &\max \lambda_q, \\ &\text{s.t. } \frac{f_q(x) - \underline{f}_q}{d_q} \geq \lambda_q \end{aligned}$$

$$\frac{g_i(x) - \underline{b}_q}{s_i} \geq \lambda_q, \quad i = 1, \ldots, m,$$

$$\frac{f_h(x) - \underline{f}_h}{d_h} \geq \beta_h^q \lambda_h^*, \quad h = 1, \ldots, q-1,$$

$$\lambda_q \geq 0, \quad x \geq \theta, \tag{6-93}$$

where $0 \leq \beta_h \leq \beta_h^q \leq \beta_h^{q-1} \leq \ldots \leq \beta_h^{h+1} \leq 1$, $h = 1, \ldots, q-1$, with β_h being the maximal permissible trade-off of f_h^* for f_q^*. Such an ordering indicates that the trade-off for more important lower order objectives are more lenient than that of the less important ones. That is, the trade-off extent decreases with decreasing order of priorities of the lower order objectives. Such a specification makes the search for a compromise solution consistent throughout the sequential procedure. Solutions obtained from later steps would not reverse or contradict to that obtained from the preceding steps. The more stringent we are on the trade-offs, the more favorable the compromise solution is to the higher order objectives.

Example 6.7. Let the following be a simple two-objective fuzzy linear programming problem in $\mathbb{R}$:

$$f_1: x \underset{\sim}{\geq} 60; \ 20$$

$$f_2: x \underset{\sim}{\leq} 10; \ 90 \tag{6-94}$$

$$g\ : x \underset{\sim}{\leq} 50; \ 70$$

where $f_1 \succ f_2$ is the priority ordering of f_1 and f_2. (See Fig. 6.5)

Based on the sequential optimization procedure, we need to solve the following two-step problems:

Step 1. Obtain $\lambda_1^* = 0.833$ by solving

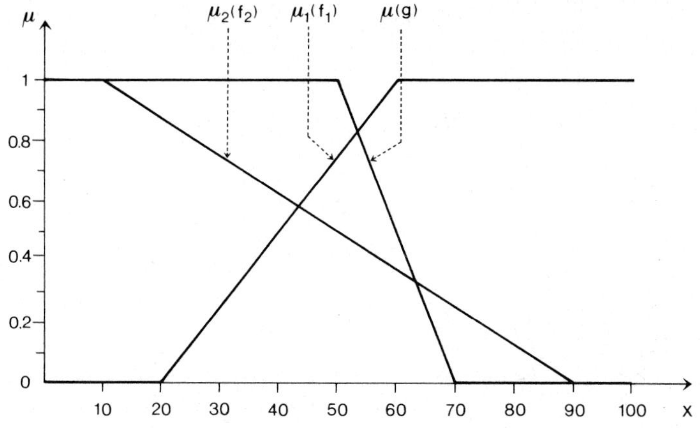

Fig. 6.5 Membership functions of two fuzzy objectives and one fuzzy constraint in (6-94)

$$\max \lambda_1,$$

$$\text{s.t.} \quad \frac{x - 20}{40} \geq \lambda_1$$

$$\frac{70 - x}{20} \geq \lambda_1 \qquad (6\text{-}95)$$

$$\lambda_1 \geq 0, \; x \geq 0.$$

Step 2. Obtain λ_2^* by solving

$$\max \lambda_2,$$

$$\text{s.t.} \quad \frac{90 - x}{80} \geq \lambda_2$$

$$\frac{70 - x}{20} \geq \lambda_2 \qquad (6\text{-}96)$$

$$\frac{x - 20}{40} \geq (0.833)\beta_1^2$$

$$\lambda_2 \geq 0, \; x \geq 0.$$

The result of the two-step optimization problem for selected values of β_1^2 is tabulated in Table 6.2 and depicted in Fig. 6.6.

Table 6.2 Compromise solutions with varying trade-off specifications

Trade-off Coefficients	Compromise Solutions		
(β_1^2)	x*	λ_1^*	λ_2^*
0.0	20.000	0.000	0.875
0.1	23.332	0.083	0.833
0.2	26.664	0.166	0.792
0.3	29.996	0.250	0.750
0.4	33.328	0.333	0.708
0.5	36.660	0.417	0.667
0.6	39.992	0.500	0.625
0.7	43.324	0.583	0.583
0.8	46.656	0.666	0.542
0.9	49.988	0.750	0.500
1.0	53.333	0.833	0.458

Obviously, not all trade-off specifications are in favor of the more important objective f_1. The critical trade-off coefficient is 0.7. At this level of trade-off, the degrees of satisfaction of both objectives are identical, i.e. 0.583. It is in fact the point at which f_1 and f_2 are of equal importance and simultaneously optimized. For $\beta_1^2 > 0.7$, the degree of satisfaction of f_1 is higher than that of f_2. For $\beta_1^2 < 0.7$, the opposite is true. Thus, to guarantee that the degree of satisfaction of f_1 would always be higher than that of f_2, the maximal permissible trade-off of f_1 for f_2 should be set at a value greater than 0.7 in the search for a compromise. That is, the condition, $0.7 = \beta_1 < \beta_1^2 \leq 1$, should be enforced.

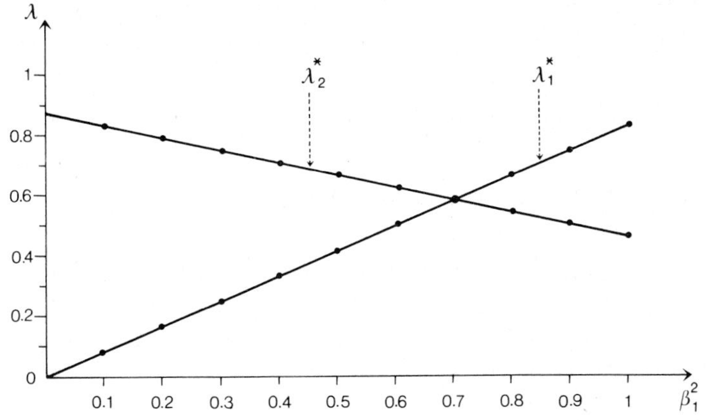

Fig. 6.6 Trade-offs of objective-achievements in (6-94)

Sometimes, instead of determing λ_i^* , decisionmakers may find it easier to determine the trade-off on the basis of how much of f_h^* they are willing to trade for f_q^* in (6-85). Under this situation, we can substitute

$$\frac{f_h(x) - \underline{f}_h}{d_h} \geq \beta_h^q \lambda_h^*, \tag{6-97}$$

in (6-93) in step q by

$$f_h(x) \geq \alpha_h^q f_h^*, \ h = 1, \ldots, q-1, \tag{6-98}$$

where $0 \leq \alpha_h \leq \alpha_h^m \leq \alpha_h^{m-1} \leq \ldots \leq \alpha_h^{h+1} \leq 1, \quad h = 1, \ldots, q-1,$ with α_h being the maximal permissible trade-off of f_h^* for f_q^*.

Remark. There are other methods in formulating the interactive optimization process (see Leung, 1985c). Some are more demanding in the acquisition of trade-off information. Though the processes are noniterative, they become iterative if any one or all of the steps are repeated throughout the compromising process.

6. 4. 2. Conflict Resolution through Fuzzy Ideal Displacement

So far, conflict resolution involving multiple objectives has been examined under the assumption that the technological environment is static. In most real-life settings, however, conflict resolution is a dynamic process. Determination of a compromise solution is subjected to new information and changes in the cognition or valuation of the conflict environment. Often, initial technological environment may fail to yield a satisfactory compromise. Modification of the aspiration levels of our objectives, tigtening or relaxation of constraints, deletion or addition of objectives or constraints would generate new alternatives and triger a new round of conflict resolution process.

If conflict is resolved via the concept of a fuzzy ideal solution as discussed in subsection 6.3.3, changes in the technological environment would displace the fuzzy ideal solution and the corresponding fuzzy compromise solution. Therefore, conflict resolution may involve several displacements of the fuzzy ideal solution (Leung, 1982c; 1983b; 1987c).

In general, any changes in the fuzzy or precise objectives, and/or fuzzy or precise constraints may displace the ideal solution. To shorten our discussion, I only examine briefly the situation involving precise objectives and fuzzy constraints, i.e. problems in subsection 6.3.3. However, other situations can be examined likewise.

Changes in the technological environment can be classified into two basic categories. The first category involves changes of specification in the original set of constraints. The second category involves the addition or deletion of constraints. Of course, we may have a mixture of both.

Category 1. This category consists of three basic changes of constraint specifications:

(a) Changes due to the shifting of target capacities. This only involves the shifting of previously selected target values of the fuzzy constraints while keeping the length of the tolerance interval constant. The predetermined availabilities of resources may turn out to be impractical, suboptimal, or impossible to resolve conflicts. Though the tolerance intervals remain the same, shifting of targets may displace the fuzzy ideal solution and generate new alternatives for compromising.

Given a fuzzy constraint

$$g_i(x) \lesssim \underline{b}_i;\ \bar{b}_i, \text{ with } d_i = (\bar{b}_i - \underline{b}_i), \tag{6-99}$$

for example, the changes

$$g_i(x) \lesssim \underline{\underline{b}}_i;\ \bar{\bar{b}}_i, \text{ with } \underline{b}_i \neq \underline{\underline{b}}_i,\ \bar{b}_i \neq \bar{\bar{b}}_i \text{ and } d_i = d_i' = (\bar{\bar{b}}_i - \underline{\underline{b}}_i), \tag{6-100}$$

would shift the target capacities from $\underline{b}_i$ to $\underline{\underline{b}}_i$ and $\bar{b}_i$ to $\bar{\bar{b}}_i$ while keeping the length of the original tolerance interval d_i the same. Such a shift may displace the original fuzzy ideal solution and the corresponding compromise solutions.

For instance, if the constraint, $x_1 \lesssim 5;\ 6$, in (6-71) in example 6.6 is changed to: $x_1 \lesssim 6;\ 7$, the original fuzzy ideal solution $\{(f_1, f_2) | 14 \leq f_1 \leq 16,\ 12 \leq f_2 \leq 16\}$ would be displaced to $\{(f_1, f_2) | 15 \leq f_1 \leq 17,\ 12 \leq f_2 \leq 16\}$.

(b) Changes due to the lengthening or shortening of tolerance intervals. Only the lengths of the tolerance intervals change in this situation. During the conflict resolution process, some tolerances may turn out to be too tight or too loose. To achieve a more appropriate compromise, loosening or tightening of tolerance intervals is then necessary. Such changes may displace the original fuzzy ideal solution and lead to a new compromise solution.

Given the fuzzy constraint in (6-99), the changes

$$g_i(x) \lesssim \underline{\underline{b}}_i,\ \bar{\bar{b}}_i, \text{ with } d_i \lesseqgtr d_i' = (\bar{\bar{b}}_i - \underline{\underline{b}}_i), \tag{6-101}$$

may displace the original fuzzy ideal solution and the associated compromise solutions.

For example, if the constraint: $x_2 \lesssim 6;\ 8$, in (6-71) in example 6.6 is changed to: $x_2 \lesssim 6;\ 7$, the original fuzzy ideal solution $\{(f_1, f_2) | 14 \leq f_1 \leq 16,\ 12 \leq f_2 \leq 16\}$ would be displaced to $\{(f_1, f_2) | 14 \leq f_1 \leq 17,\ 12 \leq f_2 \leq 14\}$.

(c) Changes due to the modification of coefficients. Such changes would lead to the variations of slopes. To resolve conflicts, decisionmakers may want to try out different technological production coefficients in the iterative process. This again may displace the fuzzy ideal solution and requires no further elaboration here.

Category 2. In this category, displacement of the fuzzy ideal solution is due to the deletion of existing constraints or inclusion of additional constraints. During the conflict resolution process, decisionmakers may realize that an alternative conduced by certain constraints is undesirable or inferior and its deletion may improve the final compromise. On the other hand, new alternatives may surface during the iterative negotiation process and its inclusion may enhance the conflict resolution capabilities. Either one of these changes may displace the original fuzzy ideal solution and the corresponding compromise solutions.

For example, the deletion of the constraint: $x_2 \underset{\sim}{\leq} 6; 8$, in (6-71) in example 6.6 displaces the original fuzzy ideal solution from $\{(f_1, f_2) | 14 \leq f_1 \leq 16, 12 \leq f_2 \leq 16\}$ to $\{(f_1, f_2) | 14 \leq f_1 \leq 16, 18 \leq f_2 \leq 20\}$. However, the deletion of the constraint: $2x_1 + 3x_2 \underset{\sim}{\geq} 8; 6$, has no influence on the fuzzy ideal solution.

Remark. Similar changes can be made on the fuzzy or precise objectives. They would have similar effects on the fuzzy ideal solution and the corresponding compromise solutions. Allowing such changes to take place, our conflict resolution becomes dynamic and flexible. Of course, to make the dynamic conflict resolution time dependent, the fuzzy objective functions and constraints can be formulated as functions of time. The analysis then involves the solution of fuzzy dynamic systems. Employing the concept of displaced ideal solution, value-based conflict resolution involving multiple objectives and multiple decisionmaking units can also be analyzed (Leung, 1983d).

CHAPTER 7

DYNAMICS AND OPTIMAL CONTROLS OF FUZZY SPATIAL SYSTEMS

7. 1. ON FUZZY SPATIAL SYSTEMS

So far, our discussion of spatial analysis and planning under imprecision has essentially been static in nature. Human spatial behavior, however, changes over time. With changes in the objective environment and our cognition and valuation, subdivision of space, for instance, may vary over time. Our preference for and the utilities of specific commodities also vary with the ever changing markets. Spatial partial and general equilibriums under imprecision are thus dynamic in character. To chart a long term plan, decisionmaking processes are generally multistaged with fuzzy objectives and fuzzy constraints set for each stage along the planning horizon. Through the optimal control of the dynamic process, we can select the best course of action. Therefore, spatial analysis and planning under imprecision would not be complete unless fuzzy spatial dynamics are considered. I focus in this chapter on the characterization of fuzzy dynamic spatial systems and their optimal controls. My intention is to outline a general framework for such an analysis. Dynamics of specific spatial problems such as regionalization, spatial partial and general equilibrium are however not analyzed here. I hope that interested researchers would perform more indepth analyses.

To simplify our investigation, only discrete-time dynamic systems at a point in space are analyzed. To shed some more light on the subject, a brief examination of the continuous-time situation is also made.

A general formulation of fuzzy dynamic spatial systems is provided in section 7.2. Emphasis is placed on systems with imprecise states, imprecise state-transitions, and precise or imprecise controls.

In section 7.3, our discussion is extended to the optimal controls of systems with precise and imprecise dynamics. Multistage spatial decisionmaking processes with fuzzy objectives and fuzzy constraints are analyzed. Stochasticity is also incorporated into the fuzzy dynamic systems.

In section 7.4, a remark is made on the analysis of fuzzy spatial dynamics. Time-varying and continuous-time fuzzy spatial systems are briefly introduced. Some technical problems of multistage decisionmaking processes under imprecision are also mentioned.

7.2. A GENERAL FRAMEWORK OF FUZZY DYNAMIC SPATIAL SYSTEMS

In general, dynamic systems consist of inputs, outputs, and state-transitions as three major components. Conventionally, these components are considered to be precise. Due to imperfect information, systems complexities, and human problem solving capabilities, inputs and outputs may not be precise, and state-transitions from stage to stage may only be known with imprecision. Instead of dynamic systems with a high level of precision, decisionmakers in fact operate within a more or less imprecise spatial system with fuzzy inputs, and/or outputs, and/or transitions. Through this simplification of the otherwise complicated dynamic processes, human intelligence, however, is able to make a relatively high standard of decision.

To analyze such a dynamic decisionmaking process, formalization of fuzzy dynamic systems is a necessity. In subsection 7.2.1, a formal representation of conventional dynamic systems is briefly outlined. A general framework for analyzing fuzzy dynamic systems is then proposed.

In subsection 7.2.2, systems with fuzzy states and fuzzy dynamics is analyzed. The purpose is to model the temporal dynamics of such a system.

In subsection 7.2.3, fuzzy inputs are incorporated into

the fuzzy dynamic process. The set of recurrence equations of the purely fuzzy dynamic system is derived.

7.2.1. A Representation of Conventional and Fuzzy Dynamic Systems

Let $S = (\underline{U}, \underline{X}, \underline{Y}, f, g)$ be a system in which $\underline{U}$ is the input (control) space, $\underline{X}$ is the state space, $\underline{Y}$ is the output space, $f: \underline{X} \times \underline{U} \longrightarrow \underline{X}$ is a state-transition function, and $g: \underline{X} \times \underline{U} \longrightarrow \underline{Y}$ is a output function.

Conventional dynamic systems require $\underline{U}$, $\underline{X}$, $\underline{Y}$, f, and g be precise. If $\underline{U}$, $\underline{X}$, and $\underline{Y}$ are finite, i.e. a finite set, and f and g are time-invariant with no uncertainties, then S is a time-invariant finite-state deterministic system.

Let $u_t \varepsilon \underline{U}$, $x_t \varepsilon \underline{X}$, and $y_t \varepsilon \underline{Y}$ be respectively the input, state, and output at time t, t = 0, 1, 2, ... Then, the state at time t+1 and output at time t are given respectively by the state and output equations:

$$x_{t+1} = f(x_t, u_t), \quad t = 0, 1, 2, \ldots \tag{7-1}$$

$$y_t = g(x_t, u_t), \quad t = 0, 1, 2, \ldots \tag{7-2}$$

If x_t and y_t cannot be determined with certainty by the previous state, x_t, and input, u_t, then f and g are random functions. The system S becomes a stochastic system whose state at time t+1 and output at time t are conditional probability distributions $P_{\underline{X}}(x_{t+1}| x_t, u_t)$ and $P_{\underline{Y}}(y_t| x_t, u_t)$ over $\underline{X}$ and $\underline{Y}$ respectively. In other words, x_{t+1} and y_t are conditioned on x_t and u_t through the respective conditional probability distribution functions $p_{\underline{X}}(x_{t+1}| x_t, u_t)$ and $p_{\underline{Y}}(y_t| x_t, u_t)$.

The literature on deterministic and stochastic systems is voluminous and is not reviewed here. Our emphasis is placed on the conditions under which the system S becomes fuzzy.

Obviously, with our imprecise knowledge of the transitions of states and the output function, both f and g are fuzzy. Under this situation, S is a fuzzy system whose state at time t+1, X_{t+1}, and output at time t, Y_t, are fuzzy subsets conditioned on x_t and u_t. That is, X_{t+1} and Y_t are conditioned fuzzy subsets characterized respectively by the functions $\mu_{\underline{X}}(x_{t+1}|x_t, u_t)$ and $\mu_{\underline{Y}}(y_t|x_t, u_t)$, or in short $\mu(x_{t+1}|x_t, u_t)$ and $\mu(y_t|x_t, u_t)$. Any system S with fuzzy state-transitions and fuzzy outputs is a system with fuzzy and dynamics.

Under fuzzy dynamics, we can have systems with fuzzy states and precise inputs, or with fuzzy states and fuzzy inputs. The former is a partially fuzzy dynamic system while the latter is a purely fuzzy dynamic system.

If only the states and/or inputs are fuzzy, but state-transitions and the output function are precise, then S is a fuzzy system with precise dynamics. Again, we have a partially fuzzy dynamic system.

In the following subsections, systems with fuzzy states, precise inputs, and fuzzy dynamics are first analyzed. Systems with fuzzy states, fuzzy inputs, and fuzzy dynamics are then discussed.

7.2.2. Dynamic Systems with Fuzzy States, Precise Inputs, and Fuzzy Dynamics

Let $S = (\underline{U}, \underline{X}, \underline{Y}, F, G, \mathscr{F}(\underline{X}), \mathscr{F}(\underline{Y}))$ be a fuzzy dynamic system. Let $\mathscr{F}(\underline{X})$ and $\mathscr{F}(\underline{Y})$ be fuzzy power sets of $\underline{X}$ and $\underline{Y}$ whose elements are fuzzy states and fuzzy outputs respectively. Let

$$F: \mathscr{F}(\underline{X}) \times \underline{U} \longrightarrow \mathscr{F}(\underline{X}), \qquad (7\text{-}3)$$

$$G: \mathscr{F}(\underline{X}) \times \underline{U} \longrightarrow \mathscr{F}(\underline{Y}), \qquad (7\text{-}4)$$

be the fuzzy state-transition function and fuzzy output function respectively.

Let $u_t \varepsilon \underline{U}$, $X_t \varepsilon \mathscr{F}(\underline{X})$, $Y_t \varepsilon \mathscr{F}(\underline{Y})$ be respectively the precise input, fuzzy state, and fuzzy output at time t, t = 0, 1, 2,... Then the state and output equations are:

$$X_{t+1} = F(X_t, u_t), \ t = 0, 1, 2, \ldots \qquad (7\text{-}5)$$

$$Y_t = G(X_t, u_t), \ t = 0, 1, 2, \ldots \qquad (7\text{-}6)$$

Let $x_t \varepsilon \underline{X}$ and $y_t \varepsilon \underline{Y}$. Then, the state at time t+1, X_{t+1}, and output at time t, Y_t, are fuzzy subsets obtained by:

$$\mu_{X_{t+1}}(x_{t+1}) = \sup_{x_t \varepsilon \underline{X}} \min[\mu_{X_t}(x_t), \mu_F(x_{t+1} | x_t, u_t)], \tag{7-7}$$

$$\mu_{Y_t}(y_t) = \sup_{x_t \varepsilon \underline{X}} \min[\mu_{X_t}(x_t), \mu_G(y_t | x_t, u_t)]. \tag{7-8}$$

That is, X_{t+1} and Y_t are conditioned fuzzy subsets through μ_F and μ_G respectively.

Leaving out $\underline{X}$ in "$x_t \varepsilon \underline{X}$" and all the subscripts attached to μ (unless specified otherwise, this simplification format is used throughout), (7-7) and (7-8) can be simplified to:

$$\mu(x_{t+1}) = \sup_{x_t} \min[\mu(x_t), \mu(x_{t+1} | x_t, u_t)], \tag{7-9}$$

$$\mu(y_t) = \sup_{x_t} \min[\mu(x_t), \mu(y_t | x_t, u_t)]. \tag{7-10}$$

Based on (7-5) and (7-6), the state and output at the next point in time can be obtained by

$$X_{t+2} = F(X_{t+1}, u_{t+1}) = F(F(X_t, u_t), u_{t+1}), \; t = 0, 1, 2, ... \tag{7-11}$$

$$Y_{t+1} = G(X_{t+1}, u_{t+1}) = G(F(X_t, u_t), u_{t+1}), \; t = 0, 1, 2, ... \tag{7-12}$$

Therefore, the state at time t+2, X_{t+2}, and output at time t+1, Y_{t+1}, can be expressed as:

$$\mu(x_{t+2}) = \sup_{x_{t+1}} \min[\mu(x_{t+1}), \mu(x_{t+2} | x_{t+1}, u_{t+1})], \tag{7-13}$$

$$\mu(y_{t+1}) = \sup_{x_{t+1}} \min[\mu(x_{t+1}), \mu(y_{t+1} | x_{t+1}, u_{t+1})]. \tag{7-14}$$

Substituting $\mu(x_{t+1})$ in (7-13) and (7-14) by (7-9), we have

$$\mu(x_{t+2}) = \sup_{x_{t+1}} \min\left\{\sup_{x_t} \min[\mu(x_t), \mu(x_{t+1} | x_t, u_t)], \mu(x_{t+2} | x_{t+1}, u_{t+1})\right\}, \tag{7-15}$$

$$\mu(y_{t+1}) = \sup_{x_{t+1}} \min\left\{\sup_{x_t} \min[\mu(x_t), \mu(x_{t+1}|x_t, y_t)], \right.$$
$$\left. \mu(y_{t+1}|x_{t+1}, u_{t+1})\right\}. \tag{7-16}$$

Based on the distributivity property of sup and min over one another, (7-15) and (7-16) become

$$\mu(x_{t+2}) = \sup_{x_t} \sup_{x_{t+1}} \min[\mu(x_t), \mu(x_{t+1}|x_t, u_t), \mu(x_{t+2}|x_{t+1}, u_{t+1})], \tag{7-17}$$

$$\mu(y_{t+1}) = \sup_{x_t} \sup_{x_{t+1}} \min[\mu(x_t), \mu(x_{t+1}|x_t, u_t), \mu(y_{t+1}|x_{t+1}, u_{t+1})]. \tag{7-18}$$

If the time frame of the temporal dynamics is finite, let N, $N < \infty$, be the specified and fixed termination time of the fuzzy dynamic system with $t = 0, 1, 2, \ldots, N-1$. Following the above line of reasoning, for $t = 0$, the state X_1 and output Y_0 are obtained as:

$$\mu(x_1) = \sup_{x_0} \min[\mu(x_0), \mu(x_1|x_0, u_0)], \tag{7-19}$$

$$\mu(y_0) = \sup_{x_0} \min[\mu(x_0), \mu(y_0|x_0, u_0)]. \tag{7-20}$$

At $t = 1$, the state and output are given by:

$$\mu(x_2) = \sup_{x_0} \sup_{x_1} \min[\mu(x_0), \mu(x_1|x_0, u_0), \mu(x_2|x_1, u_1)], \tag{7-21}$$

$$\mu(y_1) = \sup_{x_0} \sup_{x_1} \min[\mu(x_0), \mu(x_1|x_0, u_0), \mu(y_1|x_1, u_1)], \tag{7-22}$$

Finally, at $t = N-1$,

$$\mu(x_N) = \sup_{x_0} \ldots \sup_{x_{N-2}} \sup_{x_{N-1}} \min[\mu(x_0), \mu(x_1|x_0, u_0), \ldots, \mu(x_{N-1}|x_{N-2}, u_{N-2}), \mu(x_N|x_{N-1}, u_{N-1})], \tag{7-23}$$

$$\mu(y_{N-1}) = \sup_{x_0} \ldots \sup_{x_{N-2}} \sup_{x_{N-1}} \min[\mu(x_0),\ \mu(x_1|x_0,\ u_0),\ \ldots,\ \mu(x_{N-1}|x_{N-2},\ u_{N-2}),\ \mu(y_{N-1}|x_{N-1},\ u_{N-1})]. \qquad (7\text{-}24)$$

Thus, for the state sequence $(x_0, \ldots, x_{N-2}, x_{N-1})$, (7-23) and (7-24) can be rewritten as:

$$\mu(x_N) = \sup_{x_0,\ldots,x_{N-2},x_{N-1}} \min[\mu(x_0),\ \mu(x_1|x_0,\ u_0),\ \ldots,\ \mu(x_{N-1}|x_{N-2},\ u_{N-2}),\ \mu(x_N|x_{N-1},\ u_{N-1})], \qquad (7\text{-}25)$$

$$\mu(y_{N-1}) = \sup_{x_0,\ldots,x_{N-2},x_{N-1}} \min[\mu(x_0),\ \mu(x_1|x_0,\ u_0),\ \ldots,\ \mu(x_{N-1}|x_{N-2},\ u_{N-2}),\ \mu(y_{N-1}|x_{N-1},\ u_{N-1})]. \qquad (7\text{-}26)$$

In general, dynamic systems with fuzzy states and fuzzy dynamics over a finite time horizon can be represented by the following set of recurrence equations:

$$\mu(x_{N-w+1}) = \sup_{x_{N-w}} \min[\mu(x_{N-w}),\ \mu(x_{N-w+1}|x_{N-w},\ u_{N-w})], \quad w = 1, \ldots, N, \qquad (7\text{-}27)$$

$$\mu(y_{N-w}) = \sup_{x_{N-w}} \min[\mu(x_{N-w}),\ \mu(y_{N-w}|x_{N-w},\ u_{N-w})], \quad w = 1, \ldots, N. \qquad (7\text{-}28)$$

If $\underline{U}$, $\underline{X}$, and $\underline{Y}$ are finite, dynamic systems with fuzzy states and fuzzy dynamics can be represented in a more compact form. Let $\underline{U} = \{u_1, \ldots, u_k, \ldots, u_\ell\}$, $\underline{X} = \{x_1, \ldots, x_i, \ldots, x_m\}$, and $\underline{Y} = \{y_1, \ldots, y_j, \ldots, y_n\}$. At time t, let $\underline{X}_t = [\mu(x_1), \ldots, \mu(x_i), \ldots, \mu(x_m)]$ and $Y_t = [\mu(y_1), \ldots, \mu(y_j), \ldots, \mu(y_n)]$ be the fuzzy state and fuzzy output respectively.

For each element $u_k \in \underline{U}$, let the following be the fuzzy state-transition matrix:

$$R(x_{t+1}|x_t,u_t=u_k) = \begin{bmatrix} \mu(x_1|x_1,u_k)\ldots\mu(x_1|x_i,u_k)\ldots\mu(x_1|x_m,u_k) \\ \vdots \qquad\qquad \vdots \qquad\qquad \vdots \\ \mu(x_i|x_1,u_k)\ldots\mu(x_i|x_i,u_k)\ldots\mu(x_i|x_m,u_k) \\ \vdots \qquad\qquad \vdots \qquad\qquad \vdots \\ \mu(x_m|x_1,u_k)\ldots\mu(x_m|x_i,u_k)\ldots\mu(x_m|x_m,u_k) \end{bmatrix}. \tag{7-29}$$

Let the following be the output matrix

$$Q(y_t|x_t,u_t=u_k) = \begin{bmatrix} \mu(y_1|x_1,u_k)\ldots\mu(y_1|x_i,u_k)\ldots\mu(y_1|x_m,u_k) \\ \vdots \qquad\qquad \vdots \qquad\qquad \vdots \\ \mu(y_j|x_1,u_k)\ldots\mu(y_j|x_i,u_k)\ldots\mu(y_j|x_m,u_k) \\ \vdots \qquad\qquad \vdots \qquad\qquad \vdots \\ \mu(y_n|x_1,u_k)\ldots\mu(y_n|x_i,u_k)\ldots\mu(y_n|x_m,u_k) \end{bmatrix}. \tag{7-30}$$

Then, the state at time t+1, (7-9), and output at time t, (7-10), can be expressed as:

$$X_{t+1} = R(x_{t+1}|x_t,\ u_k) \circ X_t', \tag{7-31}$$

$$Y_t = Q(y_t|x_t,\ u_k) \circ X_t', \tag{7-32}$$

where $\circ \equiv$ max-min composition such that

$$\mu(x_{t+1}) = \max_{x_i \in \underline{X}} \min[\mu_R(x_i|x_i,\ u_k),\ \mu(x_i)], \tag{7-33}$$

$$\mu(y_t) = \max_{x_i \in \underline{X}} \min[\mu_Q(y_j|x_i,\ u_k),\ \mu(x_i)], \tag{7-34}$$

and X_t' is the transpose of X_t.

Therefore, at the termination time N, (7-23) and (7-24) become

$$X_N = R(x_N|x_{N-1},\ u_{k,N-1}) \circ R(x_{N-1}|x_{N-2},\ u_{k,N-2}) \circ \ldots \circ R(x_1|x_0,\ u_{k,0}) \circ X_0', \tag{7-35}$$

$$Y_{N-1} = R(y_{N-1}| x_{N-1}, u_{k,N-1}) \circ R(y_{N-2}| x_{N-2}, u_{k,N-2}) \circ \ldots \circ R(y_0| x_0, u_{k,0}) \circ X_0'. \qquad (7\text{-}36)$$

Consequently, the set of recurrence equations, (7-27) and (7-28), becomes

$$X_{N-w+1} = R(x_{N-w+1}| x_{N-w}, u_{k,N-w}) \circ X_{N-w}', \quad w = 1, \ldots, N, \qquad (7\text{-}37)$$

$$Y_{N-w} = R(y_{N-w}| x_{N-w}, u_{k,N-w}) \circ X_{N-w}', \quad w = 1, \ldots, N. \qquad (7\text{-}38)$$

Example 7.1. Assume that a region has been experiencing economic-decay and the regional authority wants to stem and possibly reverse the process. Let $\underline{X} = \{x_1, x_2, x_3\}$ be a state space whose element x_i, $i = 1, 2, 3$, indicates various severities of the economic-decay situation. The ordering of severity is $x_3 \succ x_2 \succ x_1$. Let $X_0 = (\mu_{X_0}(x_1), \mu_{X_0}(x_2), \mu_{X_0}(x_3)) = (0.1, 0.4, 0.9)$ be the initial state. Obviously, the region is in a state of highest severity since $\mu_{X_0}(x_3) = 0.9$ is the largest among the three.

To rectify the situation, the authority intends to invest in millions (a) nothing, (b) \$10, or (c) \$100. Let the input space be $\underline{U} = \{u_1, u_2, u_3\} = \{\$0, \$10, \$100\}$ and the input at time $t = 0$ can be $u_0 = \$0$, or $u_0 = \$10$, or $u_0 = \$100$. To simplify, let us omit the output (it can, for example, be employment level).

For different inputs, let the state-transition matrices be depicted in Fig. 7.1, 7.2, and 7.3.

$$R(x_{t+1}| x_t, u_t = u_1) = \begin{array}{c|ccc} x_{t+1} \backslash x_t & x_1 & x_2 & x_3 \\ \hline x_1 & 0.7 & 0 & 0 \\ x_2 & 0.6 & 0.6 & 0 \\ x_3 & 0.2 & 0.8 & 1 \end{array}$$

Fig. 7.1 State-transition matrix conditioned on u_1

$$R(x_{t+1}|x_t,\ u_t = u_2) = \begin{array}{c|ccc} x_{t+1} \backslash x_t & x_1 & x_2 & x_3 \\ \hline x_1 & 0.5 & 0.7 & 0.3 \\ x_2 & 0 & 0.4 & 0.6 \\ x_3 & 0 & 0 & 0.5 \end{array}$$

Fig. 7.2 State-transition matrix conditioned on u_2

$$R(x_{t+1}|x_t,\ u_t = u_3) = \begin{array}{c|ccc} x_{t+1} \backslash x_t & x_1 & x_2 & x_3 \\ \hline x_1 & 0.2 & 0.8 & 0.7 \\ x_2 & 0 & 0.2 & 0.9 \\ x_3 & 0 & 0 & 0.1 \end{array}$$

Fig. 7.3 State-transition matrix conditioned on u_3

Therefore, the respective states at time t = 1 are:

$$X_1 = R(x_1|x_0,\ u_0 = \$0) \circ X_0' = (0.1,\ 0.4,\ 0.9),$$

$$X_1 = R(x_1|x_0,\ u_0 = \$10) \circ X_0' = (0.4,\ 0.6,\ 0.5),$$

$$X_1 = R(x_1|x_0,\ u_0 = \$100) \circ X_0' = (0.7,\ 0.8,\ 0.1).$$

Thus, with zero input, the state at time t = 1, X_1, remains unchanged. The state, X_1, however, improves a little with an input of 10 million dollars. Lastly, an input of 100 million dollars leads to a better regional economic performance. The state at time t = 1 moves further away from X_3 and approaches closer to x_2 and x_1. That is, we can stem more effectively the economic downfall and reverse it to a less severe state of decay.

7.2.3. Dynamic Systems with Fuzzy States, Fuzzy Inputs and Fuzzy Dynamics

Let $S = (\underline{U}, \underline{X}, \underline{Y}, F, G, \mathscr{F}(\underline{U}), \mathscr{F}(\underline{X}), \mathscr{F}(\underline{Y}))$ be a fuzzy dynamic system. Let $\mathscr{F}(\underline{U})$, $\mathscr{F}(\underline{X})$, and $\mathscr{F}(\underline{Y})$ be fuzzy power sets of $\underline{U}$, $\underline{X}$, and $\underline{Y}$ whose elements are fuzzy inputs, fuzzy states, and fuzzy outputs respectively. Let

$$F: \mathscr{F}(\underline{X}) \times \mathscr{F}(\underline{U}) \longrightarrow \mathscr{F}(\underline{X}), \tag{7-39}$$

$$G: \mathscr{F}(\underline{X}) \times \mathscr{F}(\underline{U}) \longrightarrow \mathscr{F}(\underline{Y}), \tag{7-40}$$

be the fuzzy state-transition function and fuzzy output function respectively.

Let $U_t \varepsilon \mathscr{F}(\underline{U})$, $X_t \varepsilon \mathscr{F}(\underline{X})$, and $Y_t \varepsilon \mathscr{F}(\underline{Y})$ be respectively the fuzzy input, fuzzy state, and fuzzy output at time t, t = 0, 1, 2, ... Then, the state and output equations are:

$$X_{t+1} = F(X_t, U_t), \quad t = 0, 1, 2, \ldots \tag{7-41}$$

$$Y_t = G(X_t, U_t), \quad t = 0, 1, 2, \ldots \tag{7-42}$$

Let $u_t \varepsilon \underline{U}$, $x_t \varepsilon \underline{X}$, $y_t \varepsilon \underline{Y}$, and $(x_t, u_t) \varepsilon \underline{X} \times \underline{U}$. Then the state at time t+1, X_{t+1}, and output at time t, Y_t, are conditioned fuzzy subsets given by

$$\mu(x_{t+1}) = \sup_{x_t} \sup_{u_t} \min[\mu(x_t, u_t), \mu(x_{t+1}|x_t, u_t)], \tag{7-43}$$

$$\mu(y_t) = \sup_{x_t} \sup_{u_t} \min[\mu(x_t, u_t), \mu(y_t|x_t, u_t)]. \tag{7-44}$$

If x_t and u_t are non-interactive, then (7-43) and (7-44) become

$$\mu(x_{t+1}) = \sup_{x_t} \sup_{u_t} \min[\mu(x_t), \mu(u_t), \mu(x_{t+1}|x_t, u_t)], \tag{7-45}$$

$$\mu(y_t) = \sup_{x_t} \sup_{u_t} \min[\mu(x_t), \mu(u_t), \mu(y_t|x_t, u_t)], \tag{7-46}$$

or equivalently,

$$\mu(x_{t+1}) = \sup_{x_t} \min\left\{\mu(x_t), \sup_{u_t} \min[\mu(u_t), \mu(x_{t+1}|x_t, u_t)]\right\}, \tag{7-47}$$

$$\mu(y_t) = \sup_{x_t} \min\left\{\mu(x_t),\ \sup_{u_t} \min[\mu(u_t),\ \mu(y_t | x_t,\ u_t)]\right\}. \tag{7-48}$$

Based on (7-41) and (7-42), the state and output at the next point in time can be obtained by

$$X_{t+2} = F(X_{t+1},\ U_{t+1}) = F(F(X_t,\ U_t),\ U_{t+1}),\ t = 0,\ 1,\ 2,... \tag{7-49}$$

$$Y_{t+1} = G(X_{t+1},\ U_{t+1}) = G(F(X_t,\ U_t),\ U_{t+1}),\ t = 0,\ 1,\ 2,... \tag{7-50}$$

That is, the state at time t+2, X_{t+2}, and output at time t+1, Y_{t+1}, can be obtained by:

$$\mu(x_{t+2}) = \sup_{x_{t+1}} \sup_{u_{t+1}} \min[\mu(x_{t+1},\ u_{t+1}),\ \mu(x_{t+2} | x_{t+1},\ u_{t+1})], \tag{7-51}$$

$$\mu(y_{t+1}) = \sup_{x_{t+1}} \sup_{u_{t+1}} \min[\mu(x_{t+1},\ u_{t+1}),\ \mu(y_{t+1} | x_{t+1},\ u_{t+1})], \tag{7-52}$$

For x_{t+1} and u_{t+1} non-interactive, (7-51) and (7-52) become

$$\mu(x_{t+2}) = \sup_{x_{t+1}} \sup_{u_{t+1}} \min[\mu(x_{t+1}),\ \mu(u_{t+1}),\ \mu(x_{t+2} | x_{t+1}, u_{t+1})], \tag{7-53}$$

$$\mu(y_{t+1}) = \sup_{x_{t+1}} \sup_{u_{t+1}} \min[\mu(x_{t+1}),\ \mu(u_{t+1}),\ \mu(y_{t+1} | x_{t+1}, u_{t+1})]. \tag{7-54}$$

Substituting $\mu(x_{t+1})$ in (7-53) and (7-54) by (7-45), we obtain

$$\mu(x_{t+2}) = \sup_{x_{t+1}} \sup_{u_{t+1}} \min\Big\{\sup_{x_t} \sup_{u_t} \min[\mu(x_t),\ \mu(u_t),\ \mu(x_{t+1} | x_t,\ u_t)],\ \mu(u_{t+1}),\ \mu(x_{t+2} | x_{t+1},\ u_{t+1})\Big\}, \tag{7-55}$$

$$\mu(y_{t+1}) = \sup_{x_{t+1}} \sup_{u_{t+1}} \min\Big\{\sup_{x_t} \sup_{u_t} \min[\mu(x_t), \mu(u_t), \mu(y_t|x_t, u_t)], \mu(u_{t+1}), \mu(y_{t+1}|x_{t+1}, u_{t+1})\Big\}. \qquad (7\text{-}56)$$

By the distributivity property of sup and min over one another, (7-55) and (7-56) become

$$\mu(x_{t+2}) = \sup_{x_t} \sup_{x_{t+1}} \sup_{u_t} \sup_{u_{t+1}} \min[\mu(x_t), \mu(x_{t+1}|x_t, u_t), \mu(x_{t+2}|x_{t+1}, u_{t+1}), \mu(u_t), \mu(u_{t+1})], \qquad (7\text{-}57)$$

$$\mu(y_{t+1}) = \sup_{x_t} \sup_{x_{t+1}} \sup_{u_t} \sup_{u_{t+1}} \min[\mu(x_t), \mu(y_t|x_t, u_t), \mu(y_{t+1}|x_{t+1}, u_{t+1}), \mu(u_t), \mu(u_{t+1})]. \qquad (7\text{-}58)$$

By the same token, at the termination time, the state and output are obtained respectively by:

$$\mu(x_N) = \sup_{x_0} \dots \sup_{x_{N-2}} \sup_{x_{N-1}} \sup_{u_0} \dots \sup_{u_{N-2}} \sup_{u_{N-1}} \min [\mu(x_0), \mu(x_1|x_0, u_0), \dots, \mu(x_{N-1}|x_{N-2}, u_{N-2}), \mu(x_N|x_{N-1}, u_{N-1}), \mu(u_0), \dots, \mu(u_{N-2}), \mu(u_{N-1})], \qquad (7\text{-}59)$$

$$\mu(y_{N-1}) = \sup_{x_0} \dots \sup_{x_{N-2}} \sup_{x_{N-1}} \sup_{u_0} \dots \sup_{u_{N-2}} \sup_{u_{N-1}} \min [\mu(x_0), \mu(y_0|x_0, u_0), \dots, \mu(y_{N-2}|x_{N-2}, u_{N-2}), \mu(y_{N-1}|x_{N-1}, u_{N-1}), \mu(u_0), \dots, \mu(u_{N-2}), \mu(u_{N-1})]. \qquad (7\text{-}60)$$

Let $(x_0, \dots, x_{N-2}, x_{N-1})$ and $(u_0, \dots, u_{N-2}, u_{N-1})$ be respectively the state and input sequences. Then, the final state and output in (7-59) and (7-60) can be rewritten as:

$$\mu(x_N) = \sup_{\substack{x_0,\ldots,x_{N-2},x_{N-1}\\ u_0,\ldots,u_{N-2},u_{N-1}}} \min[\mu(x_0),\ \mu(x_1|x_0,\ u_0),\ \ldots,\ \mu(x_{N-1}|x_{N-2},\ u_{N-2}),\ \mu(x_N|x_{N-1},\ u_{N-1}),\ \mu(u_0),\ \ldots,\ \mu(u_{N-2}),\ \mu(u_{N-1})], \qquad (7\text{-}61)$$

$$\mu(y_{N-1}) = \sup_{\substack{x_0,\ldots,x_{N-2},x_{N-1}\\ u_0,\ldots,u_{N-2},u_{N-1}}} \min[\mu(x_0),\ \mu(y_0|x_0,\ u_0),\ \ldots,\ \mu(y_{N-2}|x_{N-2},\ u_{N-2}),\ \mu(y_{N-1}|x_{N-1},\ u_{N-1}),\ \mu(u_0),\ \ldots,\ \mu(u_{N-2}),\ \mu(u_{N-1})]. \qquad (7\text{-}62)$$

Therefore, dynamic systems with fuzzy states, fuzzy inputs, and fuzzy dynamics over a finite time horizon can be characterized by the following set of recurrence equations:

$$\mu(x_{N-w+1}) = \sup_{x_{N-w}} \sup_{u_{N-w}} \min[\mu(x_{N-w}),\ \mu(u_{N-w}),\ \mu(x_{N-w+1}|x_{N-w},\ u_{N-w})],\ w = 1,\ \ldots,\ N, \qquad (7\text{-}63)$$

$$\mu(y_{N-w}) = \sup_{x_{N-w}} \sup_{u_{N-w}} \min[\mu(x_{N-w}),\ \mu(u_{N-w}),\ \mu(y_{N-w}|x_{N-w},\ u_{N-w})],\ w = 1,\ \ldots,\ N. \qquad (7\text{-}64)$$

For $\underline{U}$, $\underline{X}$, and $\underline{Y}$ finite, the set of recurrence equations becomes:

$$X_{N-w+1} = U_{N-w} \circ R(x_{N-w+1}|x_{N-w},\ u_{N-w}) \circ X'_{N-w},\ w = 1,\ \ldots,\ N, \qquad (7\text{-}65)$$

$$Y_{N-w} = U_{N-w} \circ Q(y_{N-w}|x_{N-w},\ u_{N-w}) \circ X'_{N-w},\ w = 1,\ \ldots,\ N. \qquad (7\text{-}66)$$

where U_{N-w}, X_{N-w+1}, Y_{N-w} are respectively the fuzzy input, fuzzy state, and fuzzy output at time w; R and Q are respectively the fuzzy transition and fuzzy output matrices; and X'_{N-w}

is the transpose of X_{N-w}.

Example 7.2. Consider example 7.1. Let the state space $\underline{X}$, input space $\underline{U}$, state at time $t = 0$, X_0, and the state-transition matrices in Fig. 7.1, 7.2, and 7.3 be the same. In place of the precise inputs \$0, \$10, and \$100 (in millions), let the input at time $t = 0$ be "*high* regional investment" which is defined by the membership function $U_0 = [\mu_{U_0}(u_1), \mu_{U_0}(u_2), \mu_{U_0}(u_3)] = [\mu(\$0), \mu(\$10), \mu(\$100)] = (0, 0.2, 0.9)$. By (7-45), the state X_1 at time $t = 1$ is given by:

$$
\begin{aligned}
\mu_{X_1}(x_1) &= \max\Big\{\min[\mu_{X_0}(x_i), \mu_{U_0}(u_j), \mu_R(x_1|x_i, u_j)], \\
&\qquad i = 1, 2, 3;\ j = 1, 2\Big\} \\
&= \max\Big\{\min[\mu_{X_0}(x_1), \mu_{U_0}(u_1), \mu_R(x_1|x_1, u_1)], \\
&\qquad \min[\mu_{X_0}(x_1), \mu_{U_0}(u_2), \mu_R(x_1|x_1, u_2], \\
&\qquad \min[\mu_{X_0}(x_1), \mu_{U_0}(u_3), \mu_R(x_1|x_1, u_3], \\
&\qquad \min[\mu_{X_0}(x_2), \mu_{U_0}(u_1), \mu_R(x_1|x_2, u_1], \\
&\qquad \min[\mu_{X_0}(x_2), \mu_{U_0}(u_2), \mu_R(x_1|x_2, u_2], \\
&\qquad \min[\mu_{X_0}(x_2), \mu_{U_0}(u_3), \mu_R(x_1|x_2, u_3], \\
&\qquad \min[\mu_{X_0}(x_3), \mu_{U_0}(u_1), \mu_R(x_1|x_3, u_1], \\
&\qquad \min[\mu_{X_0}(x_3), \mu_{U_0}(u_2), \mu_R(x_1|x_3, u_2], \\
&\qquad \min[\mu_{X_0}(x_3), \mu_{U_0}(u_3), \mu_R(x_1|x_3, u_3]\Big\} \\
&= \max\{\min[0.1, 0, 0.7], \min[0.1, 0.2, 0.5], \\
&\qquad \min[0.1, 0.9, 0.2], \min[0.4, 0, 0], \\
&\qquad \min[0.4, 0.2, 0.7], \min[0.4, 0.9, 0.8], \\
&\qquad \min[0.9, 0, 0], \min[0.9, 0.2, 0.3], \\
&\qquad \min[0.9, 0.9, 0.7]\} \\
&= \max\{0, 0.1, 0.1, 0, 0.2, 0.4, 0, 0.2, 0.7\} = 0.7
\end{aligned}
$$

By the same token, $\mu_{X_1}(x_2) = 0.9$ and $\mu_{X_1}(x_3) = 0.4$. Thus, the state at time $t = 1$ is $X_1 = (0.7, 0.9, 0.4)$, an improvement from X_0.

Apparently, dynamic systems with fuzzy states, fuzzy inputs, and fuzzy dynamics generalize that with precise inputs. For example, (7-37) and (7-38) are respectively special cases of (7-65) and (7-66). Moreover, dynamic systems under precision is a special case within the imprecision framework. Though our analysis has been on systems with fuzzy dynamics, systems with fuzzy states, fuzzy inputs, and precise dynamics can likewise be analyzed. Without repetition, such a case is left for the discussion of optimal controls in subsection 7.3.1.

7. 3. MULTISTAGE SPATIAL DECISION PROBLEMS WITH FUZZY GOALS AND FUZZY CONSTRAINTS

In the preceding section, a general description of fuzzy dynamic systems is given. Often, our interest is to steer a dynamic system towards a goal along an optimal path within a fixed period of time. A common decisionmaking problem involves a multistage process with fuzzy constraints imposed on the controls (inputs) at each point in time (stage) and a fuzzy goal set for the final state. Obviously, it is an optimal control problem of a fuzzy dynamic system and is a subject of analysis in this section.

Since state transitions (systems dynamics) can be precise, fuzzy, or stochastic, we then have different optimal control problems, albeit more or less similar in format. In subsection 7.3.1, multistage optimal control problems with precise dynamics, fuzzy constraints, and fuzzy goals are analyzed. In subsection 7.3.2, our analysis is extended to systems with fuzzy dynamics, fuzzy constraints, and fuzzy goals. In subsection 7.3.3, multistage optimal control problems with stochastic dynamics, fuzzy constraints and fuzzy goals are examined.

7. 3. 1. Multistage Decisionmaking Processes with Fuzzy Goals, Fuzzy Constraints, and Precise Dynamics

Let $S = (\underline{U}, \underline{X}, f)$ be a time-invariant finite-state deterministic system. Let $\underline{U} = \{u_1, \ldots, u_k, \ldots, u_\ell\}$ and $X = \{x_1, \ldots, x_i, \ldots, x_m\}$ be the finite control (input) and state spaces respectively. Let $f\colon \underline{X} \times \underline{U} \longrightarrow \underline{X}$ be the time-invariant state-transition function. Let $u_t \in \underline{U}$ and $x_t \in \underline{X}$ be respectively the precise control and state at time t, $t = 0, 1, 2, \ldots$.

Let N, $N < \infty$, be a prespecified and fixed termination time. Then, the state equation is given by:

$$x_{t+1} = f(x_t, u_t),\ t = 0, 1, 2, \ldots, N-1. \tag{7-67}$$

Assume that at time t, the control u_t, due to imprecision, is subjected to a fuzzy constraint C_t defined by the membership function

$$\begin{aligned} &\mu_{C_t}\colon \underline{U} \longrightarrow [0, 1], \\ &u_t \longrightarrow \mu_{C_t}(u_t),\ t = 0, 1, 2, \ldots, N-1. \end{aligned} \tag{7-68}$$

For the whole planning horizon, the controls are then subjected to a sequence of fuzzy constraints $(\mu_{C_0}, \mu_{C_1}, \ldots, \mu_{C_{N-1}})$.

Let there be a fuzzy goal G_N set for the termination time with the following defining membership function.

$$\begin{aligned} &\mu_{G_N}\colon \underline{X} \longrightarrow [0, 1], \\ &x_N \longrightarrow \mu_{G_N}(x_N). \end{aligned} \tag{7-69}$$

Given the initial state x_0, the multistage optimal control problem is to find an optimal sequence of controls $(u_0^*, u_1^*, \ldots, u_{N-1}^*)$ so that the fuzzy goal is maximized and the sequence of fuzzy constraints are satisfied (see Fig. 7.4 for a schematic representation). Here, the outputs of the system are essentially equated with the states.

Formally, the multistage decisionmaking process can be treated as a fuzzy optimization process over time (Bellman and Zadeh, 1970). The decision space D over $\underline{U} \times \underline{U} \times \ldots \times \underline{U}$ is $D = C_0 \cap C_1 \cap \ldots \cap C_{N-1} \cap G_N$ which is defined by the membership function

$$\begin{aligned} \mu_D(u_0, u_1, \ldots, u_{N-1}) = \min[&\mu_{C_0}(u_0), \mu_{C_1}(u_1), \ldots, \\ &\mu_{C_{N-1}}(u_{N-1}), \mu_{G_N}(x_N)]. \end{aligned} \tag{7-70}$$

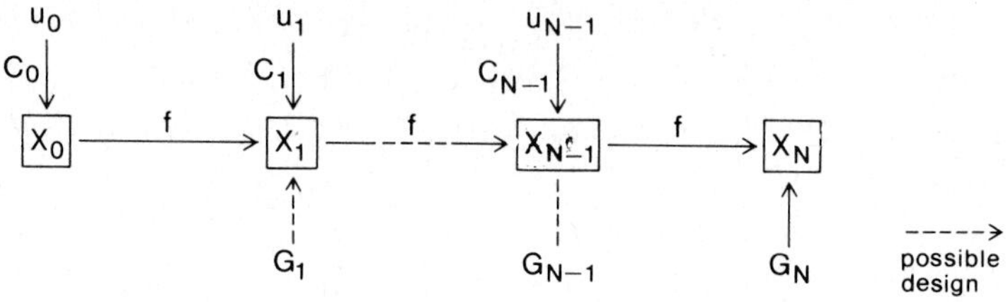

Fig. 7.4 Schematic representation of a multistage decisionmaking problem with fuzzy goal(s) and fuzzy constraints

The problem then is to find an optimal control sequence $(u_0^*, u_1^*, \ldots, u_{N-1}^*)$ such that

$$\mu_D(u_0^*, u_1^*, \ldots, u_{N-1}^*)$$

$$= \max_{u_0, u_1, \ldots, u_{N-1}} \min$$

$$[\mu_{C_0}(u_1), \mu_{C_1}(u_1), \ldots, \mu_{C_{N-1}}(u_{N-1}), \mu_{G_N}(x_N)], \qquad (7\text{-}71)$$

with $x_N = f(f(\ldots f(f(x_0, u_0), u_1, \ldots, u_{N-1}))$ being a function of $x_0, u_0, u_1, \ldots, u_{N-1}$. Since $x_N = f(x_{N-1}, u_{N-1})$, by (7-67), and max and min are mutually distributive, then (7-71) can be rewritten as:

$$\mu_D(u_0^*, u_1^*, \ldots, u_{N-1}^*)$$

$$= \max_{u_0, u_1, \ldots, u_{N-2}} \min\Big\{\mu_{C_0}(u_0), \mu_{C_1}(u_1), \ldots, \mu_{C_{N-2}}(u_{N-2}),$$

$$\max_{u_{N-1}} \min[\mu_{C_{N-1}}(u_{N-1}), \mu_{G_N}(f(x_{N-1}, u_{N-1}))]\Big\}. \qquad (7\text{-}72)$$

Following the same line of reasoning, (7-72), with respect to u_{N-2}, can be expressed as:

$$\mu_D(u_0^*, u_1^*, \ldots, u_{N-1}^*)$$

$$= \max_{u_0, u_1, \ldots, u_{N-3}} \min\Big\{\mu_{C_0}(u_0), \mu_{C_1}(u_1), \ldots, \mu_{C_{N-3}}(u_{N-3}),$$

$$\max_{u_{N-2}} \min\left\{\mu_{C_{N-2}}(u_{N-2}),\ \max_{u_{N-1}} \min [\mu_{C_{N-1}}(u_{N-1}),\ \mu_{G_N}(f(x_{N-1},\ u_{N-1}))]\right\}. \tag{7-73}$$

Let

$$\mu_{G_{N-1}}(x_{N-2}) = \max_{u_{N-1}} \min[\mu_{C_{N-1}}(u_{N-1}),\ \mu_{G_N}(f(x_{N-1},\ u_{N-1}))]. \tag{7-74}$$

Then (7-73) can be rewritten as:

$$\mu_D(u_0^*,\ u_1^*,\ \ldots,\ u_{N-1}^*) = \max_{u_0, u_1, \ldots, u_{N-3}} \min\left\{\mu_{C_0}(u_0),\ \mu_{C_1}(u_1),\ \ldots,\ \mu_{C_{N-3}}(u_{N-3}),\ \max_{u_{N-2}} \min[\mu_{C_{N-2}}(u_{N-2}),\ \mu_{G_{N-1}}(x_{N-1})]\right\}. \tag{7-75}$$

Repeating this backward iteration, Bellman and Zadeh (1970) obtain the following set of recurrence equations:

$$\mu_{G_{N-w}}(x_{N-w}) = \max_{u_{N-w}} \min[\mu_{C_{N-w}}(u_{N-w}),\ \mu_{G_{N-w+1}}(x_{N-w+1})],$$

with

$$x_{N-w+1} = f(x_{N-w},\ u_{N-w}),\ w = 1,\ 2,\ \ldots,\ N. \tag{7-76}$$

Therefore, the optimal control sequence $(u_0^*,\ u_1^*, \ldots,\ u_{N-1}^*)$ is obtained by the successive maximizing values of u_{N-w} in (7-76).

Let σ_{N-w}: $\underline{X} \longrightarrow \underline{U}$ be a policy function, with $u_{N-w}^* = \sigma_{N-w}(x_{N-w})$, $w = 1,\ 2,\ \ldots,\ N$. That is, the control u_{N-w}^* depends on the current state x_{N-w}. Due to its specific iterative format, the solution of (7-76) can be determined by conventional dynamic programming method.

Example 7. 3. (Bellman and Zadeh, 1970)

Let $\underline{U} = \{u_1,\ u_2\}$ and $\underline{X} = \{x_1,\ x_2,\ x_3\}$ be the respective control and state space in a N-stage, $N = 2$, decisionmaking process. Let the fuzzy constraints at $t = 0$ and $t = 1$ be defined respectively by:

$$\mu_{C_0}(u_1) = 0.7,\ \mu_{C_0}(u_2) = 1, \text{ and, } \mu_{C_1}(u_1) = 1,\ \mu_{C_2}(u_2) = 0.6.$$

Let a fuzzy goal G_2 be set for $t = 2$ and is defined by

$$\mu_{G_2}(x_1) = 0.3,\ \mu_{G_2}(x_2) = 1,\ \mu_{G_2}(x_3) = 0.8.$$

Assume that the state-transition is precise and is given in Fig. 7.5.

$$f(x_t,\ u_t) = \begin{array}{c|ccc} u_t \backslash x_t & x_1 & x_2 & x_3 \\ \hline u_1 & x_1 & x_3 & x_1 \\ u_2 & x_2 & x_1 & x_3 \end{array}$$

Fig. 7.5 Precise state-transition matrix

Then by (7-74), the fuzzy goal induced at $t = 1$ can be obtained in matrix form as follows:

$$[\mu_{G_1}(x_1),\ \mu_{G_1}(x_2),\ \mu_{G_1}(x_3)]$$

$$=[\mu_{C_1}(u_1),\ \mu_{C_1}(u_2)] \circ \begin{bmatrix} \mu_{G_2}(f(x_1,u_1)) & \mu_{G_2}(f(x_2,u_1)) & \mu_{G_2}(f(x_3,u_1)) \\ \mu_{G_2}(f(x_1,u_2)) & \mu_{G_2}(f(x_2,u_2)) & \mu_{G_2}(f(x_3,u_2)) \end{bmatrix}$$

$$=[\mu_{C_1}(u_1),\ \mu_{C_1}(u_2)] \circ \begin{bmatrix} \mu_{G_2}(x_1) & \mu_{G_2}(x_3) & \mu_{G_2}(x_1) \\ \mu_{G_2}(x_2) & \mu_{G_2}(x_1) & \mu_{G_2}(x_3) \end{bmatrix}$$

$$=[1,\ 0.6] \circ \begin{bmatrix} 0.3 & 0.8 & 0.3 \\ 1 & 0.3 & 0.8 \end{bmatrix} = (0.6,\ 0.8,\ 0.6).$$

Let σ_t be the policy function. Then, the optimal solution is

$$\sigma_1(x_1) = u_2,\ \sigma_1(x_2) = u_1,\ \sigma_1(x_3) = u_2.$$

By the same token, at t = 0, we have

$$[\mu_{G_0}(x_1),\ \mu_{G_0}(x_2),\ \mu_{G_0}(x_3)]$$

$$= [\mu_{C_0}(u_1),\ \mu_{C_0}(u_2)] \circ \begin{bmatrix} \mu_{G_1}(x_1) & \mu_{G_1}(x_3) & \mu_{G_1}(x_1) \\ \mu_{G_1}(x_2) & \mu_{G_1}(x_1) & \mu_{G_1}(x_3) \end{bmatrix}$$

$$= [0.7,\ 1] \circ \begin{bmatrix} 0.6 & 0.6 & 0.6 \\ 0.8 & 0.6 & 0.6 \end{bmatrix} = (0.8,\ 0.6,\ 0.6).$$

The optimal solution is

$$\sigma_0(x_1) = u_2,\ \sigma_0(x_2) = u_1 \text{ or } u_2,\ \sigma_0(x_3) = u_1 \text{ or } u_2.$$

Thus, at t = 0, if the state is x_2, the optimal decision is (u_2, u_1) with $\mu_{G_2} = 0.8$. If the state is x_2, the optimal decision is (u_1, u_2) or (u_2, u_2) with $\mu_{G_2} = 0.6$. If the state is x_3, the optimal decision is (u_1, u_2) or (u_2, u_2) with $\mu_{G_2} =$ 0.6.

The dynamic programming approach discussed above is based on a backward-iteration procedure proposed by Bellman and Zadeh (1970). We solve the optimal control problem in (7-71) by solving recursively backward the set of recurrence equations in (7-76) from the final stage to the initial stage. It, however, may be regarded by some decisionmakers as counterintuitive. People may want to proceed from the initial stage to the final stage.

The system in (7-71) can actually be obtained by proceeding forward from the initial stage to the final stage. At time t = 1, the optimal control problem is

$$\mu_D(u_0^*) = \max_{u_0} \min[\mu_{C_0}(u_0),\ \mu_{G_1}(x_1)]. \tag{7-77}$$

At time t = 2, the optimal control problem becomes

$$\mu_D(u_0^*, u_1^*)$$

$$= \max_{u_0} \max_{u_1} \min[\mu_{C_0}(u_0), \mu_{C_1}(u_1), \mu_{G_2}(x_2)]$$

$$= \max_{u_0} \max_{u_1} \min[\mu_{C_0}(u_0), \mu_{C_1}(u_1), \mu_{G_2}(f(f(x_0, u_0), u_1))]. \quad (7\text{-}78)$$

By the same token, at time t = N, we have

$$\mu_D(u_0^*, u_1^*, \ldots, u_{N-1}^*)$$

$$= \max_{u_0} \max_{u_1} \ldots \max_{u_{N-1}} \min[\mu_{C_0}(u_0), \mu_{C_1}(u_1), \ldots, \mu_{C_{N-1}}(u_{N-1}),$$

$$\mu_{G_N}(f(f(\ldots f(f(x_0, u_0), u_1, \ldots, u_{N-1}))]. \quad (7\text{-}79)$$

Thus, for the control sequence $(u_0, u_1, \ldots, u_{N-1})$, (7-79) becomes (7-71) and the solution of the N-stage decision problem can be obtained by a branch-and-bound technique (for the solution algorithm see Chang and Pavlidis, 1977; Kacprzyk, 1978; 1979).

Remark. (a) The decision space D in (7-70) is formulated on the assumption that the fuzzy goal is only set for the final stage. In some situations, however, a fuzzy goal may be set for each stage, starting from stage 1 (see Fig. 7.4). The decision space then is $D = C_0 \cap G_1 \cap C_1 \cap G_2 \cap \ldots \cap C_{N-1} \cap G_N$ defined by

$$\mu_D(u_0, u_1, \ldots, u_{N-1})$$

$$= \min[\mu_{C_0}(u_0), \mu_{G_1}(x_1), \mu_{C_1}(u_1), \mu_{G_2}(x_2), \ldots,$$

$$\mu_{C_{N-1}}(u_{N-2}), \mu_{G_N}(x_N)]. \quad (7\text{-}80)$$

(b) Similar to the discussion in chapter 5, min- is not the only operator for decision-space construction. Depending on our purpose, algebraic-product, convex combination, and max-, for example, are operators which can be used to combine fuzzy constraints and fuzzy goals of all stages. The formulations are straight forward and are not elaborated here.

Example 7.4. (An Application to the Policy Analysis of a Regional Development Problem)

Kacprzyk (1983) applied the multistage approach to identify optimal alternatives for a development problem in a predominantly agricultural region. The original formulation, as reported, was rather involving and complicated and a largely simplified version was given as an illustration.

The region under study is predominantly agricultural with 450,000 acres of arable land and a population size of 120,000. Due to huge out-migration of young people, the region has been experiencing severe economic decay. The authority believes that the continuous out-migration is due to the perception of a poor quality of life in the region. To stem out-migration is to improve the perception of life quality through an adequate appropriation of development fund to the region.

To simplify, a three-year development plan with respect to four life quality indicators is considered. Let

$x_{1t} \triangleq$ average subsidies per acre,

$x_{2t} \triangleq$ per capita sanitation (water and sewage treatment) expenditures,

$x_{3t} \triangleq$ per capita health care expenditures, and

$x_{4t} \triangleq$ expenditures for paved roads (new roads and maintenance),

be four life quality indicators measured in dollars.

Let the regional development outlays be partitioned as follows:

(a) 5 percent for subsidies,

(b) 25 percent for sanitation,

(c) 45 percent for health care, and

(d) 25 percent for infrastructure.

Let u_t, $t = 0, 1, 2$, be the only control variable.

Let the initial values of the life quality indicators be: $x_0^1 = \$0.5$, $x_0^2 = \$15$, $x_0^3 = \$27$, and $x_0^4 = \$1{,}700{,}000$.

Rather than determining the optimal policy, only two prespecified policies are considered and compared (it is quite a common practice in real-life planning):

Policy 1: $u_0 = u_1 = u_2 = \$8,000,000$, and
Policy 2: $u_0 = \$7,500,000$, $u_1 = \$8,000,000$, $u_2 = \$8,500,000$.

Tables 7.1 and 7.2 depict respectively the consecutive values of the life quality indicators under policies 1 and 2. To compare the two trajectories, fuzzy constraints and fuzzy goals are set for each stage. Let the constraint be the fuzzy "less than or equal to" type depicted in (5-7), and the goal be the fuzzy "greater than or equal to" type depicted in (5-40). (The actual functions were not given). Table 7.3 depicts the permissible minimal and maximal values of the fuzzy constraints and fuzzy goals.

Let min- be the operator for constructing the decision space D. The two policies can then be assessed as follows:

(a) Policy 1.

$$\mu_D(u_0, u_1, u_2) = \mu_D(8,000,000,\ 8,000,000,\ 8,000,000)$$

$$= \min\Big\{\min[\mu_{C_0}(8,000,000),\ \min(\mu_{G_1^1}(0.88),\ \mu_{G_1^2}(16.7),\ \mu_{G_1^3}(30),$$

$$\mu_{G_1^4}(2,000,000))],\ \min[\mu_{C_1}(8,000,000),\ \min(\mu_{G_2^1}(0.88),$$

$$\mu_{G_2^2}(16.7),\ \mu_{G_2^3}(30),\ \mu_{G_2^4}(2,000,000))],\ \min[\mu_{C_2}(8,000,000),$$

$$\min(\mu_{G_3^1}(0.88),\ \mu_{G_3^2}(16.7),\ \mu_{G_3^3}(30),\ \mu_{G_3^4}(2,000,000))]\Big\}$$

$$= \min\{\min[0.5,\ \min(1,\ 1,\ 1,\ 1)],$$
$$\min[0.8,\ \min(0.9,\ 0.85,\ 1,\ 1)],$$
$$\min[1, \min(0.52,\ 0.28,\ 0.5,\ 0.33)]\}$$
$$= \min\{0.5,\ 0.8,\ 0.28\} = 0.28.$$

(b) Policy 2.

$$\mu_D(u_0, u_1, u_2) = \mu_D(7,500,000,\ 8,000,000,\ 8,500,000)$$

$$= \min\Big\{\min[\mu_{C_0}(7,500,000),\ \min(\mu_{G_1^1}(0.83),\ \mu_{G_1^2}(15.6),\ \mu_{G_1^3}(28.1),$$

$$\mu_{G_1^4}(1,875,000))],\ \min[\mu_{C_1}(8,000,000),\ \min(\mu_{G_2^1}(0.88),$$

$\mu_{G_2^2}(16.7),\ \mu_{G_2^3}(30),\ \mu_{G_2^4}(2{,}000{,}000))],\ \min[\mu_{C_2}(8{,}500{,}000),$

$\min(\mu_{G_3^1}(0.94),\ \mu_{G_3^2}(17.7),\ \mu_{G_3^3}(31.9),\ \mu_{G_3^4}(2{,}125{,}000))]\}$

$$= \min\{\min[1, \min(0.92, 0.8, 0.55, 0.75)], \min[0.8, \min(0.9, 0.85, 1, 1)], \min[0.75, \min(0.76, 0.68, 1, 1)]\}$$

$$= \min\{0.55, 0.8, 0.68\} = 0.55.$$

Thus, policy 2 is better than policy 1. Though Kacprzyk did not give a complete description of the whole problem and some of the specifications were unaccounted for, we more or less can get a rough idea on how the whole multistage regional development policy under imprecision was or can be analyzed.

Table 7.1 Consecutive values ($) of the life quality indicators under policy 1

t	u_t	x_t^1	x_t^2	x_t^3	x_t^4
0	8,000,000				
1	8,000,000	0.88	16.7	30	2,000,000
2	8,000,000	0.88	16.7	30	2,000,000
3		0.88	16.7	30	2,000,000

Table 7.2 Consecutive values ($) of the life quality indicators under policy 2

t	u_t	x_t^1	x_t^2	x_t^3	x_t^4
0	7,500,000				
1	8,000,000	0.83	15.6	28.1	1,875,000
2	8,500,000	0.88	16.7	30.0	2,000,000
3		0.94	17.7	31.9	2,125,000

Table 7.3 Specifications of fuzzy goals and fuzzy constraints

	Permissible Values of a Constraint (C_t)		Permissible Values of a Goal (G_t^i, i = 1, 2, 3, 4)							
			G_t^1		G_t^2		G_t^3		G_t^4	
t	minimal ($\underline{u}_t$)	maximal ($\bar{u}_t$)	$\underline{x}_t^1$	$\bar{x}_t^1$	$\underline{x}_t^2$	$\bar{x}_t^2$	$\underline{x}_t^3$	$\bar{x}_t^3$	$\underline{x}_t^4$	$\bar{x}_t^4$
0	7,500,000	8,500,000								
1	7,750,000	9,000,000	0.60	0.85	14	16.0	27	29	1,800,000	1,900,000
2	8,000,000	10,000,000	0.70	0.90	15	17.0	28	30	1,900,000	2,000,000
3			0.75	1.00	16	18.5	29	31	1,950,000	2,100,000

7. 3. 2. Multistage Decisionmaking Processes with Fuzzy Goals, Fuzzy Constraints, and Fuzzy Dynamics

The multistage decisionmaking problems discussed in subsection 7.3.1 only involves precise dynamics. Though the goals and constraints are fuzzy, the states, controls, and state-transition functions are precise. Within a complex system, states, controls, and state-transitions are most likely fuzzy. We focuses on this class of purely fuzzy multistage decisionmaking processes in this subsection.

Let $S = (\underline{U}, \underline{X}, F, \mathscr{F}(\underline{U}), \mathscr{F}(\underline{X}))$ be a time-invariant finite-state fuzzy system. Let $\underline{U} = \{u_1, \ldots, u_k, \ldots, u_\ell\}$ and $\underline{X} = \{x_1, \ldots, x_i, \ldots, x_m\}$ be the finite control and state spaces respectively. Let $\mathscr{F}(\underline{U})$ and $\mathscr{F}(\underline{X})$ be the fuzzy power sets of $\underline{U}$ and $\underline{X}$. Let $F\colon \mathscr{F}(\underline{X}) \times \mathscr{F}(\underline{U}) \longrightarrow \mathscr{F}(\underline{X})$ be the time-invariant fuzzy state-transition function such that

$$X_{t+1} = F(X_t, U_t), \; t = 0, 1, 2, \ldots, N-1. \qquad (7\text{-}81)$$

where $U_t \in \mathscr{F}(\underline{U})$, X_t and $X_{t+1} \in \mathscr{F}(\underline{X})$ are fuzzy controls at time t, and fuzzy states at time t and t+1 respectively.

Let N, $N < \infty$, be the termination time. If $x_t \in \underline{X}$ and $u_t \in \underline{U}$ are non-interactive, then by (7-45) and (7-47), the state equation for X_{t+1} is

$$\mu_{X_{t+1}}(x_{t+1})$$

$$= \max_{x_t \in \underline{X}} \min\left\{\mu_{X_t}(x_t), \max_{u_t \in \underline{U}} \min[\mu_{U_t}(u_t), \mu_F(x_{t+1} | x_t, u_t)]\right\}. \qquad (7\text{-}82)$$

To simplify, $\underline{X}$ in "$x_t \in \underline{X}$", $\underline{U}$ in "$u_t \in \underline{U}$", and all the subscripts attached to μ are assumed to be understood and are left out henceforth.

Assume that at each time t, the fuzzy control U_t is subjected to a fuzzy constraint C_t defined by the membereship function in (7-68). Moreover, the final state X_N is evaluated is terms of a fuzzy goal G_N defined by the membership function in (7-69).

Since both the controls and states are fuzzy and are jointly distributed over $\underline{X} \times \underline{U}$ with

$$\mu(x_t, u_t) = \min[\mu(x_t), \mu(u_t)], \quad t = 0, 1, 2, \ldots, N-1, \tag{7-83}$$

then the fuzzy constraint and the fuzzy goal should be jointly evaluated and form a fuzzy relation R_t in $\underline{X} \times \underline{U}$ defined by the membership function:

$$\mu_{R_t}(x_t, u_t) = \min[\mu_{C_t}(u_t), \mu_{G_{t+1}}(x_{t+1})],$$

$$t = 0, 1, 2, \ldots, N-1, \tag{7-84}$$

with R_t describing how well the fuzzy constraint and fuzzy goal are satisfied by U_t and X_{t+1}.

Thus, for a fuzzy control U_t and the resulting fuzzy state X_{t+1}, Baldwin and Pilsworth (1982) define the truthfulness of constraint and goal satisfaction as

$$T(u_t, R_t, X_{t+1}) = U_t \circ R_t \circ X_{t+1}, \quad t = 0, 1, 2, \ldots, N-1, \tag{7-85}$$

which is defined by the membership function

$$T(u_t, R_t, X_{t+1})$$
$$= \max_{x_{t+1}} \min\left\{\max_{u_t} \min[\mu(u_t), \mu_{R_t}(x_t, u_t)], \mu(x_{t+1})\right\}$$
$$= \max_{x_{t+1}} \min\left\{\max_{u_t} \min[\mu(u_t), \mu_{C_t}(u_t), \mu_{G_{t+1}}(x_{t+1})], \mu(x_{t+1})\right\}. \tag{7-86}$$

By the distributivity of max and min over one another, (7-86) becomes

$$T(u_t, R_t, X_{t+1}) = \min\left\{\max_{u_t} \min[\mu(u_t), \mu_{C_t}(u_t)],\ \max_{x_{t+1}} \min[\mu(x_{t+1}), \mu_{G_{t+1}}(x_{t+1})]\right\},$$

$$t = 0, 1, 2, \ldots, N-1. \tag{7-87}$$

Example 7. 5. (Baldwin and Pilsworth, 1982)

Let $\underline{X} = \{x_1, x_2, x_3\}$ and $\underline{U} = \{u_1, u_2\}$. Let the fuzzy constraint at $t = 0$ be: $\mu_{C_0}(u_1) = 0.3$, $\mu_{C_0}(u_2) = 0.8$, and the fuzzy goal at time $t = 1$ be: $\mu_{G_1}(x_1) = 0.5$, $\mu_{G_2}(x_2) = 0.9$,

$\mu_{G_3}(x_3) = 0.2.$

Assume that the state-transitions (Fig. 7.6, and 7.7) are imprecise.

$$R(x_{t+1}|x_t,\ u_t = u_1) = \begin{array}{c|ccc} x_{t+1} \backslash x_t & x_1 & x_2 & x_3 \\ \hline x_1 & 1 & 0.7 & 0.2 \\ x_2 & 0.7 & 1 & 0.8 \\ x_3 & 0.3 & 0.3 & 1 \end{array},$$

Fig. 7.6 State-transition matrix conditioned on u_1

$$R(x_{t+1}|x_t,\ u_t = u_2) = \begin{array}{c|ccc} x_{t+1} \backslash x_t & x_1 & x_2 & x_3 \\ \hline x_1 & 0.7 & 1 & 1 \\ x_2 & 1 & 0.7 & 0.3 \\ x_3 & 0.3 & 0.3 & 0.2 \end{array}.$$

Fig. 7.7 State-transition matrix conditioned on u_2

Let the initial state be $X_0 = (0.6,\ 0.1,\ 0.8)$ and the initial input be $U_0 = (0.2,\ 0.9)$ and $U_0 = (0.7,\ 0.5)$. Then, for $U_0 = (0.2,\ 0.9)$, the state at time $t = 1$, X_1, is given by

$$\begin{aligned}
\mu_{X_1}(x_1) &= \max\Big\{\min[\mu_{X_0}(x_i),\ \mu_{U_0}(u_j),\ \mu_{C_0}(u_j),\ \mu_R(x_1|x_i,\ u_j)], \\
&\qquad i = 1,\ 2,\ 3;\ j = 1,\ 2\Big\} \\
&= \max\{\min[0.6,\ 0.2,\ 0.3,\ 1],\ \min[0.6,\ 0.9,\ 0.8,\ 0.7], \\
&\qquad \min[0.1,\ 0.2,\ 0.3,\ 0.7],\ \min[0.1,\ 0.9,\ 0.8,\ 1], \\
&\qquad \min[0.8,\ 0.2,\ 0.3,\ 0.2],\ \min[0.8,\ 0.9,\ 0.8,\ 1]\} \\
&= \max\{0.2,\ 0.6,\ 0.1,\ 0.1,\ 0.2,\ 0.8\} = 0.8.
\end{aligned}$$

Similarly, $\mu_{X_1}(x_1) = 0.8$ and $\mu_{X_1}(x_3) = 0.3$. That is, $X_1 = (0.8, 0.8, 0.3)$.

If $U_0 = (0.7, 0.5)$, by the same token, $X_1 = (0.6, 0.7, 0.7)$.

Therefore, for $U_0 = (0.2, 0.9)$, we have, by (7-87),

$$T(U_0, R_t, X_1)$$

$$=\min\{[(0.3, 0.8)\circ(0.2, 0.9)],[(0.8, 0.8, 0.3)\circ(0.5, 0.9, 0.2)]\}$$
$$=\min\{0.8, 0.8\} = 0.8.$$

For $U_0 = (0.7, 0.5)$, we have

$$T(U_0, R_t, X_1)$$

$$=\min\{[(0.3, 0.8)\circ(0.7, 0.5)],[(0.6, 0.6, 0.7)\circ(0.5, 0.9, 0.2)]\}$$
$$=\min\{0.5, 0.5\} = 0.5.$$

Obviously, $U_0 = (0.2, 0.9)$ is an optimal solution.

Apparently, for a sequence of fuzzy controls $U_0, U_1, \ldots, U_{N-1}$, the truth function of the N-stage decisionmaking process is given by $T(U_0, U_1, \ldots, U_{N-1}, R, X)$. The fuzzy relation $R \subset \underline{U} \times \underline{U} \times \ldots \times \underline{U} \times \underline{X}$ is defined by

$$R(u_0, u_1, \ldots, u_{N-1}, x_N)$$
$$= \min[\mu_{C_0}(u_0), \mu_{C_1}(u_1), \ldots, \mu_{C_{N-1}}(u_{N-1}), \mu_{G_N}(x_N)], \quad (7\text{-}88)$$

and the expression of T is given in theorem 7.1 (Baldwin and Pilsworth, 1982).

Theorem 7.1. Let $U_0, U_1, \ldots, U_{N-1}$ be a sequence of fuzzy controls in a N-stage decisionmaking process. Let X_N be the final state. If $C_0, C_1, \ldots, C_{N-1}$ is a sequence of fuzzy constraints imposed on $U_0, U_1, \ldots, U_{N-1}$ respectively, and G_N is a fuzzy goal set for X_N, then the truth function of the process is given by

$$T(U_0, U_1, \ldots, U_{N-1}, R, X_N)$$

$$= \min\Big\{\max_{u_0} \min[\mu(u_0), \mu_{C_0}(u_0)], \max_{u_1} \min[\mu(u_1), \mu_{C_1}(u_1)], \ldots,$$

$$\max_{u_{N-1}} \min[\mu(u_{N-1}), \mu_{C_{N-1}}(u_{N-1})],$$

$$\max_{x_N} \min[\mu(x_N), \mu_{G_N}(x_N)]\Big\}. \quad (7\text{-}89)$$

Therefore, with a fuzzy goal imposed on the final state and a sequence of fuzzy constraints imposed on the fuzzy controls, the optimal control of the N-stage decisionmaking process is to determine the control sequence $(U_0^*, U_1^*, \ldots, U_{N-1}^*)$ such that

$$\mu_D(u_1^*, u_2^*, \ldots, u_{N-1}^*)$$

$$= \max_{U_0,U_1,\ldots,U_{N-1}} \min\Big\{\max_{u_0} \min[\mu(u_0), \mu_{C_0}(u_0)],$$

$$\max_{u_1} \min[\mu(u_1), \mu_{C_1}(u_1)], \ldots,$$

$$\max_{u_{N-1}} \min[\mu(u_{N-1}), \mu_{C_{N-1}}(u_{N-1})],$$

$$\max_{x_N} \min[\mu(x_N), \mu_{G_N}(x_N)]\Big\}, \quad (7\text{-}90)$$

where X_N can be expressed as a function of $X_0, U_0, U_1, \ldots, U_{N-1}$ via (7-81).

Treating (7-90) as a defining function of the decision space D, then by the distributivity property of max and min over one another, (7-90) can be rewritten as

$$\mu_D(u_0^*, u_1^*, \ldots, u_{N-1}^*)$$

$$= \max_{U_0,U_1,\ldots,U_{N-1}} T(U_0, U_1, \ldots, U_{N-1}, R, X_N)$$

$$= \max_{U_0,U_1,\ldots,U_{N-2}} \min\Big\{\max_{u_0} \min[\mu(u_0), \mu_{C_0}(u_0)],$$

$$\max_{u_1} \min[\mu(u_1), \mu_{C_1}(u_1)], \ldots,$$

$$\max_{U_{N-1}} \left\{ \max_{u_{N-1}} \min[\mu(u_{N-1}), \mu_{C_{N-1}}(u_{N-1})] \right\},$$

$$\max_{x_N} \min[\mu(x_N), \mu_{G_N}(x_N)] \Big\}. \qquad (7\text{-}91)$$

Let

$$\mu_{G_{N-1}}(X_{N-1}) = \min\left\{ \max_{U_{N-1}} \left\{ \max_{u_{N-1}} \min[\mu(u_{N-1}), \mu_{C_{N-1}}(u_{N-1})] \right\}, \right.$$

$$\left. \max_{x_N} \min[\mu(x_N), \mu_{G_N}(x_N)] \right\}, \qquad (7\text{-}92)$$

where X_N can be expressed as a function of X_0, U_0, U_1, ..., U_{N-1} via (7-81), and $\mu_{G_{N-1}}(X_{N-1})$ only depends on the fuzzy control U_{N-1}.

Thus, (7-91) becomes

$$\mu_D(u_0^*, u_1^*, \ldots, u_{N-1}^*)$$

$$= \max_{U_0, U_1, \ldots, U_{N-2}} \min\left\{ \max_{u_0} \min[\mu(u_0), \mu_{C_0}(u_0)], \right.$$

$$\max_{u_1} \min[\mu(u_1), \mu_{C_1}(u_1)], \ldots,$$

$$\left. \max_{u_{N-2}} \min[\mu(u_{N-2}), \mu_{C_{N-2}}(u_{N-2})], \mu_{G_{N-1}}(X_{N-1}) \right\}.$$

$$(7\text{-}93)$$

Apparently, the multistage decisionmaking process with fuzzy dynamics can be characterized by

$$\mu_{G_{N-w}}(X_{N-w}) = \min\left\{ \max_{u_{N-w}} \left\{ \max_{u_{N-w}} \min[\mu(u_{N-w}), \mu_{C_{N-w}}(u_{N-w})] \right\}, \right.$$

$$\left. \mu_{G_{N-w+1}}(X_{N-w+1}) \right\},$$

with,

$$X_{N-w+1} = F(X_{N-w}, u_{N-w}), \; w = 1, 2, \ldots, N. \qquad (7\text{-}94)$$

Obviously, serious dimensionality is encountered in solving recursively (7-94) for the optimal sequence U_0^*, U_1^*, ..., U_{N-1}^*. A "fuzzy interpolation" technique is proposed by Baldwin and Pilsworth (1982) to mimic human problem solving under such a situation. Its effectiveness, however, still needs empirical supports.

7.3.3. Multistage Decisionmaking Processes with Fuzzy Goals, Fuzzy Constraints, and Stochastic Dynamics

As discussed in subsection 2.5 of chapter 2, randomness and imprecision can exist simultaneously within a multistage decisionmaking process. Fuzzy goals and constraints, for example, can be incorporated into a stochastic dynamic system. Our discussion in this subsection only focuses on the optimal control of such a system. It is, however, not the only case possible.

Let $S = (\underline{U}, \underline{X}, p)$ be a time-invariant finite-state stochastic system. Let $\underline{U} = \{u_1, \ldots, u_k, \ldots, u_\ell\}$ and $\underline{X} = \{x_1, x_i, \ldots, x_m\}$ be respectively the finite control and state spaces. Let the transition of states be characterized by a Markov-chain conditional probability function such that the probability of attaining state x_{t+1} at time t is $p(x_{t+1}|x_t, u_t)$, with $x_t, x_{t+1} \varepsilon \underline{X}$ and $u_t \varepsilon \underline{U}$.

At each control stage t, let the precise control u_t be subjected to a fuzzy constraint C_t defined by (7-68). Let N, $N < \infty$, be the termination time with a fuzzy goal G_N set for the final state x_t via (7-69).

Thus, the difference between the present system and the system discussed in subsection 7.3.1 is that the latter is a deterministic dynamic system with fuzzy goal and fuzzy constraints while the present is a stochastic system with fuzzy goal and fuzzy constraints.

Since the system under control is now stochastic, the N-stage problem then is to find an optimal control sequence $(u_0^*, u_1^*, \ldots, u_{N-1}^*)$ so that the probability of attaining the fuzzy goal at time N is maximized and the fuzzy constraints C_0, $C_1, \ldots, C_{N-1}$ are satisfied. Parallel to conventional stochastic programming formulations, to maximize the probability of attaining a goal is to maximize the expectation of the random variable characterizing the goal.

Since the fuzzy goal G_N can be regarded as a fuzzy event defined by μ_{G_N}, (2-150), then the conditional probability of this event given x_{N-1} and u_{N-1} is given by

$$P(G_N|x_{N-1}, u_{N-1}) = E[\mu_{G_N}] = \sum_{x_N} \mu_{G_N}(x_N)\ p(x_N|x_{N-1}, u_{N-1}). \tag{7-95}$$

Since $E[\mu_{G_N}]$ is a function of x_{N-1} and u_{N-1}, then $E[\mu_{G_N}(x_N)]$ can be treated similarly as $\mu_{G_N}(x_N)$ is treated in the deterministic system.

Thus, the decision space D of the N-stage decision problem is given by

$$\mu_D(u_0, u_1, \ldots, u_{N-1})$$
$$= \min\left\{\mu_{C_0}(u_0), \mu_{C_1}(u_1), \ldots, \mu_{C_{N-1}}(u_{N-1}), E[\mu_{G_N}(x_N)]\right\}, \tag{7-96}$$

and the problem, as shown by Bellman and Zadeh (1970), is to find an optimal control sequence $(u_0^*, u_1^*, \ldots, u_{N-1}^*)$ such that

$$\mu_D(u_0^*, u_1^*, \ldots, u_{N-1}^*)$$
$$= \max_{u_0,u_1,\ldots,u_{N-1}} \min\left\{\mu_{C_0}(u_0), \mu_{C_1}(u_1), \ldots, \mu_{C_{N-1}}(u_{N-1}), E[\mu_{G_N}(x_N)]\right\}, \tag{7-97}$$

where x_N is a function of x_0, u_0, u_1, ..., u_{N-1}.

Employing the same arguments in subsection 7.3.1, (7-97) can be rewritten as

$$\mu_D(u_0^*, u_1^*, \ldots, u_{N-1}^*)$$
$$= \max_{u_0,u_1,\ldots,u_{N-2}} \min\left\{\mu_{C_0}(u_0), \mu_{C_1}(u_1), \ldots, \mu_{C_{N-2}}(u_{N-2}), \max_{u_{N-1}} \min\left\{\mu_{C_{N-1}}(u_{N-1}), E[\mu_{G_N}(f(x_{N-1}, u_{N-1}))]\right\}\right\}$$
$$= \max_{u_0,u_1,\ldots,u_{N-2}} \min\left\{\mu_{C_0}(u_0), \mu_{C_1}(u_1), \ldots, \mu_{C_{N-3}}(u_{N-3}), \max_{u_{N-2}} \min\left\{\mu_{C_{N-2}}(u_{N-2}), \max_{u_{N-1}} \min\left\{\mu_{C_{N-1}}(u_{N-1}), E[\mu_{G_N}(f(x_{N-1}, u_{N-1}))]\right\}\right\}\right\}. \tag{7-98}$$

Let

$$E[\mu_{G_{N-1}}(x_{N-1})]$$

$$= \max_{u_{N-1}} \min\left\{\mu_{C_{N-1}}(u_{N-1}),\ E[\mu_{G_N}(f(x_{N-1},\ u_{N-1}))]\right\}. \qquad (7\text{-}99)$$

Then, (7-98) can be expressed as

$$\mu_D(u_0^*,\ u_1^*,\ \ldots,\ u_{N-1}^*)$$

$$= \max_{u_0, u_1, \ldots, u_{N-3}} \min\left\{\mu_{C_0}(u_0),\ \mu_{C_1}(u_1),\ \ldots,\ \mu_{C_{N-3}}(u_{N-3}),\ \max_{u_{N-2}} \min\left\{\mu_{C_{N-2}}(u_{N-2}),\ E[\mu_{G_{N-1}}(x_{N-1})]\right\}\right\}. \qquad (7\text{-}100)$$

Therefore, the optimal control sequence $(u_0^*,\ u_1^*,\ \ldots,\ u_{N-1}^*)$ is obtained by the successive maximizing values of u_{N-w} in the following set of recurrence equations:

$$\mu_{G_{N-w}}(x_{N-w}) = \max_{u_{N-w}} \min\left\{\mu_{C_{N-w}}(u_{N-w}),\ E[\mu_{G_{N-w+1}}(x_{N-w+1})]\right\},$$

with,

$$E[\mu_{G_{N-w+1}}(x_{N-w+1})]$$

$$= \sum_{x_{N-w+1}} \mu_{G_{N-w+1}}(x_{N-w+1})\ p(x_{N-w+1} | x_{N-w},\ u_{N-w}),$$

$$w = 1,\ 2,\ \ldots,\ N. \qquad (7\text{-}101)$$

Let $\sigma_{N-W}\colon \underline{X} \longrightarrow \underline{U}$ be a policy function. Then, the optimal control u_{N-w}^* is given by $u_{N-w}^* = \sigma_{N-w}(x_{N-w})$, $w = 1,\ 2,\ \ldots,\ N$.

Example 7. 6. (Bellman and Zadeh, 1970)

Let the number of stages, control space, state space, fuzzy constraints, and fuzzy goal in the dynamic system be the same as those specified in example 7.3. That is, $N = 2$; $\underline{U} = \{u_1,\ u_2\}$; $\underline{X} = \{x_1,\ x_2,\ x_3\}$; $\mu_{C_0}(u_1) = 0.7$, $\mu_{C_0}(u_1) = 1$; $\mu_{C_1}(u_1) = 1$, $\mu_{C_2}(u_2) = 0.6$, and $\mu_{G_2}(x_1) = 0.3$, $\mu_{G_2}(x_2) = 1$, $\mu_{G_2}(x_3) = 0.8$.

With respect to u_1 and u_2, let the conditional probability functions be:

$$p(x_{t+1}|x_t,\ u_t = u_1) = \begin{array}{c|ccc} x_{t+1} \backslash x_t & x_1 & x_2 & x_3 \\ \hline x_1 & 0.8 & 0 & 0.8 \\ x_2 & 0.1 & 0.1 & 0.1 \\ x_3 & 0.1 & 0.9 & 0.1 \end{array}, \qquad (7\text{-}102)$$

$$p(x_{t+1}|x_t,\ u_t = u_2) = \begin{array}{c|ccc} x_{t+1} \backslash x_t & x_1 & x_2 & x_3 \\ \hline x_1 & 0.1 & 0.8 & 0.1 \\ x_2 & 0.9 & 0.1 & 0 \\ x_3 & 0 & 0.1 & 0.9 \end{array}. \qquad (7\text{-}103)$$

By (7-101), the expectation of the fuzzy goal G_2 can be obtained in matrix form by:

$$(E[\mu_{G_2}(x_1)],\ E[\mu_{G_2}(x_2)],\ E[\mu_{G_2}(x_3)])$$

$$=[\mu_{G_2}(x_1), \mu_{G_2}(x_2), \mu_{G_2}(x_3)] \cdot \begin{bmatrix} p(x_1|x_1,u_1) & p(x_1|x_2,u_1) & p(x_1|x_3,u_1) \\ p(x_2|x_1,u_1) & p(x_2|x_2,u_1) & p(x_2|x_3,u_1) \\ p(x_3|x_1,u_1) & p(x_3|x_2,u_1) & p(x_3|x_3,u_1) \end{bmatrix}$$

$$=(0.3,\ 1,\ 0.8) \cdot \begin{bmatrix} 0.8 & 0 & 0.8 \\ 0.1 & 0.1 & 0.1 \\ 0.1 & 0.9 & 0.1 \end{bmatrix} = (0.42,\ 0.82,\ 0.42).$$

The expectation of G with respect to u can likewise be obtained. The result is tabulated as:

$$E[\mu_{G_2}] = \begin{array}{c|ccc} u_1 \backslash x_1 & x_1 & x_2 & x_3 \\ \hline u_1 & 0.42 & 0.82 & 0.42 \\ u_2 & 0.93 & 0.42 & 0.75 \end{array}$$

Based on (7-101), the fuzzy goal at t = 1 is obtaianed as:

$$\mu_{G_1}(x_1) = 0.6,\ \mu_{G_1}(x_2) = 0.82,\ \mu_{G_1}(x_3) = 0.6.$$

With respect to the policy function σ_t, the optimal solution is

$$\sigma_1(x_1) = u_2, \ \sigma_1(x_2) = u_1, \ \sigma_1(x_3) = u_2.$$

By the same token, at t = 0, we have

$$E[\mu_{G_1}] = \begin{array}{c|ccc} u_0 \backslash x_0 & x_1 & x_2 & x_3 \\ \hline u_1 & 0.62 & 0.62 & 0.62 \\ u_2 & 0.80 & 0.62 & 0.60 \end{array};$$

and,

$$\mu_{G_0}(x_1) = 0.8, \ \mu_{G_0}(x_2) = 0.62, \ \mu_{G_0}(x_3) = 0.62;$$

and,

$$\sigma_0(x_1) = u_1, \ \sigma_0(x_2) = u_1, \ \sigma_0(x_3) = u_1.$$

7. 4. CONCLUDING REMARKS ON THE ANALYSIS OF FUZZY DYNAMIC SPATIAL SYSTEMS

Some basic notions of fuzzy dynamic spatial systems have been discussed in this chapter. The optimal controls of multistage time-invariant decisionmaking processes with prespecified and fixed termination time have also been investigated. The analysis, albeit brief, can be extended to solve multistage decision problems involving implicitly specified termination time, fuzzy termination time, and infinite termination time (Kacprzyk, 1983). It has been demonstrated that stochasticity and imprecision may exist simultaneously and can be analyzed under the fuzzy set framework. Moreover, dynamic systems under imprecision generalize the conventional dynamic systems.

Though our analysis is on time-invariant dynamic systems, fuzzy time-varying systems can also be analyzed. If $S = (\underline{U}, \underline{X}, \underline{Y}, F_t, G_t, \mathscr{F}(\underline{U}), \mathscr{F}(\underline{X}), \mathscr{F}(\underline{Y}))$ is a fuzzy time-varying dynamic system, then the state and output equations, parallel to (7-41) and (7-42), are given by

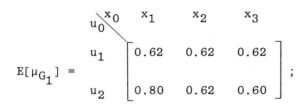

$$X_{t+1} = F_t(X_t, U_t), \ t = 0, 1, 2, \ldots \tag{7-104}$$

$$Y_t = F_t(X_t, U_t), \ t = 0, 1, 2, \ldots \tag{7-105}$$

where $U_t \varepsilon \mathscr{F}(\underline{U})$, $X_t \varepsilon \mathscr{F}(\underline{X})$, and $Y_t \varepsilon \mathscr{F}(\underline{Y})$ are respectively fuzzy input, fuzzy state, and fuzzy output at time t, and

$$F_t: \ \mathscr{F}(\underline{X}) \times \mathscr{F}(\underline{U}) \longrightarrow \mathscr{F}(\underline{X}), \tag{7-106}$$

$$G_t: \ \mathscr{F}(\underline{X}) \times \mathscr{F}(\underline{U}) \longrightarrow \mathscr{F}(\underline{Y}), \tag{7-107}$$

are respectively the fuzzy time-varying state-transition and output functions.

Since F_t and G_t vary with time, the analysis of time-varying systems is more complicated. Multistage decision problems involving such systems would lead to a rather demanding task computationally.

So far, our analysis only focuses on discrete-time fuzzy dynamic systems. The time frame, of course, can also be continuous. Similarly, states and outputs can take on real numbers as their values. Under this situation, the state and output equations are given respectively by:

$$\frac{dx_t}{dt} = \phi(x_t, u_t), \tag{7-108}$$

$$y_t = \psi(x_t, u_t), \tag{7-109}$$

where the state-transition function ϕ and the output function ψ are real-valued fuzzy function with $u_t \varepsilon \underline{U} \subseteq \mathbb{R}$ (or $\mathbb{R}^n$ for multiple inputs), $x_t \varepsilon \underline{X} \subseteq \mathbb{R}$ (or $\mathbb{R}^n$), and $y_t \varepsilon \underline{Y} \subseteq \mathbb{R}$ (or $\mathbb{R}^n$).

If u_t, x_t, and y_t are fuzzy numbers, and if some are fast-evolving while the others are slow-evolving, we may experience catastrophic changes in the system. Fuzzy catastrophe then needs to be analyzed.

The analysis of continuous-time fuzzy dynamic systems involves differentiation and integration of fuzzy functions. It is perehaps an interesting topic for further investigation.

In solving a fuzzy dynamic spatial system, dimensionality is a critical problem which needs to be handled with care.

Under imprecision, the dimensionality problem encountered in conventional dynamic programming problems is compounded. Efficient algorithms, not yet fully developed, are required to solve such systems.

CHAPTER 8

CONCLUSION

8. 1. SUMMARY

Fundamentals of spatial analysis and planning under imprecision have been laid out and discussed in this monograph. Imprecision, long evaded by spatial scientists, has been treated as another source of uncertainty within spatial systems. Such uncertainty is directly related to systems complexity, human perception, thought, and behavior. It has been demonstrated that our perception, thought, decision, and behavior over space and time can be restalled through fuzzy set theory with analytical elegance and empirical relevance. The models constructed appear to be natural, flexible, and humanized. Though the idea is not to fuzzify precise systems, fuzzy set models often generalize models under certainty. More importantly, problems which cannot be analyzed or solved by the conventional methods can be effectively handled within the fuzzy set framework. Thus, from the philosophical, theoretical, and practical points of view, fuzzy set spatial models are more embracing and revealing. Together with randomness, we can construct a more complete picture of uncertainty.

Throughout the monograph, from the fundamental issues of spatial analysis, namely regional concepts and regionalization, to the complex issues of spatial dynamics, imprecision has been systematically unravelled and analyzed. New concepts have also been proposed. It is quite persuasive that uncertainty due to imprecision is prevailing and human valuation should be a focal point of analysis within complex systems.

Since we are at the early stage of applying fuzzy set theory to spatial analysis and planning under imprecision, the scope and depth of the present monograph is incomplete and limited. Further research effort is thus crucial to the development of spatial models by which human behavior over space and time can be systematically described, analyzed and predicted.

While fuzzy set research has branched out in various directions, my proposal for further research is restricted only to areas concerning spatial analysis and planning. The purpose is to outline brief directions along which present effort should continue and further research might follow. Basic empirical issues of fuzzy set theory are first discussed. Issues in spatial analysis and planning are then addressed.

8. 2. DIRECTIONS FOR FURTHER RESEARCH

In subsection 8.2.1, I concentrate the investigation on the evaluation of some basic empirical issues of fuzzy set theory. It deals with the determination of membership functions defining fuzzy subsets. Though our discussion is not space specific, the issues are relevant to spatial analysis and planning.

In subsection 8.2.2, the psycholinguistic realities of fuzzy connectives are discussed. Operators such as min- and max- are examined from the empirical point of view.

In subsection 8.2.3, directions for further research in spatial analysis and planning are proposed. The purpose is to stimulate a full-scale investigation of spatial analysis and planning under imprecision through fuzzy set theory.

Due to the limitation of space, depth is traded off for breadth. In place of a detailed discussion, I only high light the main points in each argument. Thorough investigation of the issues is left for future research.

8. 2. 1. Empirical Aspects of the Membership Functions

The successful application of fuzzy-set spatial models,

similar to other operational mathematical models, depends to a large extent on whether or not they are appropriate representations of the real world. The real world we are modelling is complex spatial systems embedded with human perceptions and thoughts. The importance of the fuzzy set approach is that it brings forth the notion of imprecision which constitutes uncertainty within spatial systems. Though we process fuzzy information and think in imprecise terms, our decisionmaking processes could be analyzed in a formal manner.

The basis of such an analysis is that imprecise concepts can be formally represented. They can be conceptualized as fuzzy subsets defined by precise membership functions. In our discussion, the membership functions and the operations on fuzzy subsets are specified a priori. Though they are theoretically sound and exhibit satisfactory empirical relevance, it is still necessary to investigate how well they represent what they are intended to represent. In other words, it is imperative to investigate in what way membership functions can be appropriately determined on the empricial basis.

In regional classification, for example, the characterizing functions of the regions need to be attained before spatial units can be classified. In the grouping for regions, it is necessary to first determine the membership functions defining similarity or distance between spatial units. In the analysis of spatial choice, membership functions defining fuzzy preferences and utilities need to be obtained prior to the determination of choice or behavior. In spatial optimization, specifications of satisfaction functions associated with the fuzzy objectives and fuzzy constraints are prerequisite for the optimal search of alternatives. Such an empirical procedure is in fact no different from bringing any operational models, fuzzy or not, into practice. It is by no means unique, as some critics think, to fuzzy set models.

Some researchers believe that subjectivity and imprecision is non-experimental. Therefore, membership functions characterizing fuzzy subsets cannot be empirically determined. Countering such a belief is another group of researchers who think subjectivity and imprecision can be empirically unravelled in most cases. The empirical aspect becomes even more important when group behavior is the subject of analysis. For most researchers, it would be more comforting to know that systematic procedures can be formulated to determine an appropriate membership function defining a fuzzy concept. This is the area where research effort has been lacking and further research is required.

To determine a membership function, it is essential to determine the grade of membership of an element in the universe of discourse on which the fuzzy subset is defined. Methods employed have to be objective in the estimation sense. Under some circumstances, statistical procedures can be formulated to derive membership functions from experiments. Repeated experimentation generally gives a reliable and representative membership function. There are two basic methods for determining the membership function on the experimental basis. Other methods are essentially their variants.

Method 1. Each experiment consists of four basic components:

(a) The universe of discourse $\underline{X}$;

(b) A fixed element x_0 in $\underline{X}$;

(c) A precise response to the statement "x is A", with A being the fuzzy concept defined in $\underline{X}$. It indicates the confirmation of x_0 being A. For example, we can assign 1 for a positive answer and 0 for a negative answer;

(d) A condition S consisting of objective factors, physical and/or psychological, under which A is being classified.

Thus, in each trial a definite answer of whether or not x_0 is A is obtained. In n trials, the grade of membership of x_0 being A can be defined as the frequency of x_0 being considered as A:

$$\mu_A(x_0) \triangleq \text{frequency of } x_0 \text{ being considered as A}$$

$$= \frac{\text{the number of positive responses to "}x_0 \text{ is A"}}{n}. \quad (8\text{-}1)$$

The idea then is to count how many respondents consider x_0 as A and we equate the frequency with the grade of membership of x_0 being A.

Based on various empirical results (see for example Kochen and Badre, 1975; Kochen, 1975; Hersh and Caramazza, 1976; Hersh, Caramazza, and Brownell, 1979), this method appears to be quite reliable in unravelling a representative membership function for a group. Justifications for such an experimental procedure and its derivatives can be found, for example, in Nowakowska (1977), Norwich and Turksen (1981, 1982).

Slightly different procedures have also been employed to obtain membership functions for an individual. In place of asking a "yes" or "no" answer, the degree of belief is sought and is equated with the grade of membership of x_0 being A. (See for example Kochen and Badre, 1975; Dreyfuss, Kochen, Robinson, and Badre, 1975; Macvicar-Whelan, 1978; and Leung, 1981a, b, c). The scales obtained by this method, however, may be distorted by a number of response biases (Cronbach, 1950). Methods based on ordinal-scaled responses have also been developed with some successes (Thole, Zimmermann, and Zysno, 1979).

Method 2. Since an interval is an explication of a fuzzy concept, then a more novel way to estimate the membership function is through the method of set-valued statistics (Wang, 1985). The theory is based on the notion of falling random sets. Instead of a point, the result of each experiment is a random set (an ordinary set) on the universe of discourse. A fuzzy subset, i.e. the membership function, can be treated as a fuzzy projection of the random sets.

Each experiment consists of four basic components:

(a) The universe of discourse $\underline{X}$;

(b) A fixed element x_0 in $\underline{X}$;

(c) A movable ordinary subset $\underline{A}_*$ in $\underline{X}$ which is connected to a fuzzy subset A corresponding to a fuzzy concept α. The determination of $\underline{A}_*$ is a precise classification of α. It is an approximate extension of α, $\underline{A}_* = [70\%, 100\%]$ (in agricultural employment), for example, is a precise interval explicating the fuzzy concept $\alpha \triangleq$ "*predominantly agricultural*".

(d) A condition S consisting of objective factors, physical or psychological, under which α is classified. It restricts the movement of $\underline{A}_*$.

Thus, in each trial whether or not x_0 is in $\underline{A}_*$ is precise. In n trials, the grade of membership of x_0 to A can be defined as the frequency of x_0 in $\underline{A}_*$:

$$\mu_A(x_0) \triangleq \text{frequency of } x_0 \text{ in } \underline{A}_*$$

$$= \frac{\text{the number of times of "} x_0 \in \underline{A}_* \text{"}}{n}. \qquad (8\text{-}2)$$

The idea then is to count how many times x_0 is within the interval $\underline{A}_*$ explicating the fuzzy concept. The frequency is then equated with the grade of membership of x_0 in A. It has also been demonstrated that as n increases, $\mu_A(x_0)$ tends to be stable.

Technically, let Ω be the field consisting of all the objective factors. Let $w \in \Omega$ be governed by a probability law P. Let $\mathscr{A}$ be a system of subsets of Ω. Let $\underline{A}^* = \underline{A}^*(w)$. Then, $(\Omega, \mathscr{A}, P)$ is a probability space.

Under certain conditions for estimation, $\underline{A}^*$ can be treated as a random set from Ω into $P(\underline{X})$. Let

$$C(x_0) \triangleq \{\underline{A} \mid x_0 \in \underline{A},\ A \in \mathscr{A}\}. \qquad (8\text{-}3)$$

Assume that C(x), for any $x \in \underline{X}$, is estimable. Then, the fuzzy subset A is the fuzzy projection of the random set A* defined by:

$$\mu_A(x) \triangleq P(A^*(w) \in C(x)). \qquad (8\text{-}4)$$

(See Fig. 8.1 for a simple illustration)

Example 8.1. To obtain the membership function defining "young", Zhang and his associates (Zhang, 1981) employed method 2 on 129 respondents in Wuhan, China. The universe of discourse is age: $\underline{X} = [0, 100]$. The intervals $\underline{A}_*$ explicating "young"

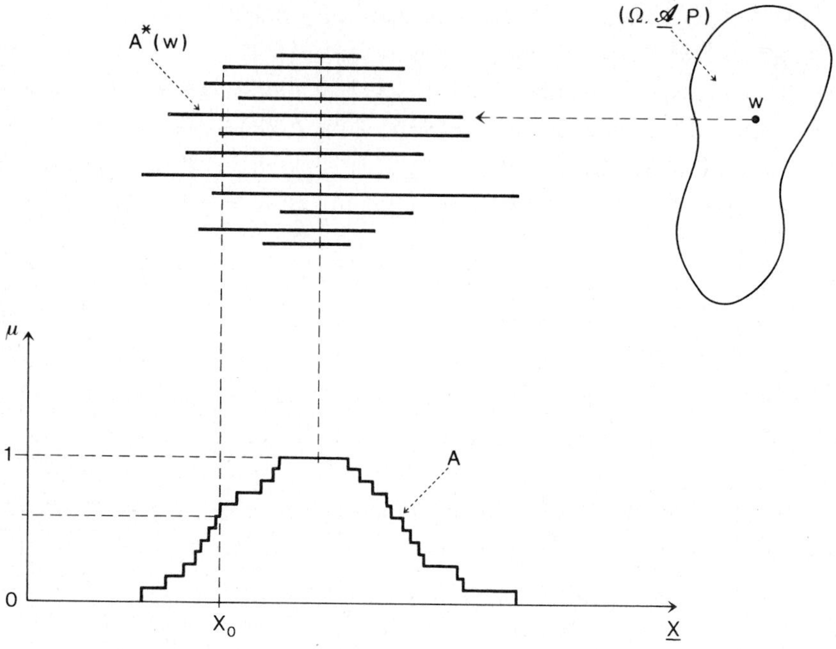

Fig. 8.1 Determination of a membership function through the fuzzy projection of random sets

obtained from the 129 respondents are tabulated in Fig. 8.2.

For a fixed age, say $x_0 = 27$, the stability of its frequency of being regarded as "young" increases as sample size becomes larger (see Fig. 8.3). Since it stablizes around 0.78, then $\mu_{young}(27) = 0.78$.

To derive the membership function, $\underline{X}$ is divided into age groups. Frequencies are then calculated with respect to the mid points of the groups (Fig. 8.4a) and the graph of the membership function is obtained accordingly (Fig. 8.4b).

Therefore, set-valued statistics, in addition to its experimental merits, also offers a statistical theory which can enhance the empirical relevance of fuzzy set theory. It not

18-25	17-30	17-28	18-25	16-35	14-25	18-30	18-35	18-35	16-25
15-30	18-35	17-30	18-25	18-25	18-35	20-30	18-30	16-30	20-35
18-30	18-30	15-25	18-30	15-28	16-28	18-30	18-30	16-30	18-35
18-25	18-25	16-28	18-30	16-30	16-28	18-35	18-35	17-27	16-28
15-28	16-30	19-28	15-30	15-26	17-25	15-36	18-30	17-30	18-35
16-35	15-25	15-25	18-28	16-30	15-28	18-35	18-30	17-28	18-35
15-28	18-30	15-25	15-25	18-30	16-24	15-25	16-32	15-27	18-35
16-25	18-28	16-28	18-30	18-35	18-30	18-30	17-30	18-30	18-35
16-30	18-35	17-25	15-30	18-25	17-30	14-25	18-26	18-29	18-35
18-28	18-30	18-25	16-35	17-29	18-25	17-30	16-28	18-30	16-28
15-30	15-35	15-30	20-30	20-30	16-25	17-30	15-30	18-30	16-30
18-28	18-35	16-30	15-30	18-35	18-35	18-30	17-30	16-35	17-30
15-25	18-35	15-30	15-25	15-30	18-30	17-25	18-29	18-28	

Fig. 8. 2 Age intervals $\underline{A}_*$ explicating "young"

n	10	20	30	40	50	60	70
Number of times	6	14	23	31	39	47	53
Frequency of belonging	0.60	0.70	0.77	0.78	0.78	0.78	0.76

n	80	90	100	110	120	129
Number of times	62	68	76	85	95	101
Frequency of belonging	0.78	0.76	0.76	0.75	0.79	0.78

Fig. 8.3 Frequencies of considering age 27 as "young"

Age Group	Number of Times	Relative Frequency	Age Group	Number of Times	Relative Frequency
13.5-14.5	2	0.0155	25.5-26.5	103	0.7984
14.5-15.5	27	0.2093	26.5-27.5	101	0.7829
15.5-16.5	51	0.3953	27.5-28.5	99	0.7674
16.5-17.5	67	0.5194	28.5-29.5	80	0.6202
17.5-18.5	124	0.9612	29.5-30.5	77	0.5969
18.5-19.5	125	0.9690	30.5-31.5	27	0.2093
19.5-20.5	129	1.0000	31.5-32.5	27	0.2093
20.5-21.5	129	1.0000	32.5-33.5	26	0.2016
21.5-22.5	129	1.0000	33.5-34.5	26	0.2016
22.5-23.5	129	1.0000	34.5-35.5	26	0.2016
23.5-24.5	129	1.0000	35.5-36.5	1	0.0078
24.5-25.5	128	0.9922			
					13.6589

Fig. 8. 4a Frequencies by age groups

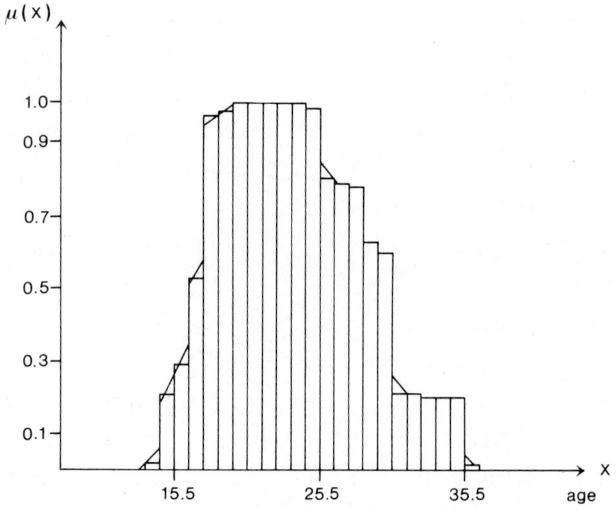

Fig. 8.4b Graph of the membership function of "young"

only shows that fuzzy set theory and probability theory can be complementary, it also demonstrate that they are two independent mathematical systems. Encouraging experimental results have also been reported in a variety of applications.

Technically, the two methods should obtain similar results, albeit the latter is much richer in theory and tends to achieve more stable estimation of μ_A in a global manner. They both attempt to provide a measure of the social value of truth. These procedures can be applied to determine membership functions defining basic fuzzy spatial concepts such as distance, direction, and connection as well as other relevant fuzzy concepts in spatial analysis and planning (Leung, 1982a). Spatial scientists, especially those whose research is experimentally oriented, should carry out more research in this area. Only through extensive empirical analysis can we determine the most appropriate procedure for unravelling imprecision and subjectivity which are crucial in shaping spatial patterns and processes.

All existing procedures, however, assume a given condition S. Very little effort has been made to determine the nature of S. Specifically, what and how are the factors governing or influencing imprecision within a specific context? How context dependent are fuzzy concepts? For example, when an individual says that a distance is *long*, is his perception of the physical distance the only determining factor? What are the major factors leading to the formation and mathematical representation of the fuzzy term "*long*"? All these problems need to be thoroughly investigated before the psycholinguistic realities of fuzzy concepts can be appropriately captured.

A major flaw of probabilistic spatial models is that researchers are too eager to work out the spatial consequences under an assumed random process. They, however, seldom justify for or fail to determine why the spatial process is random, why a human spatial behavior is random, and what guarantee that the assumed random process is the real process. Often, a random process is assumed for its mathematical convenience. An operational model is as good as how trueful we are to reality. Except for their logical structure, a model (with the exception of formal models whose merits depend solely on internal logics) without empirical relevance would not be instrumental on the practical level.

The same argument applies to fuzzy set spatial models. We have put forward the notion of imprecision and have reinstated the importance of human valuation in spatial analysis and planning. We have also formulated objective means to study subjectivity and imprecision. It appears that we are steering towards a closer approximation to the real world. The unravelling of conditions affecting imprecision and the representation of imprecision is an important step towards the realization of its empirical relevance. Determination of membership functions deserves to be a focus of further theoretical and experimental research.

8.2.2. Empirical Aspects of the Fuzzy Connectives

Another basic issue which requires further theoretical and empirical investigations is the definitions of union, intersection, and complementation of fuzzy subsets. Though Bellman and Giertz (1973) and Fung and Fu (1975) have shown theoretically that max-, min-, and $^{-}$ are the only operators under a set of assumptions, the curiosity is do human beings operate on fuzzy concepts through these operators.

Theoretically, as discussed in chapter 2, a number of operators can replace max-, min-, and $^{-}$ to obtain union, intersection, and negation of fuzzy subsets. The plurality of operators enhances the flexibility in modelling human perception, feeling, and thought but also creates confusion in selecting the appropriate operators. Besides the objective criteria proposed for the selection of appropriate operators in subsection 2.3.2, the only justification for a specific operator is through empirical verification in a specific context.

Though Hersh and Caramazza (1976) show positive experimental results, experiments done by Zimmermann (1978), Hamacher (1976), and Thole, Zimmermann and Zysno (1979), and Zimmermann and Zysno (1980, 1983) have shown that max-, min-, and $^{-}$ are not the only operators decisionmakers employ to combine fuzzy concepts or statements. The implication then is operators for "or", "and", and "not" are not unique. There may not be universal operators for these logical connectives. We may need a family of definitions with adjustable parameters to fit varying situations.

Until substantial proofs shown otherwise, max-, min-, and ¯ are useful operators for theoretical spatial analysis and can serve as a point of reference for empirical research. To make spatial and planning models more realistic and empirically relevant, further theoretical and experimental research in this area is imperative. The behavior of logical connectives in basic classes of spatial problems should also be investigated. It is hoped that a general rule for formulating operators can be ironed out in the near future.

8. 2. 3. Directions for Further Research in Spatial Analysis and Planning

In this monograph, fuzzy set frameworks for basic classes of spatial problems have been formulated. Within each class, however, there are specific problems to which further research should be conducted. Beyond these classes, there lies a wilderness for us to explore or discover. To further improve the versatility of fuzzy set theory in spatial analysis and planning, analysts should have in-depth analysis in all spatial problems where imprecision prevails. In what follows, I only outline briefly areas on which further research should be carried out.

In the area of spatial classification, fuzzy set frameworks have been proposed in chapter 3. I have compared them with the conventional frameworks at the theoretical and practical levels. Theoretically, the fuzzy set approach is more flexible, realistic, and informative. Empirical results, albeit limited, have also confirmed the claim. To be more persuasive, extensive applications of the fuzzy set frameworks is necessary. They should be applied to a variety of regionalization problems for evaluations. Furthermore, detailed comparison with the conventional methods should be made. The idea is to learn from experience under what situations one method is more appropriate than the others. An ultimate goal is perhaps to develop an automatically controlled spatial classification system which is based on human experience, valuation, and fuzzy logic and can handle various spatial classification problems. To make our conceptualization of space more general, properties of fuzzy geographical spaces should also be further investigated from the theoretical standpoint (see for example Leung, 1982a; Ponsard 1985a).

In our discussion of spatial preference, utility, and choice in chapter 4, spatial economic problems are only analyzed at the theoretical level. Specific spatial choice problems are, however, numerous and need to be analyzed. Locational problems, market area analyses, and delimitations of various geographical spheres are typical examples.

Simple locational problems such as the original Weberian problem can involve imprecision in the sources of materials, market conditions, and transport situations. As locational problems become more complex, formal treatment of imprecision becomes a necessity. Locational problems involving social values or judgements are especially imprecision prone. Though fuzzy set approach has been rudimentarily employed to solve some locational problems (Leung 1979; Bona, Inaudi, and Mauro, 1980), room for research is ample and the effort would be essential. Fuzzifying existing models should not be the target. We should focus on the characteristics of location under imprecision. Our goal is to advance new theories or models of location where systems complexity and human valuation play a prominent role. By their mathematical properties, however, it should not be surprising if the fuzzy set models turn out to be generalizations of the conventional ones.

Closely related to spatial choice and the division of space is the analysis of market area and spatial spheres of influence. Due to imprecision, Carlucci and Donati (1977), Leung (1982b), and Ponsard (1986b) have demonstrated that the mutually-exclusive property of market areas does not necessarily exist in the real world. The fuzzy set models further address to the importance of human valuation in such a space division process. While the research effort is still limited, more active research would certainly help us to better understand this type of spatial behavior.

Similar to market area analysis, delimitation of various spatial spheres also involves imprecision. For example, the delimitation of hinterlands of cities or regions, potential surfaces, utility spaces, interaction fields, cognitive spaces, and political-power surfaces, to name but a few areas, are all carried out in a complex and imprecise environment. Development of fuzzy set models would take us a step further to mimic and regulate human competition of space.

To improve the theoretical foundations of spatial economic analyses discussed in chapter 4, spatial partial and general equilibria should be investigated within a dynamic framework. Optimality and stability of the equilibrium solutions should be thoroughly scrutinized. To make the analysis more embracing, static and dynamic equilibrium analyses should be carried out under various competitions. Oligopolistic spatial economic systems, for example, would be an interesting problem for further examinations. Moreover, simultaneous solutions for location and allocation at equilibrium would be important, albeit more complicated.

With regard to the analysis of large scale interregional systems, methods discussed in chapters 5 and 6 are effective in tackling interregional decisions under imprecision. Interregional input-output analysis and optimal interregional developments, to name but two general areas, involving imprecise technological production coefficients, objectives, and/or constraints can be appropriately analyzed within the fuzzy set frameworks. While solutions of such systems still rely on conventional algorithms such as the linear programming algorithms, all-integer and mixed-integer programming algorithms, part of our research effort should be aimed at developing new algorithms to solve large scale fuzzy optimization problems. Further research effort would help us to have a firmer grasp of these problems and shed light on the paths of research in the future.

Another vexing spatial problem which I think should certainly be subjected to further pursual is the resolution of spatial conflicts under imprecision. A fuzzy set framework has been formulated in brief in chapter 6. However, more research is required before a general conflict resolution system can be constructed. Immediate efforts should be devoted to detail analysis of various types of spatial conflicts and their resolutions. Locational conflicts, center-periphery conflicts, and conflicts of spatial efficiency and equity all involve multiple conflicting objectives, multiple levels, and multiple decisionmakers. They should also be analyzed under various situations of competition (see for example Ponsard, 1985c and 1986b for an analysis under spatial duopoly and oligopoly respectively). Only through in-depth analyses of these types of spatial conflicts could a general conflict resolution framework be developed. Such studies should be important subjects for further research. They will call upon our ability in modelling complexity, uncertainty (imprecise and/or random), and human valuation in a spatial and temporal setting.

Since spatial systems are dynamic in nature. Spatial analysis under imprecision would not be complete until spatial dynamics are rigorously scrutinized. Though dynamic analysis via fuzzy set theory have been discussed in general in chapter 7, analysis of specific problems in the context of space and time has not been made. This area is certainly rich in research topics. Dynamic spatial subdivisions, locational dynamics, spatial choice dynamics, dynamic partial and general equilibriums, and dynamics of conflict management under imprecision are all important real world problems for investigation. Furthermore, control of fuzzy dynamic spatial systems with human judgement would open up a whole new area of research in the future. Similar to fuzzy optimization problems, we need more efficient algorithms to solve large problems. To be more effective, we perhaps need to analyze fuzzy dynamics within a totally new system other than the conventional state-space system.

In addition to model building, we should also investigate

the possibility of applying fuzzy set theory to spatial data analysis. In recent years, more and more researchers are not content with the assumption that spatial data are always precise (see for example the edited volume by Nijkamp, Leitner, and Wrigley, 1985). Our ability to handle imprecise data is thus crucial. The concept of a linguistic variable, the theory of possibility, and fuzzy mathematics would be essential for such an endeavor. We should also develop a spatial information system within which precise and imprecise data can be stored, retrieved, and analyzed (see Robinson, 1984; Robinson and Strahler, 1984; Robinson, Thongs, and Blaze, 1985; Rolland-May, 1986 for some initial attempts).

Moreover, developing expert systems for solving practical spatial decision problems is not a far-fetched objective. Most of the current spatial information systems are rather sophisticated data processing systems with minimal incorporation of human intelligence and experience. More intelligent systems can be constructed if human perception and decision can be successfully incorporated. Fuzzy set theory in general and fuzzy logic in particular would be instrumental in the building of such expert systems.

With my limited ability in grasping all aspects of fuzzy set theory as well as spatial analysis and planning, discussions and proposed directions for further research in this monograph should only be treated as an initial exploration. Ideas presented might not be mature. I hope that sufficient stimuli and incentives have been provided for a more significant research effort to be undertaken by geographers, spatial economists, regional scientists, fuzzy set researchers, and other social scientists. Upon the completion of the first draft of this monograph, I was informed of the publication of an interdisciplinary joint effort in fuzzy economics and spatial analysis edited by Ponsard and Fustier (1986). It is a promising sign and I hope the endeavor is just the tip of an iceberg.

Fuzzy set theory is only a method for analyzing complex and imprecise spatial systems within which human behavior is based on the way information is perceived and processed. It is highly general and flexible and should have a great potential for further development. Deeper understanding of human perception, thought, communication, and behavior in our ever changing complex environment is crucial to the improvement of the theory. A fuzzy set approach may provide a broad base for our analysis. If it turns out to be only a limited formalism of imprecision and human subjectivity -- which has not been shown to be the case -- it still would be a more appropriate framework compared to the conventional systems. My belief is we are on to a much larger theory. Fuzzy set theory is only a beginning. A more thorough development of the theory and a more systematic evaluation of its applicability to human systems in general and spatial systems in particular would most likely open to something wider and deeper.

APPENDIX 1

PROOFS OF THEOREMS IN CHAPTER 2

A1.1. Proof of Theorem 2.1.

Since

$$f_{\underline{A}_\alpha}(x) = \begin{cases} 1, & \text{if } \mu_A(x) \geq \alpha, \\ 0, & \text{if } \mu_A(x) < \alpha. \end{cases}$$

Then

$$\mu_{\bigcup_{\alpha \in [0, 1]} \alpha \cdot \underline{A}_\alpha}(x) = \sup_{\alpha \in [0, 1]} \mu_{\alpha \cdot \underline{A}_\alpha}(x)$$

$$= \sup_{\alpha \in [0, 1]} \alpha \cdot f_{\underline{A}_\alpha}(x)$$

$$= \sup_{\alpha \leq \mu_A(x)} \alpha = \mu_A(x).$$

Therefore,

$$A = \bigcup_{\alpha \in [0, 1]} \alpha \cdot \underline{A}_\alpha.$$

Q.E.D.

A1.2. Proof of Theorem 2.2.

Since

$$f_{\underline{A}_\alpha}(x) = \begin{cases} 1, & \text{if } \mu_A(x) \geq \alpha, \\ 0, & \text{if } \mu_A(x) < \alpha. \end{cases}$$

Then

$$\sup_{\alpha \in [0,\ 1]} \min[\alpha,\ f_{\underline{A}_\alpha}(x)]$$

$$= \max\left\{ \sup_{\alpha \le \mu_A(x)} \min[\alpha,\ f_{\underline{A}_\alpha}(x)],\quad \sup_{\alpha > \mu_A(x)} \min[\alpha,\ f_{\underline{A}_\alpha}(x)] \right\}$$

$$= \sup_{\alpha \le \mu_A(x)} \min[\alpha,\ 1] = \sup_{\alpha \le \mu_A(x)} \alpha = \mu_A(x).$$

Q.E.D.

A1.3. Proof of Theorem 2.3.

Let $C = *(A \times B) = A * B$. Let $x = x_1 * x_2 \in \underline{X}$, $y \in \underline{X}$. By (2-82) in definition 2.21, we have

$$\mu_{A \times B}(x,\ y) = \min[\mu_A(x),\ \mu_B(y)].$$

Based on the extension principle of Cartesian products of fuzzy subsets in definition 2.21, we have

$$\mu_C(y) = \mu_{*(A \times B)}(y) = \mu_{A\ *\ B}(y)$$

$$= \sup_{(x_1, x_2) \in *^{-1}(y)} \min[\mu_A(x_1),\ \mu_B(x_2)]$$

$$= \sup_{y = x_1 * x_2} \min[\mu_A(x_1),\ \mu_B(x_2)].$$

Therefore, when $*$ is $+$, $-$, $\cdot$, or $\div$ in $\underline{X}$ it can be extended to operate on fuzzy subsets in $\underline{X}$.

Q.E.D.

APPENDIX 2

PROOFS OF THEOREMS IN CHAPTER 3

A2.1. Proof of Therem 3.1. (see also Zadeh, 1965)

Let $M_F = \sup_y \mu_F(y)$, $M_G = \sup_y \mu_G(y)$, and $M = \sup_y \min[\mu_F(y), \mu_G(y)]$. To prove the theorem, we only need to consider two cases.

Case 1. $M = \min(M_F, M_G)$.

Assume that $M_F < M_G$. Then $M_F = M$ and $\mu_F(y) \leq M_F = M$ for all y in $\underline{\mathcal{Y}}$. Apparently, there exists a hyperplane H such that $\mu_F(y) \leq M$ for all y on one side (called the positive side) of H and $\mu_G(y) \leq M$ for all y on the other side (called the negative side) of H.

We still need to show there does not exist another hyperplane H' such that $\mu_F(y) \leq M'$ on the positive side of H' and $\mu_G(y) \leq M'$ on the negative side of H'. Assume that there exists such a M' and H'. Let the core of F (the set of all $y \in \underline{\mathcal{Y}}$ having the property $M_F = M$) be on the positive side of H'. Then, $\mu_F(y) \nleq M'$ for all points on the positive side of H'. However, $\mu_F(y) \leq M'$ for all y on the negative side of H'. Therefore, $\mu_G(y) \leq M'$ on the positive side of H'. Moreover, for all y on the positive and negative sides of H' (which implies all $y \in \underline{\mathcal{Y}}$), we have

$$\sup_y \min[\mu_F(y), \mu_G(y)] \leq M'.$$

such a result contradicts

$$\sup_y \min[\mu_F(y), \mu_G(y)] = M > M'.$$

Case 2. $M < \min(M_F, M_G)$.

Since F and G are bounded and convex fuzzy subsets, then $\underline{\Omega}_F = \{y | u_F(y) > M\}$ and $\underline{\Omega}_G = \{y | u_G(y) > M\}$ are ordinary convex sets which are non-empty. Based on the separation theorem of convex sets, $\underline{\Omega}_F$ and $\underline{\Omega}_G$ are on the positive and negative sides of H respectively. Based on the properties of $\underline{\Omega}_F$ and $\underline{\Omega}_G$, it is true that $\mu_F(y) \leq M$ for all y on the positive side of H and $\mu_G(y) \leq M$ for all y on the negative side of H. Similar to case 1, we can also show that there does not exist a H' such that $M' < M$.

Q.E.D.

A 2.2. Proof of Proposition 3.1. (see also Leung, 1985a)

If F_k, $k = 1, \ldots, \ell$, is bounded and convex, then

$$\pi_{Y_k}(\lambda y_k^1 + (1-\lambda)y_k^2) = \mu_{F_k}(\lambda y_k^1 + (1-\lambda)y_k^2)$$

$$\geq \min[\mu_{F_k}(y_k^1), \mu_{F_k}(y_k^2)] = \min[\pi_{Y_k}(y_k^1), \pi_{Y_k}(y_k^2)],$$

$$y_k^1, y_k^2 \in \underline{\mathscr{Y}}_k, \lambda \in [0, 1].$$

Since

$$\pi_Y(\lambda y^1 + (1-\lambda)y^2)$$

$$= \pi_{(Y_1, \ldots, Y_k, \ldots, Y_\ell)}(\lambda(y_1^1, \ldots, y_k^1, \ldots, y_\ell^1) +$$

$$(1-\lambda)(y_1^2, \ldots, y_k^2, \ldots, y_\ell^2))$$

$$= \pi_{(Y_1, \ldots, Y_k, \ldots, Y_\ell)}(\lambda y_1^1 + (1-\lambda)y_1^2, \ldots, \lambda y_k^1 + (1-\lambda)y_k^2,$$

$$\ldots, \lambda y_\ell^1 + (1-\lambda)y_\ell^2))$$

$$= \min[\mu_{F_1}(\lambda y_1^1 + (1-\lambda)y_1^2), \ldots, \mu_{F_k}(\lambda y_k^1 + (1-\lambda)y_k^2), \ldots,$$

$$\mu_{F_\ell}(\lambda y_\ell^1 + (1-\lambda)y_\ell^2)]. \quad (2.2\text{-}1)$$

Then, by the property of F_k, k = 1, ..., ℓ, (2.2-1) becomes

$$\pi_Y(\lambda y^1 + (1-\lambda)y^2) \geq \min\Big\{\min[\mu_{F_1}(y_1^1),\ \mu_{F_1}(y_1^2)],\ \ldots,$$
$$\min[\mu_{F_k}(y_k^1),\ \mu_{F_k}(y_k^2)],\ \ldots,$$
$$\min[\mu_{F_\ell}(y_\ell^1),\ \mu_{F_\ell}(y_\ell^2)]\Big\}$$
$$= \min\Big\{\min[\mu_{F_1}(y_1^1),\ \ldots,\ \mu_{F_k}(y_k^1),\ \ldots,\ \mu_{F_\ell}(y_\ell^1)],$$
$$\min[\mu_{F_1}(y_1^2),\ \ldots,\ \mu_{F_k}(y_k^2),\ \ldots,\ \mu_{F_\ell}(y_\ell^2)]\Big\},$$
$$= \min\Big\{\mu_F(y^1),\ \mu_F(y^2)\Big\}$$
$$= \min\Big\{\pi_Y(y^1),\ \pi_Y(y^2)\Big\}.$$

The same argument holds for G.

Q.E.D.

A 2. 3. Proof of Theorem 3. 2.

The proof is straight forward. By definition 3.9, the max-min transitive closure of a fuzzy binary relation R is defined by

$$\hat{R} = R \cup R^2 \cup R^3 \cup \ldots \cup R^k \cup R^{k+1} \cup R^{k+2} \cup \ldots \qquad (2.3\text{-}1)$$

If for some k, we have $R^{k+1} = R^k$, then $R^{k+2} = R^{k+1} = R^k$. Therefore, (2.3-1) becomes

$$\hat{R} = R \cup R^2 \cup R^3 \cup \ldots \cup R^k \cup R^k \cup R^k \cup \ldots$$
$$= R \cup R^2 \cup R^3 \cup \ldots \cup R^k.$$

Q.E.D.

A 2. 4. Proof of Theorem 3. 3.

The proof is again trivial. By definition 3.11, the min-max transitive closure of a fuzzy binary relation R is defined by

$$\breve{R} = R \cap R^2 \cap R^3 \cap \ldots \cap R^k \cap R^{k+1} \cap R^{k+2} \cap \ldots \qquad (2.4\text{-}1)$$

If for some k, we have $R^{k+1} = R^k$, then $R^{k+2} = R^{k+1} = R^k$. Thus, (2.4-1) becomes

$$\breve{R} = R \cap R^2 \cap R^3 \cap \ldots \cap R^k \cap R^k \cap R^k \cap \ldots$$
$$= R \cap R^2 \cap R^3 \cap \ldots \cap R^k.$$

Q.E.D.

A2.5. Proof of Theorem 3.4.

If the fuzzy preorder is reducible, then the similitude classes have to form among themselves an antisymmetric fuzzy relation. If not, they would not be disjoint. Furthermore, reflexivity obviously follows. Thus, (3-117) and (3-118) hold.

Since the similarity of any two similitude classes is obtained as the global projection of the subrelation formed between them, then in terms of the strongest path concept, max-min transitivity, (3-119), exists among the similitude classes.

Q.E.D.

APPENDIX 3

PROOFS OF THEOREMS IN CHAPTER 4

A3.1. Proof of Theorem 4.1.

This is a fuzzy set generalization of a theorem in the classical theory of consumption (Uzawa, 1960), the proof is given in Ponsard (1981).

A3.2. Proof of Theorem 4.2. (see also Tanaka, Okuda, and Asai, 1974)

Since $\mu_D(x_i) = \min[\mu_O(x_i), \mu_C(x_i)]$, then $\sup_{x_i \in \underline{X}} \mu_D(x_i) = \sup_{x_i \in \underline{X}} [\min \mu_O(x_i), \mu_C(x_i)]$.

Applying the decomposition theorem (theorem 2.1) on μ_C, we obtain

$$\sup_{x_i \in \underline{X}} \mu_D(x_i) = \sup_{x_i \in \underline{X}} \left\{\min \mu_O(x_i), \sup_{\alpha \in [0,1]} \min[\alpha, \mu_{\underline{C}_\alpha}(x_i)]\right\}$$

$$= \sup_{x_i \in \underline{X}} \sup_{\alpha \in [0,1]} \min\left\{\alpha, \min[\mu_O(x_i), \mu_{\underline{C}_\alpha}(x_i)]\right\}$$

$$= \sup_{\alpha \in [0,1]} \sup_{x_i \in \underline{X}} \min\left\{\alpha, \min[\mu_O(x_i), \mu_{\underline{C}_\alpha}(x_i)]\right\}$$

$$= \sup_{\alpha \in [0,1]} \min\left\{\alpha, \sup_{x_i \in \underline{X}} \min[\mu_O(x_i), \mu_{\underline{C}_\alpha}(x_i)]\right\}$$

$$= \sup_{\alpha \in [0,1]} \min\left\{\alpha, \max\left(\sup_{x_i \in \underline{C}_\alpha} \min[\mu_O(x_i), \mu_{\underline{C}_\alpha}(x_i)],\right.\right.$$

$$\left.\left.\sup_{x_i \notin \underline{C}_\alpha} \min[\mu_O(x_i), \mu_{\underline{C}_\alpha}(x_i)]\right)\right\}$$

$$= \sup_{\alpha \in [0,\, 1]} \min\left\{\alpha,\ \sup_{x_i \in \underline{C}_\alpha} \min[\mu_O(x_i),\ 1]\right\}$$

$$= \sup_{\alpha \in [0,\, 1]} \min\left\{\alpha,\ \sup_{x_i \in \underline{C}_\alpha} \mu_O(x_i)\right\}.$$

Q.E.D.

A3.3. Proof of Theorem 4.3.

This is a fuzzy set generalization of a classical competitive equilibrium theorem (Debreu, 1959), the proof is given in Ponsard (1986a).

APPENDIX 4

PROOFS OF THEOREM IN CHAPTER 5

A4.1. Proof of Theorem 5.1. (see also Leung, 1985d)

If (λ^*, x^*) is a solution of the linear programming problem in (5-73), then

$$\frac{\bar{z} - \sum_k \sum_j c_j^k x_j^{k*}}{d_0} \geq \lambda^*,$$

$$\frac{\bar{b}_i^k - \sum_j a_{ij}^k x_j^{k*}}{d_i^k} \geq \lambda^*, \quad i = 1, \ldots, m; \; k = 1, \ldots, \ell.$$

Consequently,

$$\min \left\{ \frac{\bar{z} - \sum_k \sum_j c_j^k x_j^{k*}}{d_0}, \; \min_{i,k} \left[\frac{\bar{b}_i^k - \sum_j a_{ij}^k x_j^{k*}}{d_i^k} \right] \right\} \geq \lambda^*.$$

Let

$$\lambda^{**} = \min \left\{ \frac{\bar{z} - \sum_k \sum_j c_j^k x_j^{k*}}{d_0}, \; \min_{i,k} \left[\frac{\bar{b}_i^k - \sum_j a_{ij}^k x_j^{k*}}{d_i^k} \right] \right\}.$$

Since

$$\frac{\bar{z} - \sum_k \sum_j c_j^k x_j^{k*}}{d_0} \geq \lambda^{**},$$

$$\frac{\bar{b}_j^k - \sum_j a_{ij}^k x_j^{k*}}{d_i^k} \geq \lambda^{**}, \; i = 1, \ldots, m; \; k = 1, \ldots, \ell.$$

then, (λ^{**}, x^*) satisfies the constraints in (5-73) and we have $\lambda^{**} > \lambda^*$. If λ^* is optimal, then $\lambda^* = \lambda^{**}$ and

$$\lambda^* = \min\left\{\frac{\bar{z} - \sum_k \sum_j c_j^k x_j^{k*}}{d_0}, \; \min_{i,k}\left[\frac{\bar{b}_i^k - \sum_j a_{ij}^k x_j^{k*}}{d_i^k}\right]\right\}.$$

Q.E.D.

A4.2. Proof of Theorem 5.2. (see also Leung, 1985d)

$(\Longrightarrow)$ Based on theorem 5.1, (λ^*, x^*) satisfies the constraints in (5-73) and is thus a solution of (5-73). If (λ^*, x^*) is not a solution of (5-73), there exists (λ^{**}, x^{**}) such that $\lambda^{**} > \lambda^*$ and

$$\lambda^{**} \leq \frac{\bar{z} - \sum_k \sum_j c_j^k x_j^{k**}}{d_0},$$

$$\lambda^{**} \leq \frac{\bar{b}_i^k - \sum_j a_{ij}^k x_j^{k**}}{d_i^k}, \; i = 1, \ldots, m; \; k = 1, \ldots, \ell.$$

Consequently,

$$\min\left\{\frac{\bar{z} - \sum_k \sum_j c_j^k x_j^{k**}}{d_0}, \; \min_{i,k}\left[\frac{\bar{b}_i^k - \sum_j a_{ij}^k x_j^{k**}}{d_i^k}\right]\right\} \geq \lambda^{**}$$

$$> \min\left\{\frac{\bar{z} - \sum_k \sum_j c_j^k x_j^{k*}}{d_0}, \; \min_{i,k}\left[\frac{\bar{b}_i^k - \sum_j a_{ij}^k x_j^{k*}}{d_i^k}\right]\right\} = \lambda^*.$$

A contradiction to the optimality of x^*.

(⟸) Suppose (λ^*, x^*) is a solution of (5-73) but not optimal. Then there exists an alternative x^{**} such that

$$\lambda^{**} = \min\left\{ \frac{\bar{z} - \sum_k \sum_j c_j^k x_j^{k**}}{d_0},\ \min_{i,k} \left[\frac{\bar{b}_i^k - \sum_j a_{ij}^k x_j^{k**}}{d_i^k} \right] \right\}$$

$$> \min\left\{ \frac{\bar{z} - \sum_k \sum_j c_j^k x_j^{k*}}{d_0},\ \min_{i,k} \left[\frac{\bar{b}_i^k - \sum_j a_{ij}^k x_j^{k*}}{d_i^k} \right] \right\} = \lambda^*.$$

Thus,

$$\lambda^* < \lambda^{**} \leq \frac{\bar{z} - \sum_k \sum_j c_j^k x_j^{k**}}{d_0},$$

$$\lambda^* < \lambda^{**} \leq \frac{\bar{b}_i^k - \sum_j a_{ij}^k x_j^{k**}}{d_i^k},\quad i = 1, \ldots, m;\ k = 1, \ldots, \ell.$$

Q.E.D.

A4.3. Proof of Corollary 5.1. (see also Leung, 1985d)

Let $x = (x_1^1, \ldots, x_j^k, \ldots, x_n^\ell) \in \mathbb{R}^{n\ell}$.

Let

$$\mu_0(x) = \frac{\bar{z} - \sum_k \sum_j c_j^k x_j^k}{d_0},$$

$$\mu_i^k(x) = \frac{\bar{b}_i^k - \sum_j a_{ij}^k x_j^k}{d_0},\quad i = 1, \ldots, m;\ k = 1, \ldots, \ell.$$

Based on theorems 5.1 and 5.2, we have

$$\lambda^* = \min[\mu_0(x^*), \min_{i,k} \mu_i^k(x^*)] = \sup_{x \in \mathbb{R}^{n\ell}} \min\left\{\mu_0(x), \min_{i,k}[\mu_i^k(x)]\right\}.$$

By the decomposition theorem, the fuzzy constraints μ_i^k's can be respectively expressed as

$$\mu_i^k(x) = \sup_{\alpha_i^k \in [0,\ 1]} \min[\alpha_i^k,\ \mu_{\underline{C}_{\alpha_i^k}}(x)],\ i = 1,\ \ldots,\ m;$$
$$k = 1,\ \ldots,\ \ell.$$

Then,

$$\lambda^* = \sup_{x \in \mathbb{R}^{n\ell}} \min\left\{\mu_0(x),\ \min_{i,k}[\mu_i^k(x)]\right\}$$

$$= \sup_{x \in \mathbb{R}^{n\ell}} \min\left\{\mu_0(x),\ \min_{i,k}[\sup_{\alpha_i^k \in [0,1]} \min(\alpha_i^k,\ \mu_{\underline{C}_{\alpha_i^k}}(x))]\right\}$$

$$= \sup_{x \in \mathbb{R}^{n\ell}} \sup_{\alpha_i^k \in [0,1]} \min\left\{\min_{i,k} \alpha_i^k,\ \mu_0(x),\ \min_{i,k}(\mu_{\underline{C}_{\alpha_i^k}}(x))\right\}$$

$$= \sup_{\alpha_i^k \in [0,1]} \min\left\{\min_{i,k} \alpha_i^k,\ \sup_{x \in R^{n\ell}} \min[\mu_0(x),\ \min_{i,k}(\mu_{\underline{C}_{\alpha_i^k}}(x))]\right\}$$

$$= \sup_{\alpha_i^k \in [0,1]} \min\left\{\min_{i,k} \alpha_i^k,\ \max[\sup_{x \in \bigcap_{i,k} \underline{C}_{\alpha_i^k}} \min(\mu_0(x),\ \mu_{\bigcap_{i,k} \underline{C}_{\alpha_i^k}}(x)),\right.$$
$$\left.\sup_{x \notin \bigcap_{i,k} \underline{C}_{\alpha_i^k}} \min(\mu_0(x),\ \mu_{\bigcap_{i,k} \underline{C}_{\alpha_i^k}}(x))]\right\}$$

$$= \sup_{\alpha_i^k \in [0,1]} \min\left\{\min_{i,k} \alpha_i^k,\ \sup_{x \in \bigcap_{i,k} \underline{C}_{\alpha_i^k}} \min[\mu_0(x),\ 1]\right\}$$

$$= \sup_{\alpha_i^k \in [0,1]} \min\left\{\min_{i,k} \alpha_i^k,\ \sup_{x \in \bigcap_{i,k} \underline{C}_{\alpha_i^k}} \mu_0(x)\right\}.$$

Obviously,

$$\sup_{\alpha_i^k \in [0,1]} \min\left\{\min_{i,k} \alpha_i^k,\ \sup_{x \in \bigcap_{i,k} \underline{C}_{\alpha_i^k}} \mu_0(x)\right\}$$

$$= \sup_{\alpha \in [0,1]} \min\left\{\alpha,\ \sup_{x \in \underline{C}_\alpha} \mu_0(x)\right\}.$$

Therefore,

$$\lambda^* = \sup_{\alpha \in [0,1]} \mu_D(x) = \sup_{\alpha \in [0,1]} \min \left\{ \alpha, \sup_{x \in \underline{C}_\alpha} \left[\frac{\bar{z} - \sum_k \sum_j c_j^k x_j^k}{d_0} \right] \right\}.$$

Q.E.D.

A4.4. Proof of Theorem 5.3.

Since the fuzzy optimization problem has been transformed into conventional linear programming problems in (5-74) and (5-75), then theorem 5.3 is in fact the theorem of duality of conventional linear programming problems. Readers are referred to Gale, Kuhn, and Tucker (1951) for a general proof.

A4.5. Proof of Theorem 5.4.

This is in fact the existence theorem of conventional linear programming problems. For a general proof, see Gale, Kuhn, and Tucker (1951).

A4.6. Proof of Theorem 5.5.

Let the fuzzy objective, O, and the fuzzy constraints, C_i^k, $i = 1, \ldots, m$; $k = 1, \ldots, \ell$, be fuzzy subsets respectively defined by μ_O and and μ_i^k, $i = 1, \ldots, m$; $k = 1, \ldots, \ell$. Since $D = O \cap [\bigcap_{i,k} C_i^k]$. Obviously, an optimal solution exists if D is non-vacuous.

A4.7. Proof of Theorem 5.6. (see also Leung, 1985d)

Since $(\lambda^*, x_1^{1*}, \ldots, x_j^{k*}, \ldots, x_n^{\ell*})$ and $(y_0^*, y_1^{1*}, \ldots, y_i^{k*}, \ldots, y_m^{\ell*})$ are feasible solutions, then

$$\beta_0 = y_0^*(\bar{z} - \sum_k \sum_j c_j^k x_j^{k*} - d_0 \lambda^*) \geq 0,$$

$$\beta_i^k = y_i^{k*}(\bar{b}_i^k - \sum_j a_{ij}^k x_j^{k*} - d_i^k \lambda^*) \geq 0,$$

$$i = 1, \ldots, m; \; k = 1, \ldots, \ell;$$

and,

$$\gamma_0 = \lambda^*(d_0 y_0^* + \sum_k \sum_i d_i^k y_i^{k*} - 1) \geq 0,$$

$$\gamma_j^k = x_j^{k*}(c_j^k y_0^* + \sum_k \sum_i a_{ij}^k y_i^{k*}) \geq 0,$$

$$j = 1, \ldots, n; \; k = 1, \ldots, \ell.$$

Moreover,

$$(\beta_0 + \sum_k \sum_i \beta_i^k) + (\gamma_0 + \sum_k \sum_j \gamma_j^k) = \bar{z} y_0^* + \sum_k \sum_i \bar{b}_i^k y_i^{k*} - \lambda^* \geq 0.$$

Based on the Fundamental Theorem of Duality, for $(\lambda^*, x_1^{1*}, \ldots, x_j^{k*}, \ldots, x_n^{\ell*})$ and $(y_0^*, y_1^{1*}, \ldots, y_i^{k*}, \ldots, y_m^{\ell*})$ to be optimal the following relation has to be satisfied

$$\bar{z} y_0^* + \sum_k \sum_i \bar{b}_i^k y_i^{k*} - \lambda^* = 0,$$

It implies that

$$\beta_0 = 0, \; \beta_i^k = 0, \; i = 1, \ldots, m; \; k = 1, \ldots, \ell,$$

and,

$$\gamma_0 = 0, \; \gamma_j^k = 0, \; j = 1, \ldots, m; \; k = 1, \ldots, \ell.$$

Q.E.D.

APPENDIX 5

PROOFS OF THEOREMS IN CHAPTER 6

A5.1. Proof of Theorem 6.1. (see also Leung, 1985e)

Suppose $x^* = (x_1^{1*}, \ldots, x_j^{k*}, \ldots, x_n^{\ell *})$ is not an efficient solution of (6-22). Then, there exists $x^{**} = (x_1^{1**}, \ldots, x_j^{k**}, \ldots, x_n^{\ell **}) \in \underline{X}$ such that $f_h(x^{**}) \geq f_h(x^*)$, $h = 1, \ldots, q$. This implies,

$$\lambda^{**} = \min_h \left[\frac{\sum_k \sum_j c_{hj}^k \, x_j^{k**} - \underline{f}_h}{d_h} \right] \geq \min_h \left[\frac{\sum_k \sum_j c_{hj}^k \, x_j^{k*} - \underline{f}_h}{d_h} \right] = \lambda^* .$$

If $\lambda^{**} > \lambda^*$, we have a contradiction of the optimality of x^* in (6-27). Thus, x^* has to be an efficient solution of (6-22).

However, if $\lambda^{**} = \lambda^*$, we have alternate optimal solutions. Both x^* and x^{**} are efficient solutions.

Since $0 \leq \lambda^* \leq 1$, then $f_h(x^*) \in [\underline{f}_h, \overline{f}_h]$, $h = 1, \ldots, q$. Specifically, if $\lambda^* = 0$, $f_h(x^*) = \underline{f}_h$. If $\lambda^* = 1$, $f_h(x^*) = \overline{f}_h$.

Q.E.D.

A5.2. Proof of Theorem 6.2. (see also Leung, 1985e)

Since $\min d_\infty = \min \max_h \left\{ 1 - \frac{\sum_k \sum_j c_{hj}^k \, x_j^k - \underline{f}_h}{d_h} \right\}$, then (6-27) is equivalent to

$$\min_h \max \left\{ 1 - \frac{\sum_k \sum_j c^k_{hj}\, x^k_j - \underline{f}_h}{d_h} \right\},$$

$$\text{s.t. } \sum_j a^k_{ij}\, x^k_j \le b^k_i,\ i = 1, \ldots, m;\ k = 1, \ldots, \ell,$$

$$x^k_j \ge 0,\ j = 1, \ldots, n;\ k = 1, \ldots, \ell.$$

However,

$$\min_h \max \left\{ 1 - \frac{\sum_k \sum_j c^k_{hj}\, x^k_j - \underline{f}_h}{d_h} \right\}$$

$$= \min_h \max \left\{ 1 - \mu_h(x) \right\} = \max \min_h \mu_h(x).$$

Therefore, the solutions in (6-27) and (6-29) are identical.

Q.E.D.

A5.3. Proof of Theorem 6.3. (see also Leung, 1985e)

It is sufficient to show that for $x = (x^1_1, \ldots, x^k_j, \ldots, x^\ell_n) \in \underline{X}$,

$$\frac{f_h(x^*) - \underline{f}_h}{d_h} \ge \frac{f_h(x) - \underline{f}_h}{d_h},\quad h = 1, \ldots, q.$$

Since (λ^*, x^*) is optimal, then

$$\min_h \left[\frac{f_h(x^*) - \underline{f}_h}{d_h} \right] = \sup_{x^k_j} \min_h \left[\frac{\sum_k \sum_j c^k_{hj}\, x^k_j - \underline{f}_h}{d_h} \right] \ge$$

$$\min_h \left[\frac{\sum_k \sum_j c^k_{hj}\, x^k_j - \underline{f}_h}{d_h} \right]$$

$$= \min_h \left[\frac{f_h(x) - \underline{f}_h}{d_h} \right].$$

Furthermore,

$$\frac{f_h(x^*) - \underline{f}_h}{d_h} \geq \min_h \left[\frac{f_h(x^*) - \underline{f}_h}{d_h} \right] \geq \min_h \left[\frac{f_h(x) - \underline{f}_h}{d_h} \right].$$

Let

$$\frac{\underline{f}_{\underline{h}}(x) - \underline{f}_{\underline{h}}}{d_{\underline{h}}} = \min_h \left[\frac{\underline{f}_h(x) - \underline{f}_h}{d_h} \right].$$

Then,

$$\frac{f_{\underline{h}}(x^*) - \underline{f}_{\underline{h}}}{d_{\underline{h}}} \geq \min_h \left[\frac{f_{\underline{h}}(x) - \underline{f}_{\underline{h}}}{d_{\underline{h}}} \right].$$

Thus,

$$\frac{f_h(x^*) - \underline{f}_h}{d_h} \geq \frac{f_h(x) - \underline{f}_h}{d_h}, \quad h = 1, \ldots, q.$$

Apparently, $f_h(x^*) \in [\underline{f}_h, \overline{f}_h]$, $h = 1, \ldots, q$.

Q.E.D.

APPENDIX 6

PROOFS OF THEOREMS IN CHAPTER 7

A6.1. Proof of Theorem 7.1. (see also Baldwin and Pilsworth, 1982)

Based on (7-83), if all controls are precise, then

$$T(u_0, u_1, \ldots, u_{N-1}, R, X_T)$$
$$= \min[\mu_{C_0}(u_0), \mu_{C_1}(u_1), \ldots, \mu_{C_{N-1}}(u_{N-1}), \mu_{G_N}(x_N)].$$

For u_0 fuzzy with $U_0 \in \mathscr{F}(\underline{U})$, then

$$T(U_0, u_1, \ldots, u_{N-1}, R, x_N)$$
$$= \max_{u_0} \min[\mu_{C_0}(u_0), \mu_{C_1}(u_1), \ldots, \mu_{C_{N-1}}(u_{N-1}), \mu_{G_N}(x_N), \mu(u_0)].$$
$$= \min\Big\{\max_{u_0} \min[\mu(u_0), \mu_{C_0}(u_0)],$$
$$\mu_{C_1}(u_1), \ldots, \mu_{C_{N-1}}(u_{N-1}), \mu_{G_N}(x_N)\Big\}.$$

Similarly, for $U_0, U_1 \in \mathscr{F}(\underline{U})$,

$$T(U_0, u_1, u_2, \ldots, u_{N-1}, R, x_N)$$
$$= \max_{u_1} \min\Big\{\max_{u_0} \min[\mu(u_0), \mu_{C_0}(u_0)],$$
$$\mu_{C_1}(u_1), \mu_{C_2}(u_2), \ldots, \mu_{C_{N-1}}(u_{N-1}), \mu_{G_N}(x_N), \mu(u_1)\Big\}$$
$$= \min\Big\{\max_{u_0} \min[\mu(u_0), \mu_{C_0}(u_0)], \max_{u_1} \min[\mu(u_1), \mu_{C_1}(u_1)],$$
$$\mu_{C_2}(u_2), \ldots, \mu_{C_{N-1}}(u_{N-1}), \mu_{G_N}(x_N)\Big\}.$$

Finally for $U_0, U_1, \ldots, U_{N-1} \varepsilon \mathscr{F}(\underline{U})$, $X_N \varepsilon \mathscr{F}(\underline{X})$, the truth function, by extension is given by

$$T(U_0, U_1, \ldots, U_{N-1}, R, X_N)$$

$$= \min\Big\{\max_{u_0} \min[\mu(u_0), \mu_{C_0}(u_0)], \max_{u_1} \min[\mu(u_1), \mu_{C_1}(u_1)], \ldots,$$

$$\max_{u_{N-1}} \min[\mu_{N-1}(u_{N-1}), \mu_{C_{N-1}}(u_{N-1})],$$

$$\max_{x_N} \min[\mu(x_N), \mu_{G_N}(x_N)]\Big\}.$$

Q.E.D.

BIBLIOGRAPHY

Baldwin, J.F. and B.W. Pilsworth. "Dynamic Programming for Fuzzy Systems with Fuzzy Environment", Journal of Mathematical Analysis and Applications 85(1982), 1-23.

Bellman, R.E. and M. Giertz. "On the Analytic Formalism of the Theory of Fuzzy Sets", Information Sciences 5(1973), 149-156.

Bellman, R.E. and L.A. Zadeh. "Decision-making in a Fuzzy Environment", Management Sciences 17(1970), B-141-164.

Berry, B.J.L. "Grouping and Regionalization, an Approach to the Problem using Multivariate Analysis", in W.L. Garrison and D.F. Marble (eds.), Quantitative Geography, Part I: Economic and Cultural Topics, Evanston: Northwestern University Press, 1967, pp. 219-251.

Bezdek, J.C. Pattern Recognition with Fuzzy Objective Function Algorithms. New York: Academic Press, 1981.

Black, M. "Vagueness", Philosophy of Science 4(1937), 427-455.

Black, M. "Reasoning with Loose Concepts", Dialogue 2(1963), 1-12.

Black, M. Margins of Precision: Essays in Logic of Language. Ithaca: Cornell University Press, 1970.

Bona, B., D. Inaudi, and V. Mauro. "A Fuzzy Approach to Residential Location Theory", in R. Trappl, G.J. Klin, and L. Ricciardi (eds.), Progress in Cybernetics and Systems Research, Vol. III, New York: John Wiley, 1980, pp. 314-320.

Butnariu, D. "Fixed Points for Fuzzy Mappings", Fuzzy Sets and Systems 7(1982), 191-207.

Carlucci, D. and F. Donati. "Fuzzy Cluster of Demand within a Regional Service System", in M.M. Gupta, G.N. Saridis, and B.R. Gaines (eds.), Fuzzy Automata and Decision Processes, Amsterdam: North-Holland, 1977, pp. 379-385.

Cayrol, M., H. Farreny, and H. Prade. "Possibility and Necessity in a Pattern-Matching Process", in Proceedings of the 9th International Congress on Cybernetics, Namur: International Association of Cybernetics, 1980, pp. 53-65.

Chanas, S. "Parametric Programming in Fuzzy Linear Programming", Fuzzy Sets and Systems 11(1983), 243-251.

Chang, R.L.P. and T. Pavlidis. "Fuzzy Decision Tree Algorithm", IEEE Transaction on Systems, Man and Cybernetics SMC-7(1977), 28-35.

Charnes, A. and W.W. Cooper. Management Models and Industrial

Applications of Linear Programming. New York: John Wiley, 1961.
Copilowish, I.M. "Border-line Cases, Vagueness, and Ambiguity", Philosophy of Science 6(1939), 181-195.
Cronbach, L.J. "Further Evidence on Response Sets and Test Design", Educational and Psychological Measurement 10(1950), 3-31.
Debreu, G. Theory of Value. Cowles Commission Monograph 17. New York: John Wiley, 1959.
Delft, A. Van and P. Nijkamp. Multi-Criteria Analysis and Regional Decision-Making. The Hague: Martinus Nijhoff, 1979.
Dreyfuss, G.R., M. Kochen, J. Robinson, and A.N. Badre. "On the Psycholinguistic Reality of Fuzzy Sets", in R. Grossman, J. San, and T. Vance (eds.), Proceedings of the Chicago Linguistic Society, No. 1, Chicago: University of Chicago Press, 1975, pp. 135-149.
Dubois, D. and H. Prade. "Operations on Fuzzy Numbers", International Journal of Systems Science 9(1978), 613-626.
Dubois, D. and H. Prade. Fuzzy Sets and Systems: Theory and Applications. New York: Academic Press, 1980.
Dyer, J.S. "Interactive Goal Programming", Management Science 19(1972), 62-70.
Ecker, J.G. and I.A. Kouada. "Finding Efficient Points for Linear Multiple Objective Programs", Management Science 19(1975), 375-377.
Fischer, M.M. "Regional Taxonomy: Some Reflections on the State of the Art", paper presented at the International Seminar on Regional Planning and Information System Requirements: Analysis Methods and Information System Requirements in Regional Planning, Cagliari, October 27-28, 1983.
Fung, L.W. and K.S. Fu. "An Axiomatic Approach to Rational Decisionmaking in a Fuzzy Environment", in L.A. Zadeh, K.S. Fu, K. Tanaka, and M. Shimura (eds.), Fuzzy Sets and their Application to Cognitive and Decision Processes, New York: Academic Press, 1975, pp. 227-256.
Gaines, B.R. "Foundations of Fuzzy Reasoning", International Journal of Man-Machine Studies 8(1976), 623-668.
Gale, D., H.W. Kuhn, and A.W. Tucker. "Linear Programming and the Theory of Games", in T.C. Koopmans (ed.), Activity Analysis of Production and Allocation, Cowles Commission Monograph 13, New York: John Wiley, 1951, Chapter 19.
Gale, S. "Inexactness, Fuzzy Sets and the Foundation of Behavioral Geography", Geographical Analysis 4(1972), 337-349.
Gale, S. "A Resolution of the Regionalization Problem and its Implications for Political Geography and Social Justice", Geografiska Annaler 58-B(1976), 1-16.
Geoffrion, A.M., J.S. Dyer, and A. Feinberg. "An Interactive Approach for Multicriteria Optimization, with an Application to the Operation of an Academic Department", Management Science 19(1972), 357-368.
Grigg, D. "Regions, Models and Classes", in R.J. Chorley and P. Haggett (eds.), Models in Geography, London: Methuen, 1967, 461-509.

Hamacher, H. On Logical Connectives of Fuzzy Statements and their Affiliated Truth-Functions. Paper presented at the Third European Meeting on Cybernetics and Systems Research, Vienna, Austria, 1976.

Hannan, E.L. "Linear Programming with Multiple Fuzzy Goals", Fuzzy Sets and Systems 6(1981), 235-248.

Hempel, C.G. "Vagueness and Logic", Philosophy of Science 6(1939), 163-180.

Hersh, H.M. and A. Caramazza. "A Fuzzy Set Approach to Modifiers and Vagueness in Natural Language", Journal of Experimental Psychology: General 105(1976), 254-276.

Hersh, H.M., A. Caramazza, and H.H. Brownell. "Effects of Context on Fuzzy Membership Functions", in M.M. Gupta, R.K. Ragade, and R.R. Yager (eds.), Advances in Fuzzy Set Theory and Applications, Amsterdam: North-Holland, 1979, pp. 389-408.

Ignizio, J.P. Goal Programming and Extensions. Lexington: Heath, 1976.

Kacprzyk, J. "A Branch-and-Bound Algorithm for the Multistage Control of a Nonfuzzy System in a Fuzzy Environment", Control and Cybernetics 7(1978), 51-64.

Kacprzyk, J. "A Branch-and-Bound Algorithm for the Multistage Control of a Nonfuzzy System in a Fuzzy Environment", Kybernetes 8(1979), 139-147.

Kacprzyk, J. Multistage Decision-making under Fuzziness. Koln: Verlag TUV Rheinland, 1983.

Kakutani, S. "A Generalization of Brouwer's Fixed Point Theorem", Duke Mathematical Journal 8(1941), 457-459.

Kaleva, O. "A Note on Fixed Points for Fuzzy Mappings", Fuzzy Sets and Systems 15(1985), 99-100.

Kaufmann, A. Introduction to the Theory of Fuzzy Subsets, Vol. 1. New York: Academic Press, 1975.

Kaufmann, A. and M.M. Gupta. Introduction to Fuzzy Arithmetic, Theory and Applications. New York: Van Nostrand Reinhold, 1985.

Kochen, M. "Applications of Fuzzy Sets to Psychology", in L.A. Zadeh, K.S. Fu, K. Tanaka, and M. Shimura (eds.), Fuzzy Sets and their Applications to Cognitive and Decision Processes, New York: Academic Press, 1975, pp. 395-408.

Kochen, M. and A.N. Badre. "On the Precision of Adjectives which denote Fuzzy Sets", Journal of Cybernetics 4(1975), 49-59.

Kolmogoroff, A.N. Foundations of Probability. (Translated by N. Morrison). New York: Chelsea, 1950.

Koopman, B.O. "The Axioms and Algebra of Intuitive Probability", Annuals of Mathematics 41(1940), 269-292.

Korner, S. "Reference, Vagueness, and Necessity", Philosophical Review 66(1957), 363-376.

Lee, S. Goal Programming for Decision Analysis. Philadelphia: Auerback, 1972.

Leung, Y. "Locational Choice, a Fuzzy Set Approach", Geographical Bulletin 15(1979), 28-34.

Leung, Y. "A Fuzzy Set Procedure for Project Selection with Hierarchical Objectives", in P.P. Wang and S.K. Chang (eds.), Fuzzy Sets: Theory and Applications to Policy Analysis and Information Systems, New York: Plenum, 1980a,

pp. 329-340.
Leung, Y. "A Fuzzy Set Analysis of Sociometric Structure", Journal of Mathematical Sociology 7(1980b), 159-180.
Leung, Y. "On the Exactness of Membership Functions in Fuzzy Sets Theory", in Proceedings of the International Conference on Policy Analysis and Information Systems, Taiwan: Tamkang University, 1981a, pp. 765-775.
Leung, Y. "An Empirical Analysis of Linguistic Hedges, I", in Proceedings of the International Conference on Policy Analysis and Information Systems, Taiwan: Tamkang University, 1981b, pp. 237-249.
Leung, Y. "An Empirical Analysis of Linguistic Hedges, II", in Proceedings of the International Conference on Policy Analysis and Information Systems, Taiwan: Tamkang University, 1981c, pp. 251-262.
Leung, Y. "Maximum Entropy Estimation with Inexact Information", in G.E. Lasker (ed.), Applied Systems and Cybernetics Vol. VI, New York: Pergamon, 1981d, pp. 2974-2979.
Leung, Y. "Approximate Characterization of some Fundamental Concepts of Spatial Analysis", Geographical Analysis 14(1982a), 19-40.
Leung, Y. "Market Area Separation in a Fuzzy Environment", in R.R. Yager (ed.), Fuzzy Set and Possibility Theory: Recent Developments, New York: Pergamon, 1982b, pp. 551-561.
Leung, Y. "Dynamic Conflict Resolution through a Theory of a Displaced Fuzzy Ideal", in M.M. Gupta and E. Sanchez (eds.), Approximate Reasoning in Decision Analysis, Amsterdam: North-Holland, 1982c, pp. 381-390.
Leung, Y. "Urban and Regional Planning with Fuzzy Information", in L. Chatterjee and P. Nijkamp (eds.), Urban and Regional Policy Analysis for Developing Countries, Hants: Gower, 1983a, pp. 231-249.
Leung, Y. "A Concept of a Fuzzy Ideal for Multicriteria Conflict Resolution", in P.P. Wang (ed.), Advances in Fuzzy Sets, Possibility Theory, and Applications, New York: Plenum, 1983b, pp. 387-403.
Leung, Y. "Fuzzy Sets Approach to Spatial Analysis and Planning -- A Nontechnical Evaluation", Geografiska Annaler 65B(1983c), 65-75.
Leung, Y. "A Value-Based Approach to Conflict Resolution Involving Multiple Objectives and Multiple Decisionmaking Units", in W. Isard and Y. Nagao (eds.), International and Regional Conflict: Analytic Approaches, Cambridge: Ballinger, 1983d, Chapter 4.
Leung, Y. "Towards a Flexible Framework for Regionalization", Environment and Planning A 16(1984a), 203-215.
Leung, Y. "Compromise Programming under Fuzziness", Control and Cybernetics 13(1984b), 203-215.
Leung, Y. Robust Programming for Spatial Optimization Problems. Occasional Paper No. 68, Hong Kong: Department of Geography, The Chinese University of Hong Kong, 1984c.
Leung, Y. "A Linguistically-based Regional Classification System", in P. Nijkamp, H. Leitner, and N. Wrigley (eds.), Measuring the Unmeasurable, Dordrecht: Martinus Nijhoff, 1985a, pp. 451-486.

Leung, Y. "Basic Issues of Fuzzy Set Theoretic Spatial Analysis", Papers of the Regional Science Association 58(1985b), 35-46.

Leung, Y. Hierarchical Programming with Fuzzy Objectives and Constraints. Occasional Paper No. 73, Hong Kong: Department of Geography, The Chinese University of Hong Kong, 1985c. [To appear in J. Kacprzyk and S.A. Orlovski (eds.), Optimization Models using Fuzzy Sets and Possibility Theory, Dordrecht: D. Reidel.]

Leung, Y. Interregional Equilibrium and Fuzzy Linear Programming. Occasional Paper No. 77, Hong Kong: Department of Geography, The Chinese University of Hong Kong, 1985d. [A modified version entitled "Interregional Equilibrium and Fuzzy Linear Programming: 1" is to appear in Environment and Planning A 19(1987).]

Leung, Y. Multiobjective Interregional Equilibrium and Fuzzy Linear Programming. Occasional Paper No. 79, Hong Kong: Department of Geography, The Chinese University of Hong Kong, 1985e. [A modified version entitled "Interregional Equilibrium and Fuzzy Linear Programming: 2" is to appear in Environment and Planning A 19(1987).]

Leung, Y. "On the Imprecision of Boundaries", Geographical Analysis 19(1987a), 125-151.

Leung, Y. Regional Input-Output Analysis with Imprecise Technical Production Coefficients. Occasional Paper No. 95, Hong Kong: Department of Geography, The Chinese University of Hong Kong, 1987b.

Leung, Y. "Concepts of a Fuzzy Ideal for Conflict Resolution", in M.G. Singh (ed.), Systems and Control Encyclopedia, New York: Pergamon, 1987c. (To appear)

MacVicar-Welan, P.J. "Fuzzy Sets, the Concept of Height, and the Hedges", IEEE Transaction on System, Man and Cybernetics, SMC-8(1978), 507-511.

Mathieu-Nicot, B. "Fuzzy Expected Utility", Fuzzy Sets and Systems 20(1986), 163-173.

Mizumoto, M. and K. Tanaka. "Some Properties of Fuzzy Numbers", in M.M. Gupta, R.K. Ragade, and R.R. Yager (eds.), Advances in Fuzzy Set Theory and Applications, Amsterdam: North-Holland, 1979, pp. 153-164.

Narasimhan, R. "Goal Programming in a Fuzzy Environment", Decision Sciences 11(1980), 325-336.

Negoita, C.V. Management Applications of Systems Theory. Bucharest: Birkhauser Verlag, 1979.

Negoita, C.V., S. Minoiu, and E. Stan. "On Considering Imprecision in Dynamic Linear Programming", Economic Computation and Economic Cybernetics Studies and Research 3(1976), 83-96.

Negoita, C.V. and D.A. Ralescu. Applications of Fuzzy Sets to Systems Analysis. Basle: Birkhauser Verlag, 1975.

Negoita, C.V. and D.A. Ralescu. "On Fuzzy Optimization", Kybernetes 6(1977), 193-195.

Negoita, C.V. and M. Sularia. "On Fuzzy Mathematical Programming and Tolerances in Planning", Economic Computation and Economic Cybernetics Studies and Research 1(1976), 3-15.

Nijkamp, P. Theory and Application of Environmental Economics. Amsterdam: North-Holland, 1977.

Nijkamp, P. "Conflict Patterns and Compromise Solutions in Fuzzy Choice Theory", Journal of Peace Science 4(1979), 67-90.
Nijkamp, P., H. Leitner, and N. Wrigley (eds.). Measuring the Unmeasurable. Dordrecht: Martinus Nijhoff, 1985.
Norwich, A.M. and I.B. Turksen. "Measurement and Scaling of Membership Functions", in G.E. Lasker (ed.), Applied Systems and Cybernetics, Vol. VI, New York: Pergamon, 1981, pp. 13-22.
Norwich, A.M. and I.B. Turksen. "Stochastic Fuzziness", in M.M. Gupta and E. Sanchez (eds.), Approximate Reasoning in Decision Analysis, Amsterdam: North-Holland, 1982, pp. 13-22.
Nowakowska, M. "Methodological Problems of Measurement of Fuzzy Concepts in the Social Sciences", Behavioral Science 22(1977), 107-115.
Orlovsky, S.A. "On Programming with Fuzzy Constraint Sets", Kybernetes 6(1977), 197-201.
Orlovsky, S.A. "Decision-making with a Fuzzy Preference Relation", Fuzzy Sets and Systems 1(1978), 155-167.
Orlovsky, S.A. "On Formalization of a General Fuzzy Mathematical Problem", Fuzzy Sets and Systems 3(1980), 311-321.
Peirce, C.S. "Vagueness", in J.M. Baldwin (ed.), Dictionary of Philosophy and Psychology, Vol. 2, New York: Macmillan, 1902.
Pipkin, J.S. "Fuzzy Sets and Spatial Choice", Annals, Association of American Geographers 68(1978), 196-204.
Ponsard, C. "On the Imprecision of Consumer's Spatial Preferences", Papers, Regional Science Association 42(1979), 59-71.
Ponsard, C. Fuzzy Economic Spaces. Working paper No. 43, Dijon: Institute of Economics Mathematics, University of Dijon, 1980.
Ponsard C. "An Application of Fuzzy Subsets Theory to the Analysis of the Consumer's Spatial Preferences", Fuzzy Sets and Systems 5(1981), 235-244.
Ponsard, C. "Producer's Spatial Equilibrium with a Fuzzy Constraint", European Journal of Operational Research 10(1982a), 302-313.
Ponsard, C. "Partial Spatial Equilibria with Fuzzy Constraints", Journal of Regional Science 22(1982b), 159-175.
Ponsard, C. "Fuzzy Human Spaces", Sistemi Urbani 2(1985a), 191-204.
Ponsard, C. "Fuzzy Sets in Economics, Foundations of Soft Decision Theory", in J. Kacprzyk and R.R. Yager (eds.), Management Decision Support Systems using Fuzzy Sets and Possibility Theory, Koln: Verlag TUV Rheinland, 1985b, pp. 25-37.
Ponsard, C. Nash Fuzzy Equilibrium: Theory and Application to a Spatial Duopoly. Paper presented at the 7th European Congress on Operational Research. Bologna, Italy, 1985c. [To appear in European Journal of Operational Research 31(1987)]
Ponsard, C. A Theory of Spatial General Equilibrium in a Fuzzy Economy", in C. Ponsard and B. Fustier (eds.), Fuzzy Economics and Spatial Analysis, Dijon: Librairie de

l'Universite, 1986a, pp. 1-27.
Ponsard, C. "Spatial Oligopoly as Fuzzy Game", in C. Ponsard and B. Fustier (eds.), Fuzzy Economics and Spatial Analysis, Dijon: Librairie de l'Universite, 1986b, pp. 51-67.
Ponsard, C. and B. Fustier (eds.). Fuzzy Economics and Spatial Analysis, Dijon: Librairie de l'Universite, 1986.
Ponsard, C. and P. Tranqui. "Fuzzy Economic Regions in Europe", Environment and Planning A 17(1985), 873-887.
Ralescu, D.A. "Measures, Capacities, and Optimization with Inexact Constraints", 1980. (unpublihsed)
Robinson, V.B. "Modeling Inexactness in Spatial Information Systems", Modeling and Simulation, Vol. 15, Pittsburg: Instrument Society of America, 1984, pp. 157-161.
Robinson, V.B. and A.H. Strahler. "Issues in Designing Geographic Information Systems under Conditions of Inexactness", Proceedings of the Tenth International Symposium on Machine Processing of Remotely Sensed Data, West Lafayette: Purdue University, 1984, pp. 198-204.
Robinson, V.B., D. Thongs, M. Blaze. A Process for Specifying Linguistic Variables for Query of Spatial Information. Paper presented at the National Meetings of the Association of American Geographers, Detroit, 1985.
Rolland-May, C. "Fuzzy Geographical Systems", in C. Ponsard and B. Fustier (eds.), Fuzzy Economics and Spatial Analysis, Dijon: Librairie de l'Universite, 1986, pp. 71-90.
Rubin, P.A. and R. Narasimhan. "Fuzzy Goal Programming with Nested Priorities", Fuzzy Sets and Systems 14(1984), 115-129.
Russell, B. "Vagueness", Australasian Journal of Psychology and Philosophy 1(1923), 84-92.
Soyster, A.L. "Convex Programming with Set-Inclusive Constraints: Applications to Inexact Linear Programming", Operations Research 21(1973), 1154-1157.
Sugeno, M. "Fuzzy Measures and Fuzzy Integrals, a Survey", in M.M. Gupta, G.N. Saridis, B.R. Gaines (eds.), Fuzzy Automata and Decision Processes, Amsterdam: North-Holland, 1977, pp. 98-102.
Tanaka, H., T. Okuda, and K. Asai. "On Fuzzy-mathematical Programming", Journal of Cybernetics 3(1974), 37-46.
Thole, U., H.-J. Zimmermann, and P. Zysno. "On the Suitability of Minimum and Product Operators for the Intersection of Fuzzy Sets", Fuzzy Sets and Systems 2(1979), 167-180.
Uzawa, H. "Preference and Rational Choice in the Theory of Consumption", in Proceedings of a Symposium on Mathematical Methods in Social Sciences, Standford: Standford University Press, 1960.
Verdegay, J.L. "Fuzzy Mathematical Programming", in M.M. Gupta and E. Sanchez (eds.), Fuzzy Information and Decision Processes, Amsterdam: North-Holland, 1982, pp. 231-237.
Wang, Pei-Zhuang. Fuzzy Sets: Theory and Applications. Shanghai: Shanghai Science and Technology Publishing Co., 1983. (In Chinese)
Wang, Pei-Zhuang. Fuzzy Sets and Falling Random Sets. Beijing: Beijing Normal University Press, 1985. (In Chinese)

Yu, P.L. and M. Zeleny. "The Sets of all Nondominated Solutions in Linear Cases and a Multicriteria Simplex Method", Journal of Mathematical Analysis and Applications 49(1975), 430-468.

Zadeh, L.A. "Fuzzy Sets", Information and Control 8(1965), 338-353.

Zadeh, L.A. "The Concept of a Linguistic Variable and its Application to Approximate Reasoning, I", Information Sciences 8(1975a), 199-249.

Zadeh, L.A. "The Concept of a Linguistic Variable and its Application to Approximate Reasoning, II", Information Sciences 8(1975b), 301-357.

Zadeh, L.A. "The Concept of a Linguistic Variable and its Application to Approximate Reasoning, III", Information Sciences 9(1975c), 43-80.

Zadeh, L.A. "Fuzzy Sets as a Basis for a Theory of Possibility", Fuzzy Sets and Systems 1(1978), 3-28.

Zhang, Nan-lun. "The Subordination and Probability Feature of Random Phenomena", Academic Journal of Wuhan Building Material College 2(1981). (In Chinese)

Zimmermann, H.-J. "Description and Optimization of Fuzzy Systems", International Journal of General Systems 2(1976), 209-215.

Zimmermann, H.-J. "Fuzzy Programming and Linear Programming with Several Objective Functions", Fuzzy Sets and Systems 1(1978), 45-55.

Zimmermann, H.-J. Fuzzy Set Theory -- and its Applications. Boston-Dordrecht-Lancaster: Kluwer-Nijhoff, 1985.

Zimmermann, H.-J. and P. Zysno. "Latent Connectives in Human Decision Making", Fuzzy Sets and Systems (1980), 37-51.

Zimmermann, H.-J. and P. Zysno. "Decisions and Evaluations by Hierarchical Aggregation of Information", Fuzzy Sets and Systems 10(1983), 243-266.

Zionts, S. and J. Wallenius. "An Interactive Programming Method for Solving the Multiple Criteria Problem", Management Science 22(1976), 652-663.

INDEX